Nucleoside Modifications

Special Issue Editors

Mahesh K. Lakshman
Fumi Nagatsugi

MDPI

Guest Editors

Mahesh K. Lakshman
The City College and
The City University of New York
USA

Fumi Nagatsugi
Tohoku University
Japan

Editorial Office
MDPI AG
St. Alban-Anlage 66
Basel, Switzerland

This edition is a reprint of the Special Issue published online in the open access journal *Molecules* (ISSN 1420-3049) from 2014–2016 (available at: http://www.mdpi.com/journal/molecules/special_issues/Nucleoside_Modifications).

For citation purposes, cite each article independently as indicated on the article page online and as indicated below:

Author 1; Author 2; Author 3 etc. Article title. *Journal Name.* **Year.** Article number/page range.

ISBN 978-3-03842-354-6 (Pbk)
ISBN 978-3-03842-355-3 (PDF)

Table of Contents

About the Guest Editors

Mahesh Lakshman obtained the B.Sc. and M.Sc. degrees from the University of Bombay (Mumbai), and MS and Ph.D. degrees from The University of Oklahoma. He completed postdoctoral work at the National Institutes of Health (NIDDK) developing the first total chemical synthesis approaches to site-specific DNA modification with stereochemically-defined polycyclic aromatic hydrocarbon metabolite adducts. After serving for a short while in industry, he returned to academia, joining the University of North Dakota and then relocating to The City College of New York (The City University of New York system). In addition to an active research program, funded by both the National Science Foundation and the National Institutes of Health, he has held several administrative positions such as Executive Officer for The City University of New York Ph.D. Program in Chemistry and as a Vice Chair of the Department of Chemistry and Biochemistry at The City College of New York. Professor Lakshman was recently inducted as a Fellow of Royal Society of Chemistry (UK).

Fumi Nagatsugi joined the Faculty of Pharmaceutical Sciences at Kyushu University as a research assistant in 1989. She received her PhD from the university in 1996 and performed postdoctoral research at the National Institutes of Aging (NIA) in 2001–2002. She was promoted to associate professor at Kyushu University in 2003. She moved to Tohoku University in 2006 as a professor. Her research interests are chemical biology and nucleic acid chemistry.

Preface to "Nucleoside Modifications"

Nucleosides are the fundamental components of genetic material, and are present in all living organisms, and in viruses. By virtue of their ubiquity, they are highly important biomolecules. Nucleosides consist of a heterocyclic aglycone and a sugar unit. For several decades, the natural nucleoside structures have inspired the development of chemical and biochemical modifications, leading to new nucleoside-like entities via aglycone as well as saccharide modifications. As a result, modified nucleosides and nucleoside analogues have widespread utilities in biochemistry, biology, as pharmaceutical agents and as biological probes. There is a constant need for access to novel nucleoside analogues for a plethora of applications, prompting the development of new methodologies.

This book contains twelve original research articles that include such diverse topics as metal-complexation with nucleoside analogues, synthesis of flexible nucleoside analogues (fleximers), development of a new "clamp" for thioguanosine via hydrogen-bond interactions, synthesis and properties of nucleic acids containing a pyranose sugar, studies on small molecule binding to RNA via a tethering technique, synthesis of an isonucleoside developed on a dioxabicycloheptane scaffold, development of a peptide-nucleic acid for DNA crosslinking, synthesis of C-nucleoside analogues developed on a triazole linked to an isoxazolidine, development of acyclonucleoside-based RNAse A inhibitors, synthesis and studies of oligonucleosides containing acyclic analogues of polyaminonucleosides, development of new methodology for synthesis and structure–activity studies on *cladribine* and its analogues, and enzymatic ribosylation of 8-azaguanine and 2,6-diamino-8-azapurine. The book also contains two reviews on contemporary methods for modification of nucleosides. One is on the modification of purine and pyrimidine nucleosides by C–H bond activation and the other describes palladium-catalyzed modifications of unprotected nucleosides, nucleotides, and oligonucleotides.

Mahesh K. Lakshman and Fumi Nagatsugi
Guest Editors

molecules

MDPI

Review

Modification of Purine and Pyrimidine Nucleosides by Direct C-H Bond Activation

Yong Liang and Stanislaw F. Wnuk *

Department of Chemistry and Biochemistry, Florida International University, Miami, FL 33199, USA; ylian004@fiu.edu
* Correspondence: wnuk@fiu.edu; Tel.: +1-305-348-6195; Fax: +1-305-348-3772

Academic Editor: Mahesh Lakshman
Received: 15 February 2015; Accepted: 13 March 2015; Published: 17 March 2015

Abstract: Transition metal-catalyzed modifications of the activated heterocyclic bases of nucleosides as well as DNA or RNA fragments employing traditional cross-coupling methods have been well-established in nucleic acid chemistry. This review covers advances in the area of cross-coupling reactions in which nucleosides are functionalized via direct activation of the C8-H bond in purine and the C5-H or C6-H bond in uracil bases. The review focuses on Pd/Cu-catalyzed couplings between unactivated nucleoside bases with aryl halides. It also discusses cross-dehydrogenative arylations and alkenylations as well as other reactions used for modification of nucleoside bases that avoid the use of organometallic precursors and involve direct C-H bond activation in at least one substrate. The scope and efficiency of these coupling reactions along with some mechanistic considerations are discussed.

Keywords: C-H activation; cross-coupling; direct arylation; nucleosides; purines; pyrimidines

1. Introduction

Transition metal catalyzed traditional cross-coupling reactions have contributed significantly to the formation of new carbon-carbon bonds and to the synthesis of biaryl compounds. With few exceptions, the traditional Pd-catalyzed coupling reactions require two activated substrates, one is the organometallic, alkene (Heck reaction), or terminal alkyne (Sonogashira reaction) and the other is the halide or triflate [1,2]. The most often used Stille, Suzuki, Negishi, Kumada and Hiyama reactions need an organometallic (Sn, B, Zn, Mg, and Si) component and a halide or pseudohalide. Owing to the high impact of these reactions in organic synthesis, natural product synthesis and pharmaceutical applications, the 2010 Nobel Prize in Chemistry was awarded jointly to Richard F. Heck, Ei-ichi Negishi and Akira Suzuki [3]. Pd-catalyzed cross-coupling reactions are carried out under mild conditions and can be performed in the presence of most functional groups. The mechanisms in most cases follow three major steps of: (*i*) oxidative addition, (*ii*) transmetallation, and (*iii*) and reductive elimination [1,2].

Transition metal-catalyzed cross-coupling reactions which are based on direct C-H functionalization have been recently developed [4–9]. These methodologies, which eliminate the use of organometallic substrates, compete with traditional Pd-catalyzed cross-couplings in the development of new strategies for the formation of carbon-carbon bonds. These reactions require only one activated substrate (C-H activation) and sometimes even no activation is required for either substrate (double C-H activation). They are atom efficient and avoid the synthesis of often unstable activated substrates. Major challenges associated with C-H functionalization reactions include: (*i*) the need for developing regioselective activation of specific C-H bonds in the presence of other C-H bonds; (*ii*) low chemoselectivity which means it is necessary to protect sensitive functional groups before performing the coupling; and (*iii*) the necessity to work at high temperature needed to activate C-H bonds with intrinsic low activity, which often causes decomposition of the substrates. Pd and Cu are two of the most common transition metal catalyst used for the C-H functionalization.

Transition metal-catalyzed approaches towards the synthesis of base-modified nucleosides can be divided into five major categories as depicted in Figure 1. The first two approaches are based on cross-couplings between two activated components. One involves reactions between metal-activated nucleoside bases and halides (Figure 1, Path *a*) while the second employs couplings between halo (or triflate) modified nucleoside bases and organometallics (Path *b*). These approaches were extensively reviewed [10–12] and are not discussed in this account. The next two approaches are based on cross-couplings between only one activated component and require C-H activation at the second substrate. One involves reactions between C-H activated bond in nucleoside bases and halides (Path *c*), while the second employs couplings between halo-modified nucleoside bases and arenes, which, in turn, require selective C-H activation (Path *d*). The last approach involves cross-couplings between two inactivated substrates [cross-dehydrogenative coupling (CDC) reactions; Path *e*]. Direct C-H functionalization approaches (Paths *c-e*) alleviate some drawbacks associated with the synthesis of modified nucleosides employed in traditional Pd-catalyzed cross-coupling reactions (Paths *a-b*). They also avoid usage of the toxic organotin components, which are problematic during biological studies, or the sometimes unstable organoboronic substrates.

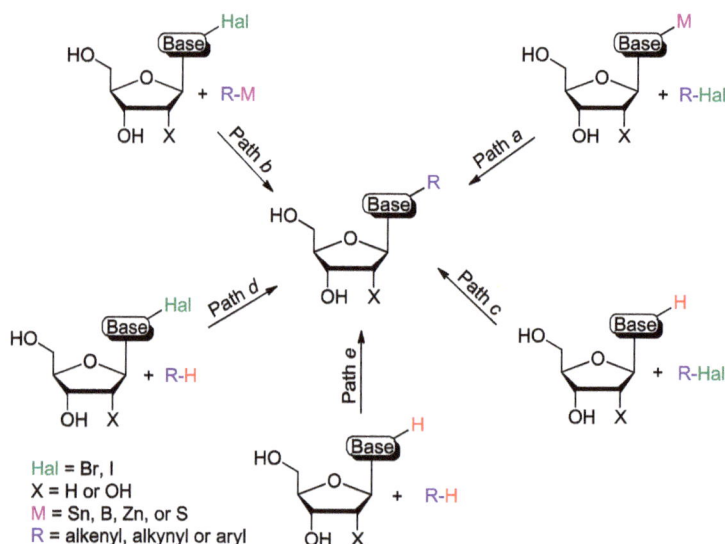

Figure 1. Transition metal catalyzed cross-coupling approaches towards the synthesis of base-modified nucleosides.

Numerous C5 or C6 modified pyrimidine nucleosides and C2 or C8 modified purine nucleosides have been synthesized in last 40 years employing the transition-metal assisted cross-coupling reactions [10]. Some of them show potent biological activity and/or are utilized as mechanistic or labelling probes (Figure 2). For example, the (*E*)-5-(2-bromovinyl)-2′-deoxyuridine (**1**, BVDU) has been found to be a highly potent and selective anti-herpes agent [13]. The bicyclic furanopyrimidine-2-one nucleoside analogues bearing an aryl side chain **2** display remarkable antiviral potency against the Varicella-Zoster virus [14]. The 5-thienyl- **3** or 5-furyluridine **4** were used as molecular beacons for oligonucleotide labeling [15–18]. The 8-pyrenyl-2′-deoxyguanosine **5** serves as a probe for the spectroscopic study of the reductive electron transfer through DNA [19,20]. Furthermore, the 8-vinyl and 8-ethynyladenosines **6** show cytotoxic activity against tumor cell lines [21], while oligodeoxynucleotides modified with the 8-alkynyl-dG possess thrombin inhibitory activity [22].

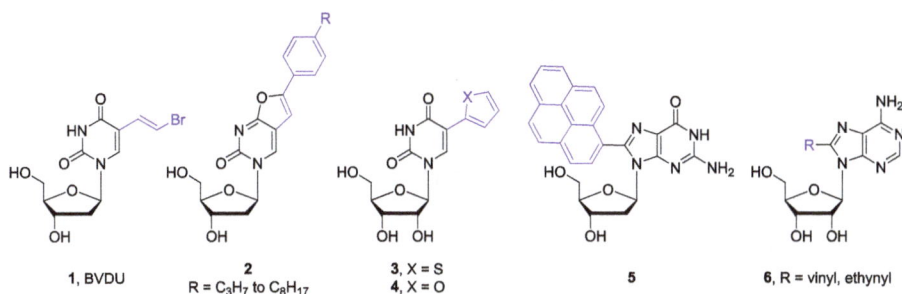

Figure 2. Selected base-modified pyrimidine and purine nucleosides.

2. Direct Activation of C8-H Bond in Purine and Purine Nucleosides

2.1. Cross-Coupling of Adenine Nucleosides with Aryl Halides

Hocek and coworkers reported the first example of direct arylation of adenosine **7** with aryl halides by selective activation of the C8-H bond which gave access to 8-arylated adenosine analogues **9**. The cross-coupling occurred in the presence of a stoichiometric amount of CuI (3 equiv.) and a catalytic load of Pd(OAc)$_2$ (5 mol %) in DMF at elevated temperature (100 °C/22 h or 150 °C/5 h) to produce **9** in 50%–68% yields (Scheme 1, Table 1 entry 1) [23]. The authors were able to improve the coupling conditions (e.g., shortening reaction time and lowering the reaction temperature), as compared to their earlier work on C8-H arylation of purines and adenines [24,25] (*vide infra*), by addition of piperidine to the reaction mixture. They assumed [23] that formation of dimethylamine, as a side product of the prolonged heating of the DMF solvent during the C8-H arylation of purines, favorable influenced the rate of the arylation reaction, which is consistent with Fairlamb's findings [26,27]. Consequently, they found that the addition of higher boiling secondary amine such as piperidine (4 equiv.) was beneficial to the coupling reactions. Couplings of **7** with aryl iodines also produced N6,8-diarylated byproducts **11** in 12%–18% yield, whereas only 8-arylated products **9** were isolated when less reactive aryl bromine were employed.

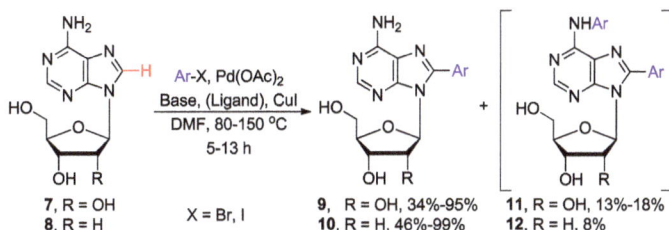

Scheme 1. Pd-catalyzed direct C8-H arylation of adenosine **7** and 2'-deoxyadenosine **8** with aryl halides.

When 2'-deoxyadenosine **8** was subjected to this direct arylation protocol desired 8-arylated products **10** were produced only after the temperature was lowered to 125 °C (31% after 5 h; entry 2). It is worth noting that this protocol was applicable to unprotected nucleosides and allowed for the first time the single-step introduction of the aryl group at the C8 position without the need to (*i*) halogenate nucleoside substrates, or (*ii*) use expensive arylboronic acids or toxic arylstannanes [10].

3

Table 1. Effect of different bases on Pd-catalyzed direct C8-H arylation of adenosine 7 and 2'-deoxyadenosine 8 with aryl halides.

Entry	Base/Ligand	Substrates	Ar	Temp. (°C)	Time (h)	Products	Yield (%)	Reference
1	Piperidine	7	4-Tol-I	150	5	9	68	[23]
2	Piperidine	8	4-Tol-I	125	5	10	31	[23]
3	Cs_2CO_3	8	Ph-I	80	13	10	84	[27]
4	Cs_2CO_3/Pyridine	7	Ph-I	120	13	9	30–95 [a]	[27]
5	Cs_2CO_3/Piperidine	8	4-Tol-I	80	15	10	85	[26]

[a] The yield depends on the substitution at the pyridine ring.

Fairlamb and coworkers independently developed a Pd-catalyzed direct C8-arylation of adenosine 7 with aryl iodine in the presence of Cs_2CO_3 as a base (instead of piperidine), $Pd(OAc)_2$ and 3 equiv. of CuI (DMF/120 °C/13 h) to give 9 in good to high yields (Table 1, entry 3) [27]. In addition to 9, small quantities (~3%) of the N6-arylated byproducts (e.g., 11) were also produced. Coupling of 7 with 0.5 equiv of 1,4-diiodobenzene yielded 1,4-di-(8-adenosinyl)benzene, albeit in low yield. The less stable 2'-deoxyadenosine 8 could also be arylated under these conditions but the synthesis required lower temperature (80 °C/13 h; 84%) to avoid substantial deglycosylation, which was observed at 120 °C. The authors also found that microwave heating was ineffective due to significant decomposition. However addition of the pyridine substituted with an EWG (e.g., 3-nitropyridine) provided 8-arylated product 9 in up to 95% yield (Table 1, entry 4). It was hypothesized that the electron-deficient pyridines can stabilize active Pd(0) species and increase the reactivity (electrophilicity) of the Pd(II) species and that their beneficial effect is substrate dependent [27]. The direct C8-H arylation of 2'-deoxyadenosine 8 with aryl iodides catalyzed by Pd-nanoparticles (60 °C, 15 h) to give access to 8-arylated product 10 (50% yield) have been recently reported [28].

Fairlamb and coworkers also demonstrated that a combination of a stoichiometric amount of Cs_2CO_3 with a substoichiometric amount of piperidine provided the best yield for Pd/Cu-mediated C8 arylation of 2'-deoxyadenosine 8 with various aryl halides (80 °C, 15 h) to give 10 in 32%–95% yields (Table 4 entry 5) [26]. They also noted that sequential direct arylation of 8 with iodo(bromo)benzene followed by Suzuki-Miyaura cross-coupling of the resulting 8-bromophenyl-2'-deoxyadenosine gave convenient access to the new class of rigid organofluorescent nucleosides (RONs) analogues [29]. The arylation conditions were also extended to the adenosine analogues modified at either the ribose or the adenine moieties. Thus, 2'-deoxy-2'-fluoroadenosine 13 gave the 8-arylated product 14 almost quantitatively (94% isolated yield) under similar conditions; probably because the 2'-fluoro substituent is known to increase the stability of the N-glycosylic bond and to favor the *syn* conformation (Scheme 2). Coupling of 2-fluoro-2'-deoxyadenosine 15 with iodobenzene also effected 8-arylation concomitant with the displacement of fluorine by piperidine to give the 2,8-disubstituted 2'-deoxyadenosine 16 (Scheme 3) [26]. The chemistry of these couplings has been discussed in recent reviews [11,30].

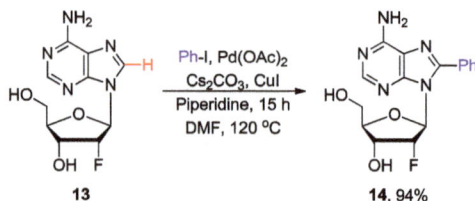

Scheme 2. Pd-catalyzed direct C8-H arylation of 2'-deoxy-2'-fluoroadenosine 13 with iodobenzene.

Scheme 3. Pd-catalyzed direct arylation of 2-fluoro-2'-deoxyadenosine **15** with iodobenzene.

These Pd-catalyzed/Cu-mediated methodologies were successfully applied to the synthesis of numerous 8-arylated purines and adenines. In 2006, Hocek and coworkers elaborated the original protocol for the efficient direct C8-H arylation of 6-phenylpurine analogues **17** using aryl iodides in the presence of Cs$_2$CO$_3$ and CuI to give **18** (Scheme 4) [24]. This route required prolonged heating at high temperature (160 °C/60 h) in DMF. Furthermore, it was essential to perform the coupling with strict exclusion of air to avoid formation of two byproducts, sometimes, in substantial yields (6%–54%). One byproduct was **19**, which was formed by double arylation at C8 position and the *ortho* position of the phenyl ring at C6. The other byproduct was the 8,8'-bispurine dimer. Various 6,8,9-trisubstituted and 2,6,8,9-tetrasubstituted purine analogues were synthesized using this approach in combination with Suzuki cross-coupling reaction and Cu-catalyzed *N*-arylation at 9 position [24,25]. These conditions (Cs$_2$CO$_3$ or piperidine) were successfully employed for the synthesis of 8,9-disubstituted adenines but 6-*N*-(di)arylated byproducts were observed [24,25]. This protocol was also applied for direct C8-H arylation of adenines anchored to solid phase via 6-*N* amino group (in the presence of piperidine as base) [31].

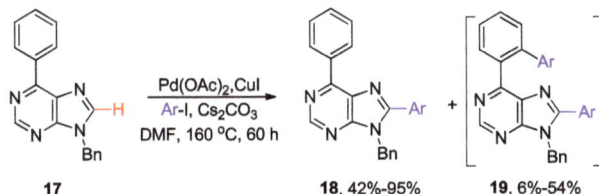

Scheme 4. Pd/Cu-mediated direct C8-H arylation of 6-phenylpurines with aryl halides.

Alami and coworkers developed a microwave-assisted direct C8 arylation of free-(NH$_2$) 9-*N*-protected adenine **20** with aryl halides catalyzed by Pd(OH)$_2$/C (Pearlman's catalyst) in the presence of CuI (Scheme 5) [32]. Reaction took only 15 min at 160 °C in NMP solvent when Cs$_2$CO$_3$ was used as base to give **21** in up to 90% yield. The application of Pd(OH)$_2$/C catalyst allowed coupling with aryl bromides and even less reactive aryl chlorides [33]. Sequential combination of C8-arylation with ArCl and 6-*N*-arylation with ArBr or ArI (using Xantphos [34] instead of CuI) provides access to disubstituted adenines **22** [33].

Scheme 5. Microwave-assisted direct C8-H arylation of 9-*N*-benzyladenine with aryl halides.

Fairlamb and coworkers reported a detailed mechanism for the direct C8-arylation of adenine ring with aryl halides mediated by Pd and Cu in the presence of Cs₂CO₃ [26,27]. The authors noted that the use of a stoichiometric amount of Cu(I) is key to the direct arylation of the adenine ring and that the process parallels the arylation of imidazole ring at the 2 position [35]. As depicted in Scheme 6, Cu(I) was proposed to assist the C-H functionalization process by an initial coordination to the adenine N7 atom. The subsequent base-assisted deprotonation leads to the formation of 8-cuprioadenine intermediate **A** or *N*-heterocyclic carbene like cuprates, which can then undergo a standard Pd(0) catalytic cycle for cross-coupling with aryl halides. This process resembles Sonogashira's reaction between alkynylcuprates and halides [26,27]. The requirement for excess of CuI was attributed to the high binding affinity of Cu(I) for both the substrate and presumably the 8-arylated product(s). The dinucleoside copper(I) complex between 7-N and 6-NH atoms of the adenine have been identified as important intermediate [26].

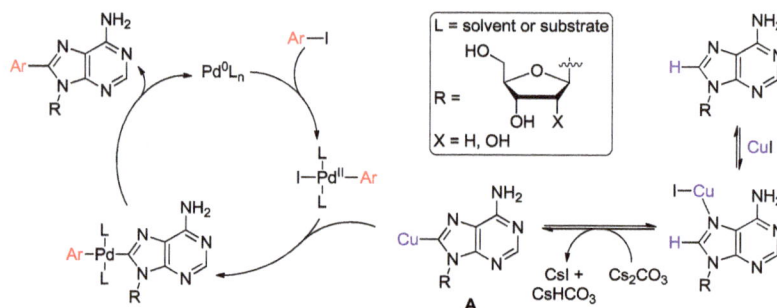

Scheme 6. Proposed mechanism for the direct arylation at C8 position of adenosine [26,27].

2.2. Cross-Coupling of Inosine and Guanine Nucleosides with Aryl Halides

Guanosine **23** was found to be a poor substrate for the direct C8 arylation as indicated by the low yield (15%) of 8-phenylated product **25** under the conditions (120 °C) which were effective for adenosine analogues (Scheme 7) [26]. Analogous arylation of 2'-deoxyguanosine **24** at lower temperature (80 °C) yielded product **26** but only in 6% yield. The authors hypothesized that in the case of guanine substrates, Cu[I]-coordination most probably occurs at sites distal to C8 hampering efficient arylation. A similar inhibitory effect, associated with the ionizable protons in the guanine moiety was observed during the Suzuki couplings with 8-haloguanine nucleosides [36]. The authors also suggested that guanine-type nucleosides are poor substrates for direct C8-H arylation due to the lack of the "templating" role of exocyclic 6-amino group present in adenine nucleosides.

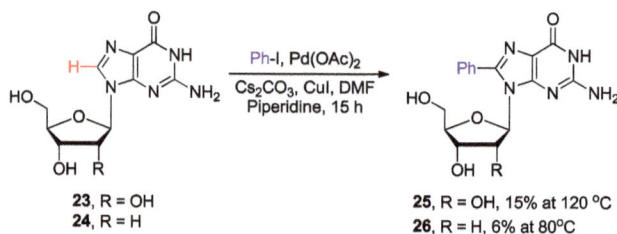

23, R = OH
24, R = H

25, R = OH, 15% at 120 °C
26, R = H, 6% at 80°C

Scheme 7. Direct C8-H arylation of guanosine **23** and 2'-deoxyguanosine **24** with iodobenzene.

The Pd-catalyzed/Cu-mediated direct C8-H arylation of inosine **27** proceeded proficiently to afford 8-phenylated product **29** in good yield (60%) at 120 °C (Scheme 8) [26]. The analogous functionalization of 2'-deoxyinosine **28**, due to the stability of the glycosylic bond, had to be carried out at lower temperature to give product **30** but in only 19% yield.

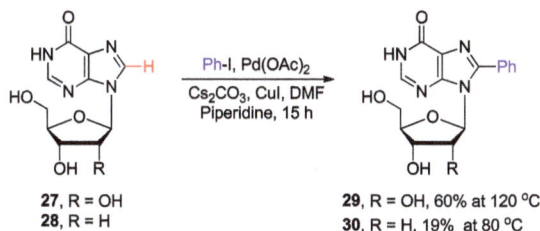

Scheme 8. Direct C8-H arylation of inosine **27** and 2'-deoxyinosine **28** with iodobenzene.

Recently, Pérez and coworkers synthesized the 8-arylated inosine analogues via a microwave-assisted Pd/Cu-catalyzed direct C8-H arylation [37]. In order to increase the solubility of the nucleoside substrate, 2',3'-*O*-isopropylideneinosine **29** was employed to couple with iodopyridines or aryl iodides, by adopting Fairlamb's protocol [26,29], to produce **30** in only 1 h at 120 °C (Scheme 9).

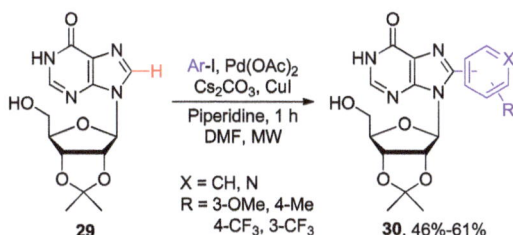

Scheme 9. Microwave-assisted direct C8-H arylation of inosine **29** with aryl iodides.

2.3. Synthesis of Fused Purines via Inter- or Intramolecular Direct C8-H Arylation

Hocek and coworkers developed a direct C8-H arylation of 9-*N*-phenylpurine **31** for the synthesis of fused purine analogues of type **32** with *e*-fusion (position 8 and 9 of purine ring). Thus, Pd-catalyzed intermolecular double direct C-H arylation of 6-methyl-9-*N*-phenylpurine **31** with 1,2-diiodobenzene gave **32** (R = CH$_3$) in modest yield (35%). Alternatively, the sequential Suzuki coupling of 9-(2-bromophenyl)adenine **33** (R = NH$_2$) with 2-bromophenylboronic acid **34** followed by intramolecular C8-H arylation also gave the desired product **32** (R = NH$_2$) in moderate to high yields which preserves the base-pairing and major groove facets of the intact adenine ring (Scheme 10) [38]. However, attempted intramolecular oxidative coupling of 8,9-diphenyladenine failed to give **32**.

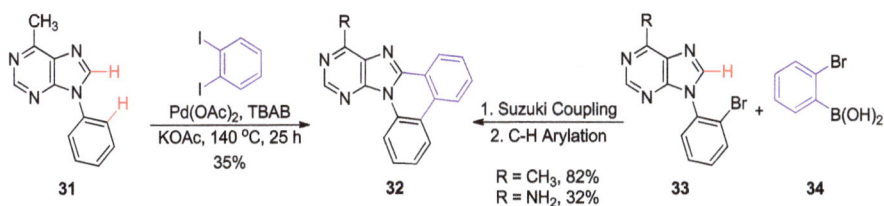

Scheme 10. Pd-catalyzed cyclization of 9-*N*-arylpurines via C8-H activation.

The purines **37** with five or six-membered e-fused rings were also synthesized by intramolecular cyclizations of 9-(2-chlorophenylalkyl)purines **35** (n = 1 or 2; X = Cl) employing conditions developed by Fagnou [39] for direct arylation with aryl halides [Pd(OAc)$_2$/tricyclohexylphoshine/K$_2$CO$_3$ in DMF] (Scheme 11) [38,40]. Domínguez and coworks reported the synthesis of five-membered ring analogue **36** by Cu-catalyzed direct C8-H arylation of 9-(2-iodophenylmethyl)purines **35** (n = 1, X = I) in 58% yield [41].

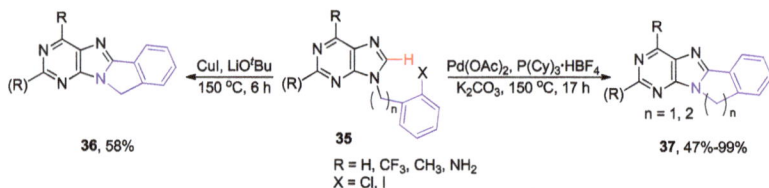

Scheme 11. Pd/Cu-catalyzed intramolecular cyclization of 9-*N*-substituted purines via C8-H activation.

The five-, six- or seven-membered *e*-fused purines **39** have been prepared by the intramolecular double C-H activation (at C8 and *ortho* position of the phenyl ring) of 9-*N*-phenylalkylpurines **38** in the presence of Pd catalyst and silver salt oxidant in high to excellent yields (Scheme 12) [42]. The seven-, eight-, or nine-membered *e*-fused purines of type **40** were prepared by the one-pot coupling of **38** with iodobenzene [42]. The reaction sequence was believed to be initiated by direct intermolecular C8-H arylation of **38** with iodobenzene followed the intramolecular cross-dehydrogenative-arylation between two phenyl rings to give product **40**.

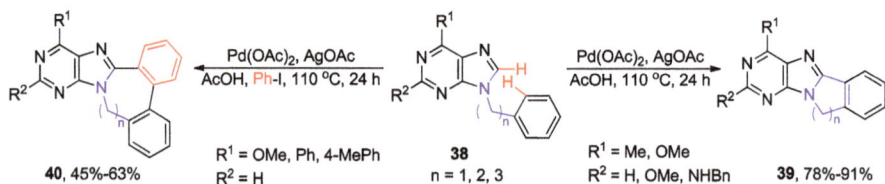

Scheme 12. Pd-catalyzed intramolecular cross-dehydrogenative arylation of 9-*N*-substituted purines via C8-H bond activation.

2.4. Miscellaneous Direct C8-H Functionalizations

You and co-workers reported intermolecular Pd/Cu-catalyzed regioselective C8-H cross-coupling of 1,3-diethyl xanthine **41** (R^1 = R^2 = Et, R^3 = H) with electron-rich furans **42a** and thiophenes **42b** [43]. For example coupling of **41** with 2-methylthiophene in the presence of catalytic amount of the copper salt gave diheteroarene product of type **43** in 96%, indicating the tolerance of the free NH group at the 9 position to the reaction conditions (Scheme 13). The differences in the electron density of two heteroarene components was believed to facilitate the reactivity and selectivity in the two metalation steps [44] of the catalytic cycle of this cross-dehydrogenative arylation reaction.

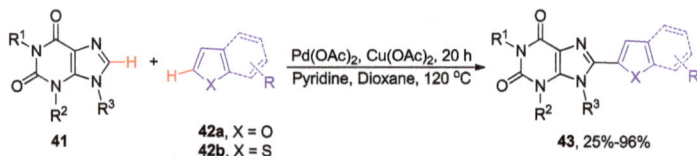

Scheme 13. Pd/Cu-catalyzed cross-dehydrogenative arylation of purines with heteroarenes.

The 8-alkenyl adenine analogues **45** have been synthesized via microwave-assisted direct C8-H alkenylation of 9-*N*-benzyladenines **20** with alkenyl bromides **44** (Scheme 14) [32]. Analogous Pd/Cu-mediated C8 alkenylations of 6-(benzylthio)-9-*N*-benzylpurines with styryl bromides provided access to 6,8,9-trisubstituted purines [45]. The optimized conditions (Pd/CuI/*t*BuOLi) were applicable for the selective alkenylation of caffeine, benzimidazole and other aromatic azole heterocycles [45,46]. These are significant developments since it was reported that 8-bromoadenosine was not a good substrate for Mizoroki-Heck reaction [47] making modification at 8 position via direct functionalization of C8-H bond a desirable transformation.

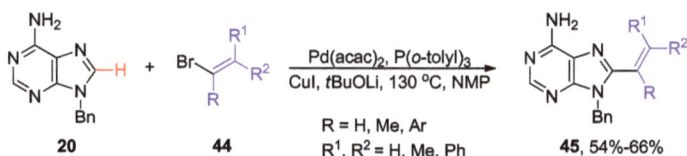

Scheme 14. Pd-catalyzed direct C8-H alkenylation of 9-*N*-benzyladenine with alkenyl halides.

Modification of biologically important 7-deazapurines by direct C-H activation have also been explored (Scheme 15). Thus, regioselective Pd-catalyzed direct C8-H arylation of the 6-phenyl-7-deazapurine analogue **46** (R^1 = Bn) with aryl halides gave corresponding 8-arylated products **46a** albeit in low to moderate yields (0%–41%) [48]. Alternatively, Ir-catalyzed C-H borylation of **46** (R = Ph, R^1 = Bn) followed by Suzuki coupling with aryl halides afforded **46a** in high yields (79%–95%). Interestingly, Ir-catalyzed C-H borylation was not successful with purines suggesting that the complexation of Ir catalyst to N7 nitrogen might be responsible for the lack of reactivity [48]. The regioselective Pd/Cu-catalyzed direct C8-H amination of the 6-phenyl-7-deazapurine analogue **46** (R^1 = Bn) with *N*-chloro-*N*-sulfonamides provided the 8-amino-7-deazapurine analogues **46b** [49]. However, subjection of the 6-chloro-7-deazapurine **46** (R^1 = Bn) to the similar coupling conditions produced a complex mixture. Remarkably, application of conditions, developed by Suna and co-workers for direct C5-H amination of uracils (see Scheme 33), to the same substrate **46** (R = Cl, R^1 = Bn) provided 7-amino-7-deazapurine analogue **46c** in 60% [50]. Cu-catalyzed direct C-H sulfenylation of 6-substituted-7-deazapurines **46** (R^1 = H) with aryl or alkyl disulfides provided 7-aryl(or alkyl)sulfanyl products **46d** (47%–96%) in addition to minor quantities of 7,8-bis(sulfanyl) byproducts [51].

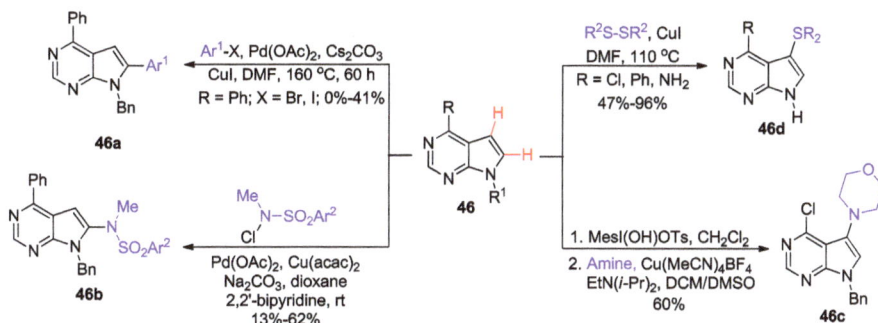

Scheme 15. Transition-metal catalyzed direct C-H activation of 7-deazapurines.

3. N1-Directed Modifications of C6-Substituted Purine Nucleosides via *ortho* C-H Bond Activation

The Cu-catalyzed direct C-H activation/intramolecular amination reaction of the 2′,3′,5′-tri-*O*-acetyl-6-*N*-aryladenosines **47** were employed for the synthesis of fluorescent polycyclic purine and purine nucleosides of type **48** (Scheme 16) [52]. It was found that addition of Ac$_2$O significantly improved the reaction rate (2 h at 80 °C) when Cu(OTf)$_2$ (5 mol %) was used as copper source and PhI(OAc)$_2$ was used as the oxidant. The 6-*N*-aryladenosines substrates **47** containing electron-withdrawing groups in the benzene ring gave better yields (~85%–92%) than those bearing electron donating groups (~45%–62%). The proposed catalytic cycle involves initial coordination of Cu(OTf)$_2$ to the 6-NH····N1 tautomer of substrate **47** at N1 position followed by electrophilic substitution yielding Cu(II) intermediate bridging N1 position of the purine ring and *ortho* position in the aryl ring. Subsequent reductive elimination then provides the fused product **48**.

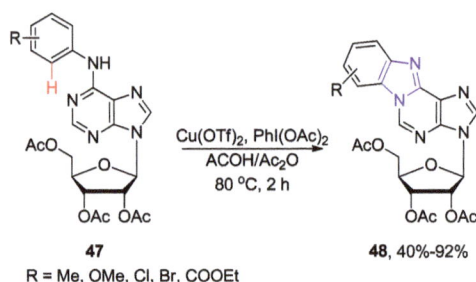

Scheme 16. Cu-catalyzed intramolecular direct *ortho* C-H activation/amination of 6-*N*-aryladenosines.

Qu and coworkers developed a Pd-catalyzed strategy for the regioselective *ortho* monophenylation of 6-arylpurine nucleosides (e.g., **49**) via (N1 purine nitrogen atom)-directed C−H activation using a large excess (30 equiv.) of iodobenzene (Scheme 17) [53]. It was essential to perform the reaction under an inert N$_2$ atmosphere at 120 °C in the presence of AcOH in order to synthesize **50** in excellent yield (85%). It was believed that AcOH might help (*i*) to overcome the poisoning of the catalyst attributed to the multiple nitrogens present in purine ring, and (*ii*) to facilitate the reductive elimination step. However, the phenylation reaction was not successful when PhI was replaced with PhBr or PhCl.

It is noteworthy that the conditions described by Qu and coworkers (Pd(OAc)$_2$/AgOAc/AcOH) are selective for the arylation at the *ortho* site of the C6-phenyl ring, while the CuI-catalyzed coupling of **49** with aryl halides in the presence of Pd(OAc)$_2$/piperidine produced exclusively C8-arylated product **51** [23], or a mixture of double arylated products **50/51** in addition to 8,8′-dimers when analogous purine substrates **49** and Pd(OAc)$_2$/Cs$_2$CO$_3$ were used [24,25].

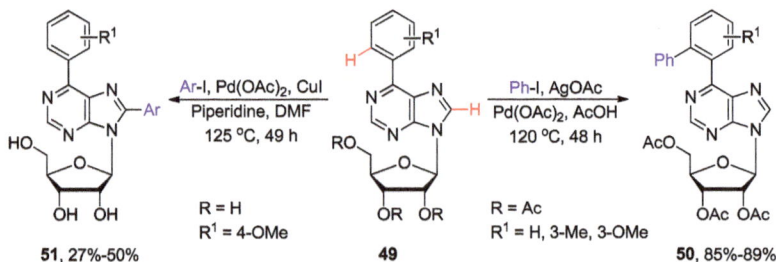

Scheme 17. Direct C-H arylation at the *ortho* position in 6-arylpurine nucleosides *vs.* C8-H arylation.

Lakshman and coworkers reported direct arylation of 6-arylpurine nucleosides of type **52** by the Ru-catalyzed C-H bond activation (Scheme 18) [54]. The coupling usually required only 2 equiv. of aryl halides and K$_2$CO$_3$ or Cs$_2$CO$_3$ as a base to give mixture of the mono and diarylated *ortho* products **53** and **54** in ratio of approximately of 2–7 to 1 with no 8-arylated byproducts detected. It is noteworthy that these conditions were applicable to acid-sensitive substrates such as 2'-deoxynucleosides as opposed to the previously described Pd-catalyzed arylation promoted by AcOH [53]. Both aryl iodides and bromides gave arylated products in good yield, but aryl iodides proceeded with higher yields and better product ratio. Also, aryl halides bearing EWG gave higher yields compared to the ones bearing EDG.

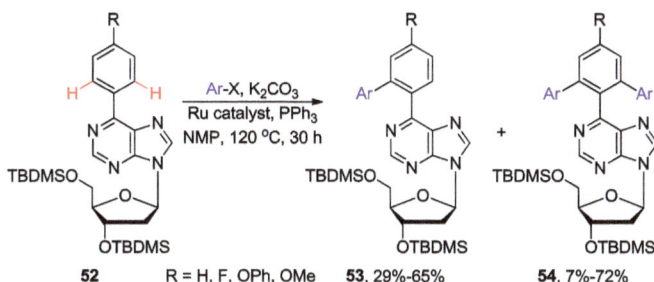

Scheme 18. Ru-Catalyzed *N*1-directed *ortho* C-H arylation of 6-phenylpurine nucleosides.

A possible mechanism for this direct *ortho*-arylation was proposed involving purinyl *N*1-directed electrophilic attack by the aryl/Ru^{IV} complex **A** on the *ortho* position of the C6-aryl ring atom of the substrate **52** (Scheme 19) [54]. Subsequent reductive-elimination of the five-membered Ru complex **B** gave product **53**. The 2-amino-6-arylpurine nucleosides were unreactive, indicating that the presence of C2-amino group is critical and often inhibits the reactivity of purine nucleoside towards C-H activation reactions.

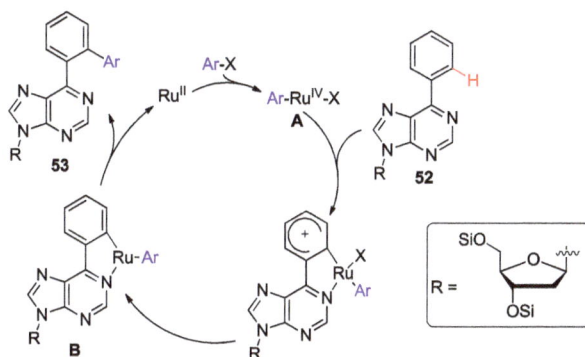

Scheme 19. Proposed mechanism for the Ru-catalyzed *N*1-directed C-H bond activation of 6-phenylpurine nucleoside [54].

The Pd-catalyzed C-H bond activation and oxidation of the silyl protected C6-aryl ribonucleosides of type **55**, as well as the 2'-deoxy counterparts (e.g., **56**), in the presence of PhI(OAc)$_2$ as a stoichiometric oxidant in MeCN provided access to monoacetoxylated products **57** or **58** in good yields (Scheme 20) [55]. Increasing the loading of PhI(OAc)$_2$ to 3 equiv gave mainly the diacetoxylated products **59** or **60**. The involvement of the *N*1-purinyl atom in this *N*-directed C-H bond activation was demonstrated by isolation and crystallographic characterization of the dimeric Pd^{II}-containing cyclopalladated C6 naphthylpurine derivative. The latter complex, together with PhI(OAc)$_2$, was shown to be effective in catalyzing oxidation of substrate **56** to **58**.

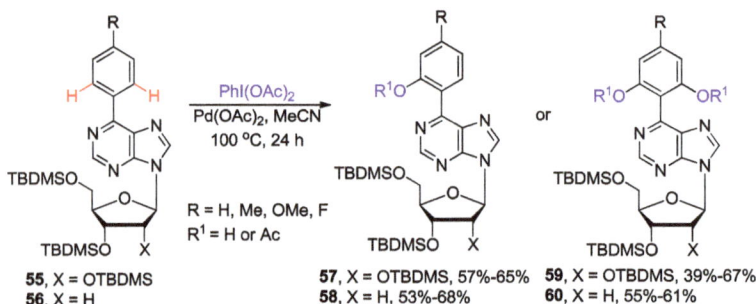

Scheme 20. Pd-Catalyzed *N*1-directed *ortho* C-H acetoxylation of 6-arylpurine nucleosides.

Chang and coworkers reported Rh-catalyzed intermolecular amidation of 6-phenylpurine nucleosides with sulfonyl azides via purinyl *N*1-assisted C-H activation (Scheme 21) [56]. Amidation of 6-arylpurine nucleoside **61** proceeded smoothly and with excellent *ortho*-selectivity to afford product **63** in 70% yield. The presence of a free amino group at the C2 position of purine substrate **62** inhibits once more the coupling efficiency to give amidation product **64** in 45% yield. No glycosylic bond cleavage was noted under the optimized conditions. It is worth noting that the coupling conditions require no additional oxidant and release N$_2$ as the only byproduct.

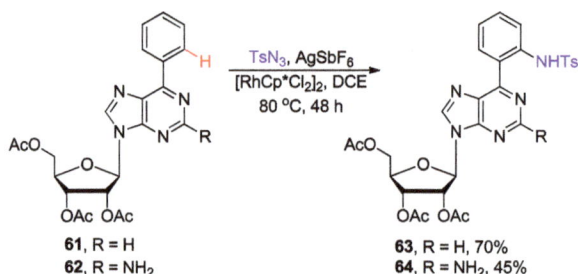

Scheme 21. Rh-catalyzed intermolecular direct *ortho* C-H amidation of 6-phenylpurine nucleosides.

4. Direct Activation of C5-H or C6-H Bond in Uracils and Uracil Nucleosides

4.1. Cross-Coupling with Aryl Halides

Direct C-H arylation of the pyrimidine nucleosides is currently limited to uracil bases. The main challenge which needed to be overcome was a regioselective activation of the C5-H or C6-H bond of the uracil ring. Also lacking are efficient conditions which could be applicable to the natural uridine and 2'-deoxyuridine analogues. Hocek and coworkers reported the regioselective C-H arylations of 1,3-dimethyluracil (**65**, DMU, R = Me) with aryl halides. Thus, arylation in the presence of Pd(OAc)$_2$ and Cs$_2$CO$_3$ mainly formed the 5-arylated uracil analogues **66**, while coupling in the presence of Pd catalyst and CuI (3 equiv.) preferentially formed the 6-arylated derivatives **67** (Scheme 22) [57,58]. Interestingly, Cu-mediated arylation in the absence of Pd catalyst gave exclusively 6-aryluracils **67** albeit in lower yields. These couplings required high temperature (160 °C) and long time (48 h) and were not applicable to the synthesis of unsubstituted uracils. Also the electron-deficient aryl bromides were poor substrates. In order to prepare the unprotected uracil derivatives, these protocols were also applied to *N*-benzyl protected uracil derivatives. Subsequent debenzylation (e.g., **67**, R = Bn) afforded efficiently 5- or 6-aryled uracil derivatives [57,58]. The experimental results indicated that different mechanisms are involved in these diverse arylation reactions [57,58].

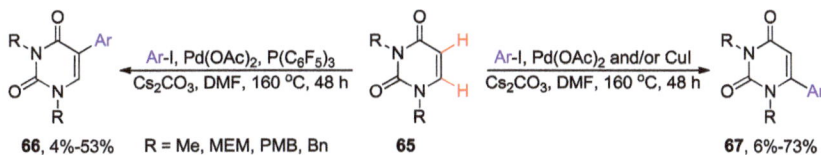

66, 4%–53%   R = Me, MEM, PMB, Bn   **65**   **67**, 6%–73%

Scheme 22. Catalyst-controlled direct C5-H or C6-H arylation of uracil analogues with aryl halides.

Kim and coworkers reported the direct arylation of 1,3-dimetyluracil **65** with aryl bromides (including electron-deficient ones) in the presence of Pd(OAc)$_2$, K$_2$CO$_3$ and PivOH (130 °C/12 h/DMF to give predominantly C5 arylated products **66** in up to 79% yield. A small amount of 6-arylated isomers **67** were also observed [59]. However, application of this condition to 1-(tetrahydrofuran-2-yl)-3-benzyluracil **68** (a substrate having a glycosylic bond) resulted in severe decomposition. Nonetheless lowering the temperature to 100 °C (12 h) allowed the isolation of the 5-phenyluracil analogue **69** as the sole product in 55% yield (Scheme 23) [59]. The intramolecular C6-H arylation of the uracil derivatives bearing a Morita-Baylis-Hillman adduct at the N1 position in the presence of TBAB and Pd(OAc)$_2$ provided a convenient access to the azepine scaffold [60].

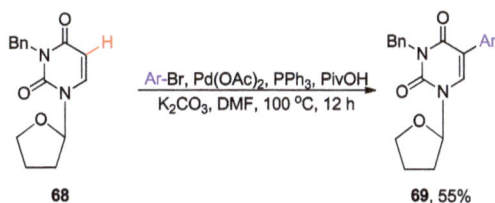

68   **69**, 55%

Scheme 23. Pd-catalyzed direct C5-H arylation of 1-(tetrahydrofuran-2-yl)-3-benzyluracil.

Chien and coworkers further explored the base-dependence of direct activation of uracil C6-H in the presence of Cu catalysts for the synthesis of 6-aryl uracil and 2-aryl-4-pyridone derivatives. The authors found that the CuBr-mediated arylation of DMU **65** (R = H) with aryl iodides (2 equiv.) in the presence of *t*-BuOLi in DMF at reflux gave the C6-arylated derivatives **71** (R^1 = H) [61]. The 5-substituted DMU analogues **70** (R^1 = Me) afforded the 5,6-disubstituted derivatives **71** (R^1 = Me) but in lower yields (Scheme 24). Both N1 and N3 positions in the uracil ring have to be protected for the reactions to proceed.

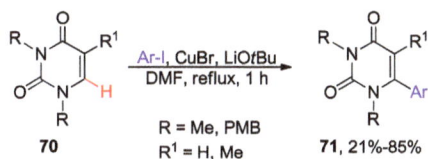

R = Me, PMB   **71**, 21%–85%
70   R^1 = H, Me

Scheme 24. Cu-mediated direct C6-H arylation of 1,3-disubstituteduracil **70** with aryl iodides.

Based on the literature reports, the mechanism for the direct arylation of uracil analogues (e.g., DMU, **65**) with aryl halides at either the C5 or C6 position in the presence of Pd or Cu/(Pd) are summarized in Scheme 24. The coupling with only Pd(OAc)$_2$ was proposed to proceed via an electrophilic metalation-deprotonation (EMD) mechanism [5,9] (path *a*, complex **A**), and thus follow the regioselectivity of substitution at the C5 position to give **66** [59]. The formation of the C5 arylated products was also suggested [57] to proceed via the concerted metalation-deprotonation (CMD) mechanism [9,62] (complex **B**). On the other hand, direct C-H arylation in the presence of CuI or in combination with Pd(OAc)$_2$ was hypothesized to proceed via an Ullmann-type mechanism [27,63], which involves the formation of a carbanion by the abstraction of the most acidic C6-H [64–66] in the uracil ring (e.g., **65**) with base (path *b*). Subsequent cupration, transmetallation and reductive elimination leading to the formation of C6 arylated uracil **67** [57,61]. Alternatively, addition of the copper center of the arylcopper(III) complex **C** to the more nucleophilic C5-position in **65** followed by either base-promoted *anti*-elimination or Heck-type *syn* β-hydrogen elimination has been also proposed for C6-arylation [61]. Moreover, Heck-type carbopalladation of the 5,6-double bond of **65** with ArPdX species has been also considered for the arylation of uracil ring with aryl halides [59].

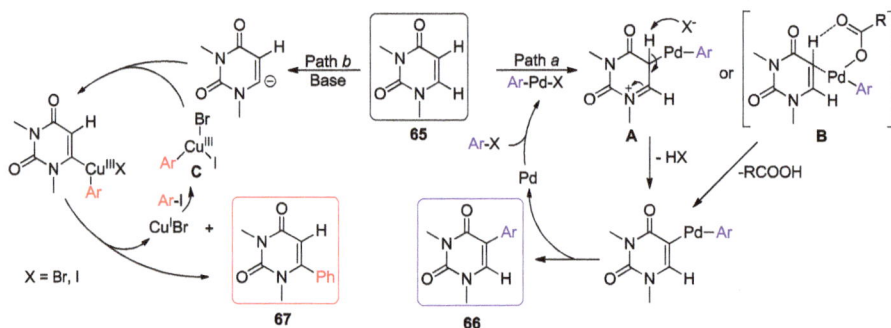

Scheme 25. Proposed mechanisms for the regioselective direct arylation of DMU [59,61].

4.2. Cross-Dehydrogenative Coupling with Arenes and Heteroarenes at C5 Position

Cross-couplings which required double C-H activation both at uracil and arene substrates have also been developed. Thus, Do and Daugulis reported a highly regioselective CuI/phenanthroline-catalyzed oxidative direct arylation of DMU **65** with arenes (e.g., 3-methylanisole **72**) to give C6-arylated uracil **73** as a sole product in 61% yield (130 °C, 48 h) (Scheme 26) [67]. The protocol used iodine as a terminal oxidant and required only a small excess of the arene (1.5–3.0 equiv.). This *overall* cross-dehydrogenative arylation was believed to proceed by *in situ* iodination of one of the coupling components (e.g. 3-methylanisole **72** to give 4-iodo-3-methylanisole **72'**) followed by Cu-catalyzed direct arylation at the most acidic bond in the DMU substrate (C6-H). Recently, an efficient (~70%–80%) and regioselective C-6 arylation of DMU **65** with arylboronic acids in the presence of Pd(OAc)$_2$ and ligand (1,10-phenanthroline) at 90 °C for 16 h has been also developed [68]. The coupling failed, however, when unsubstituted uracil or 2',3',5'-tri-*O*-acetyl-3-*N*-methyluridine was used as substrate or when the heteroaromatic boronic acids were employed [68].

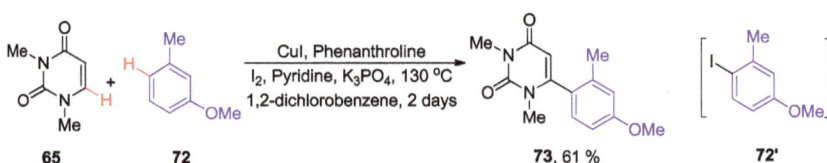

Scheme 26. Sequential iodination/direct C6-H arylation of DMU **65** with arenes.

The Pd-catalyzed cross-dehydrogenative coupling (CDC) of DMU **65** with benzene or xylenes **74** in the presence of PivOH and AgOAc at reflux was found to produce 6-aryluracil analogue **67** as the major product. Minor quantities of 5-arylated counterpart **66** and uracil dimeric byproducts were also formed (Scheme 27a) [59,69]. It is believed that the 6-arylation occurred via CMD process involving Pd^{II}(L)(OPiv) species. Deprotonation occurred at the more acid hydrogen at C6 of uracil ring, followed by a second CMD process to give the 6-arylated uracil product **67**. Interestingly, reaction of DMU **65** with mesitylene led only to the formation of 5,5- and 5,6-DMU dimers. Also, application of this protocol to 2′,3′,5′-tri-*O*-acetyl-3-*N*-benzyluridine **75** gave only the C5-C5 dimer **76** in 43% yield (Scheme 27b) [69].

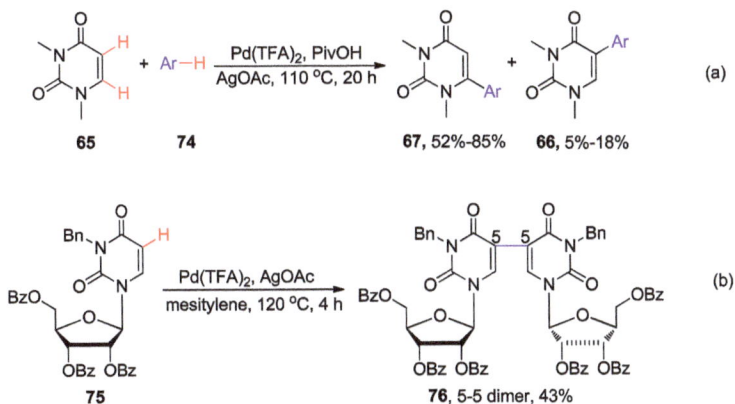

Scheme 27. Pd-catalyzed cross-dehydrogenative coupling of uracils and uracil nucleosides.

The Pd-catalyzed cross-dehydrogenative heteroarylation between 1,3-dialkyluracils **77** and pyridine-*N*-oxides **78** (3 equiv.) substrates in the presence of Ag$_2$CO$_3$ at 140 °C for 12 h gave 5-(2-pyridyl-*N*-oxide)uracils **79** in good-to-high yields (Scheme 28) [70]. As expected, the 3-substituted pyridine-*N*-oxides gave products **83** with excellent regioselectivity at the less bulky site. The coupling, however, was not compatible with either 1,3-dibenzoyluracil or unprotected uracil substrates. Reduction of *N*-oxides **79** with PCl$_3$ in toluene yielded the corresponding 5-uracil derivatives substituted with a 2-pyridyl ring. The electrophilic palladation at the C5 of uracil and the coordination of the palladium atom to *N*-oxide was believed to control the regioselectivity (at C5 of the uracil ring and C2 of pyridine oxide) of these double C-H activation cross-couplings.

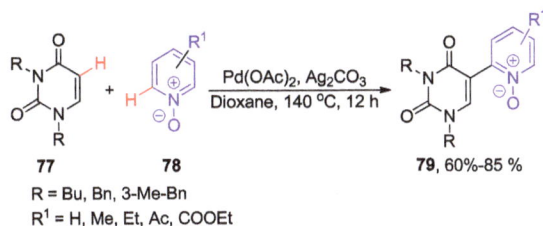

Scheme 28. Synthesis of 5-(2-pyridyl-*N*-oxide)uracils **79** via cross-dehydrogenative arylation.

4.3. Cross-Dehydrogenative Alkenylation at C5 Position

The discovery of the potent antiviral activity of *E*-5-(2-bromovinyl)-2′-deoxyuridine (**1**, BVDU) led to the exploration of synthetic routes based on the oxidative coupling of uridine nucleosides with alkenes. Such routes avoided the use of mercury that was central to Walker's synthetic approach involving the condensation of 5-mercurated 2'-deoxyuridine with methyl acrylate and radical decarboxylation-bromination sequence [71]. They also seem advantageous to other coupling protocols which employ coupling between 5-halouracil nucleosides and organometallics or methyl acrylate [10,72]. In 1987, Itahara reported the oxidative coupling of uracil nucleosides **80** with maleimides **81** which gave 5-substituted coupling products of type **82** (Scheme 29) [73]. Using stoichiometric amounts of Pd(OAc)$_2$ was however necessary. The yields for uridine and 2'-deoxyuridine substrates were lower (4%–22%) than those of the DMU substrate (~20%–50%). The yields for DMU substrate slightly increased in the presence reoxidants such as AgOAc, Na$_2$S$_2$O$_8$ or Cu(OAc)$_2$.

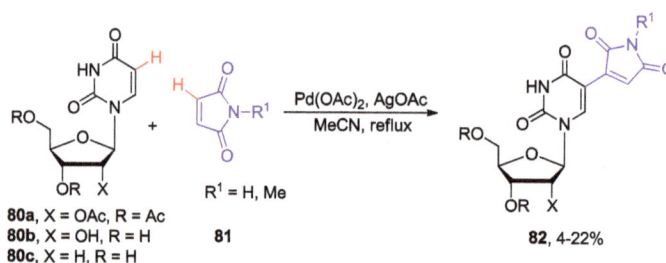

80a, X = OAc, R = Ac
80b, X = OH, R = H
80c, X = H, R = H

R^1 = H, Me

81

82, 4-22%

Scheme 29. Pd-mediated oxidative coupling of uridine and 2'-deoxyuridine with maleimides.

Also in 1987, Hirota and coworkers reported the oxidative coupling of uridine **80b** and 2'-deoxyuridine **80c** with methyl acrylate or styrene using either stoichiometric amounts of Pd(OAc)$_2$ in MeCN at ambient temperature or catalytic loading of Pd(OAc)$_2$ in the presence of *tert*-butyl perbenzoate as the reoxidant to generate 5-vinyl uridine analogues **84** (Scheme 30) [74]. The couplings proceeded stereoselectively to give the *trans* isomers. The reaction conditions were compatible with unprotected and protected uridine substrates, though, coupling of 2',3'-di-*O*-isopropylideneuridine with methyl acrylate produced also the 5',6-cyclouridine byproduct in 23% yield.

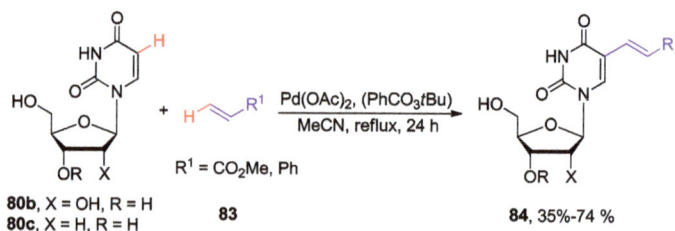

80b, X = OH, R = H
80c, X = H, R = H

R^1 = CO$_2$Me, Ph

83

84, 35%-74 %

Scheme 30. Stoichiometric and catalytic oxidative coupling of uracil nucleosides with methyl acrylate.

Yun and Georg recently reported Pd-catalyzed cross-dehydrogenative coupling of 1,3-disubstituted uracils as well as protected uridine **85** and 2'-deoxyuridine derivatives **88** with *tert*-butyl acrylate in the presence of AgOAc and PivOH in DMF at 60 °C/24 h to give 5-alkenyl products **87** and **89** in 66% and 75% yield, respectively (Scheme 31) [75]. The coupling occurred with the regio- (C5) and stereoselectivity (*E*-isomer). However, 3-*N*-methyl protection at the uracil ring was necessary. The coupling was postulated to occur via the electrophilic palladation pathway [76] at the C5 position of the uracil ring followed by deprotonation with the pivalate anion to give palladated intermediate. Coordination with the alkenes via transmetallation, and subsequent β-elimination provided 5-alkenyluracil derivatives.

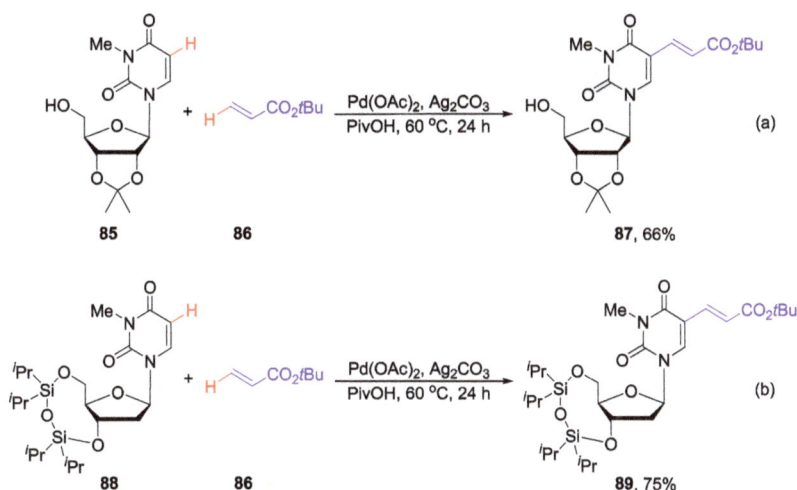

(a)

(b)

Scheme 31. Pd-Catalyzed cross-dehydrogenative alkenylation of uracil nucleosides.

4.4. Miscellaneous Direct C-H Functionalizations

Pd-catalyzed direct C-H acetoxylation at the electron-rich C5 position of uracil nucleosides with PhI(OAc)$_2$ under reasonably mild conditions (60 °C, 3–5 h) have been also developed (Scheme 32) [77]. The acyl protected uridine **80a** and silyl protected 2'-deoxyuridine **80d** were compatible with these conditions to give the corresponding 5-acetoxy products **90** in 55% and 25% yields, respectively. The reaction was proposed to proceed via oxidative electrophilic palladation at the electron rich C5 position to give the 5-palladauracil intermediate followed by oxidation to give Pd(IV) intermediate, which yielded 5-acetoxyluridine via the reductive elimination of Pd(II).

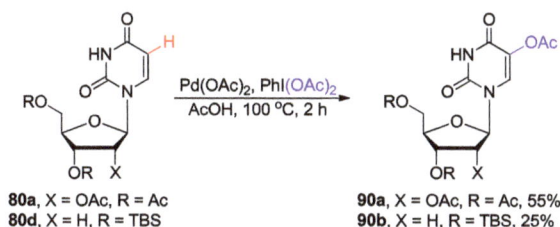

80a, X = OAc, R = Ac
80d, X = H, R = TBS

90a, X = OAc, R = Ac, 55%
90b, X = H, R = TBS, 25%

Scheme 32. Pd-catalyzed acetoxylation of uracil nucleosides.

The Cu-catalyzed intermolecular C5-H amination of DMU **65** with 4-bromoaniline in the presence of [hydroxy(tosyloxy)iodo]mesitylene has been recently developed (Scheme 33) [50]. The regioselectivity of this C5 amination was proposed to be controlled via the formation of iodonium salt intermediate **91**, which is consistent with C5 electrophilic aromatic substitution that is typical for uracil ring. Subsequent Cu-catalyzed amination gave the 5-amino product **92** in 65% overall yield.

Majumdar and coworkers reported the synthesis of pyrrolo[3,2-*d*]pyrimidine derivatives of type **94** by the intramolecular dehydrogenative coupling of the 5-amidouracils **93** via the selective activation of uracil C6-H bond in the presence of Cu(OTf)$_2$ (Scheme 34) [78]. This coupling between C$_{sp2}$-H (in the uracil ring) and C$_{sp3}$-H (in the side chain) bonds was not, however, successful when R^2 = H. The authors suggested that coupling most probably involved single electron transfer (SET) processes and might require a more stable tertiary radical on the side chain to proceed.

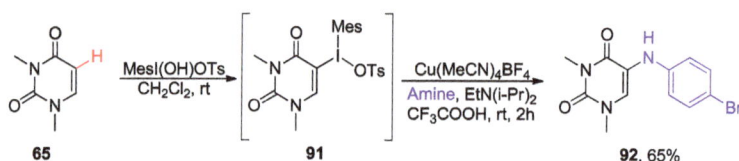

Scheme 33. Cu-catalyzed direct C5-H amination of 1,3-dimethyluracil **65**.

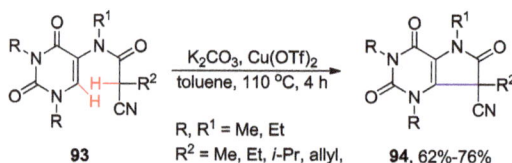

Scheme 34. Intramolecular cross-dehydrogenative cyclization at C6 of uracil ring.

Recently, C5-H trifluoromethylation of DMU **65** by means of electrophilic, nucleophilic or radical "CF$_3$" species in an effort to synthesize 6-aryl-5-(trifluoromethyl)uracils (e.g., **96**) by direct activations of both C5-H and C6-H bonds in consecutive manner has been attempted (Scheme 35) [79]. Thus, reactions of **65** with electrophilic (Umemoto's or Togni's) or nucleophilic (Rupert's) reagents in combination with Pd or Cu catalyst either failed or led to the formation of 5,5- or 5,6-dimeric products or 5-CF$_3$ product in low yields. However, radical trifluoromethylation of **65** with sodium trifluoromethanesulfinate in the presence of *tert*-butyl hydroperoxide provided 1,3-dimethyl-5-(trifluoromethyl)uracil **95** in 67% yield [79]. The subsequent direct arylation at C6-H of **95** with 4-iodotoluene in the presence of Pd(OAc)$_2$ and CuI/CsF afforded desired 5,6-disubstituted uracils **96** albeit in low yield (25%); probably because of the electron-withdrawing effect of the CF$_3$ group at the 5 position on the pyrimidine ring. This C6-H arylation was not applicable to other aryl halides and often was accompanied by cleavage of the CF$_3$ group due to hydrolysis followed by decarboxylation, especially when Cs$_2$CO$_3$ was used as the base.

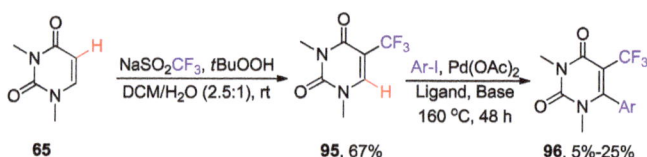

Scheme 35. Synthesis of 6-aryl-5-(trifluoromethyl)uracils by direct activation of C5-H and C6-H bonds.

5. Coupling of 5-Halouracil Nucleosides with Arenes and Heteroarenes

The lack of regioselectivity in direct activation of uracil derivatives (C5-H *vs.* C6-H) during cross-couplings with aryl halides, and fact that coupling conditions are usually unsuitable for unprotected uracils and natural nucleosides, were recently overcome by switching the halide substituents from aryl halides to uracil ring and allowing to react of 5-halouridines with arenes instead. Wnuk and coworkers found that the 5-iodouracil nucleosides **97** coupled with simple arenes or heteroaromatics **98** in the presence of Pd$_2$(dba)$_3$ and TBAF in DMF under milder condition (100 °C/1–2 h) to give the 5-arylated uracil nucleosides **99** in high yields (Scheme 36) [80].

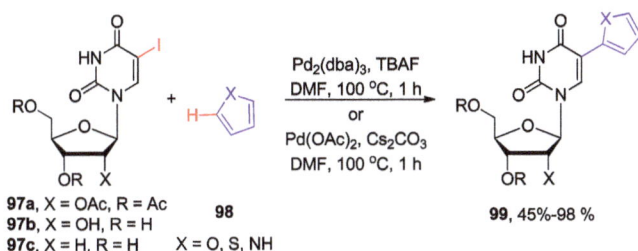

Scheme 36. Pd-catalyzed cross-coupling of 5-halouracil nucleosides with arenes and heteroarenes.

This TBAF-promoted protocol, which proceeded without the necessity of adding ligands and/or additives, worked efficiently with the natural uracils and uracil nucleosides, and was compatible with the stability of the glycosylic bond for 2'-deoxyuridine substrates (e.g., **97c**). The 5-(2-furyl, or 2-thienyl, or 2-pyrrolyl)uridine derivatives **99**, that are important RNA and DNA fluorescent probes [15–18], were synthesized in up to 98% yields without the necessity of using organometallic substrates. The arylation proceeded also when TBAF was replaced with Cs$_2$CO$_3$ base with or without the presence of PivOH [80]. The analogous coupling of 5-iodocytidines with arenes failed to afford 5-arylated products.

The fact that 3-*N*-methyl-5-iodouracil substrates, which lack the ability to tautomerize to the enol form, did not undergo these couplings with arenes indicates that the C4-alkoxide (enol form of uracil) may participate in the intramolecular processes of hydrogen abstraction as depicted in Figure 3. The mechanism pathway, after the initial oxidative-addition of palladium to C5-halogen bond, might involve electrophilic aromatic palladation assisted by C4-alkoxide (e.g., **A**) or direct proton abstraction assisted by C4-alkoxide (e.g., **B**) [80]. This would be in agreement with earlier finding [7,81] that direct arylation is facilitated by a Pd-coordinated carboxylate group, which also can assist in intramolecular proton abstraction.

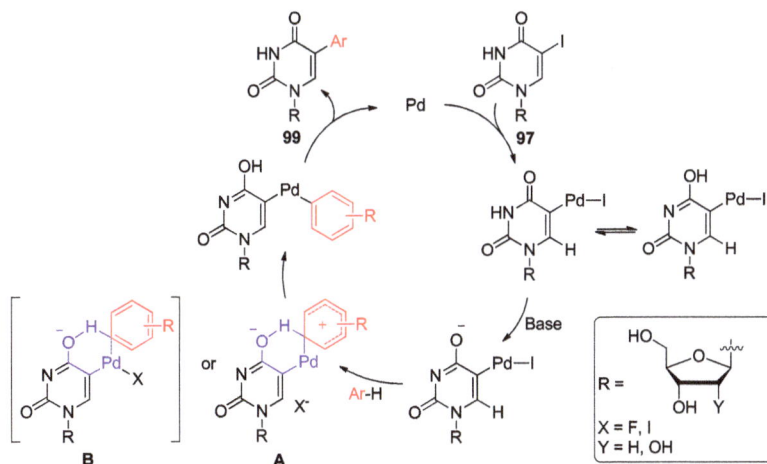

Figure 3. The plausible mechanism for the Pd-catalyzed arylation of 5-iodouridine with arenes.

6. Conclusions

The use of the Pd(cat.)/Cu(stoich) system in the presence of bases such as Cs$_2$CO$_3$ and/or piperidine in DMF effects direct regioselective arylation of purine nucleosides at the 8 position. Protected and deprotected adenosine, 2'-deoxyadenosine as well as inosine or guanosine derivatives

Molecules **2015**, *20*, 4874–4901

coupled efficiently under these conditions with aryl halides to give access to 8-arylated products in up to 99% yields. The C8-H functionalization process is proposed to occur via 8-cupriopurine intermediates or *N*-heterocyclic carbene like cuprates, which subsequently can undergo a standard Pd(0)-catalytic cycle for cross-coupling with aryl halides. The cross-dehydrogenative arylation protocols which involve C8-H bond activation have been also developed for the synthesis of purine fused ring systems. Intramolecular N1-purinyl nitrogen directed C-H bond activation was utilized for direct *ortho* modification (arylation, amination or acetoxylation) of the aryl rings at C6 position of purine nucleosides.

In the case of pyrimidine nucleosides, direct arylation protocols have only been developed for uracil bases, which, in turn, usually require protection at the 1-*N* and 3-*N* positions. The biggest challenge how to overcome regioselective C-H activation at the C5 or C6 position of uracil ring have been accomplished by employing different catalysts and ligands. For the cross-coupling with aryl halides, it was found that Pd-based catalytic systems were effective to promote C5 arylation, whereas Pd/Cu or Cu-mediated systems affected C6 arylation with good selectivity. The C5 arylation is proposed to proceed via the electrophilic or concerted metalation-deprotonation mechanisms, while the C6 arylation likely to occur via a cuprate intermediates. Pd-catalyzed cross-dehydrogenative coupling of uracil nucleosides with alkenes or arenes, in the presence of oxidants, were developed to give convenient access to 5-alkenyl/aryl derivatives. The 5-acetoxy and 5-amino derivatives were also synthesized by direct activation of C5-H bond in uracil ring. No examples of couplings involving direct C-H activation of cytosine ring have been reported.

The transition-metal catalyzed syntheses of the modified nucleobases by direct C-H bond activations have been substantially improved in the past decade. However, despite improved coupling efficiency and availability of milder reaction conditions applicable to the less stable deoxynucleosides, application of the direct C-H functionalization approaches to nucleotides and/or short (deoxy)oligonucleotides fragments still will require developing the conditions compatible both with solvent requirements for water-soluble nucleotides and stability of phosphate esters.

Acknowledgments: YL would like to thank Florida International University Graduate School (FIU) for the Dissertation Year Fellowship.

Author Contributions: Study concept and design (YL, SFW). Drafting of the manuscript and preparation of the schemes (YL). Critical review of the manuscript (YL, SFW).

Conflicts of Interest: The authors declare no conflict of interest.

References

1. Meijere, A.D.; Diederich, F. *Metal-Catalyzed Cross-Coupling Reactions*; Wiley-VCH Verlag GmbH: Weinheim, Germany, 2008.
2. Kambe, N.; Iwasaki, T.; Terao, J. Pd-catalyzed cross-coupling reactions of alkyl halides. *Chem. Soc. Rev.* **2011**, *40*, 4937–4947. [CrossRef] [PubMed]
3. The Nobel Prize in Chemistry Announcement. Available online: http://www.nobelprize.org/nobel_prizes/chemistry/laureates/2010/# (accessed on 12 March 2015).
4. Ackermann, L.; Vicente, R.; Kapdi, A.R. Transition-metal-catalyzed direct arylation of (hetero)arenes by C-H bond cleavage. *Angew. Chem. Int. Ed. Engl.* **2009**, *48*, 9792–9826. [CrossRef] [PubMed]
5. Fagnou, K. Mechanistic considerations in the development and use of azine, diazine and azole *N*-oxides in palladium-catalyzed direct arylation. *Top. Curr. Chem.* **2010**, *292*, 35–56. [PubMed]
6. Su, Y.X.; Sun, L.P. Recent progress towards transition-metal-catalyzed direct arylation of heteroarenes. *Mini Rev. Org. Chem.* **2012**, *9*, 87–117. [CrossRef]
7. Lafrance, M.; Fagnou, K. Palladium-catalyzed benzene arylation: incorporation of catalytic pivalic acid as a proton shuttle and a key element in catalyst design. *J. Am. Chem. Soc.* **2006**, *128*, 16496–16497. [CrossRef] [PubMed]

8. Liégault, B.; Lapointe, D.; Caron, L.; Vlassova, A.; Fagnou, K. Establishment of broadly applicable reaction conditions for the Palladium-catalyzed direct arylation of heteroatom-containing aromatic compounds. *J. Org. Chem.* **2009**, *74*, 1826–1834. [CrossRef] [PubMed]

9. Joo, J.M.; Touré, B.B.; Sames, D. C−H Bonds as ubiquitous functionality: A general approach to complex arylated imidazoles via regioselective sequential arylation of all three C−H bonds and regioselective *N*-alkylation enabled by SEM-group transposition. *J. Org. Chem.* **2010**, *75*, 4911–4920. [CrossRef] [PubMed]

10. Agrofoglio, L.A.; Gillaizeau, I.; Saito, Y. Palladium-assisted routes to nucleosides. *Chem. Rev.* **2003**, *103*, 1875–1916. [CrossRef] [PubMed]

11. De Ornellas, S.; Williams, T.J.; Baumann, C.G.; Fairlamb, I.J.S. (Eds.) Catalytic C-H/C-X bond functionalisation of nucleosides, nucleotides, nucleic acids, amino acids, peptides and proteins. In *C-H and C-X Bond Functionalization: Transition Metal Mediation*; The Royal Society of Chemistry: London, UK, 2013.

12. Shaughnessy, K. Palladium-catalyzed functionalization of unprotected nucleosides in aqueous media. *Molecules* **2015**, in press.

13. De Clercq, E.; Descamps, J.; de Somer, P.; Barr, P.J.; Jones, A.S.; Walker, R.T. (*E*)-5-(2-Bromovinyl)-2′-deoxyuridine: A potent and selective anti-herpes agent. *Proc. Natl. Acad. Sci. USA* **1979**, *76*, 2947–2951.

14. McGuigan, C.; Barucki, H.; Blewett, S.; Carangio, A.; Erichsen, J.T.; Andrei, G.; Snoeck, R.; de Clercq, E.; Balzarini, J. Highly potent and selective inhibition of varicella-zoster virus by bicyclic furopyrimidine nucleosides bearing an aryl side chain. *J. Med. Chem.* **2000**, *43*, 4993–4997. [CrossRef] [PubMed]

15. Greco, N.J.; Tor, Y. Simple fluorescent pyrimidine analogues detect the presence of DNA abasic sites. *J. Am. Chem. Soc.* **2005**, *127*, 10784–10785. [CrossRef] [PubMed]

16. Srivatsan, S.G.; Tor, Y. Using an emissive uridine analogue for assembling fluorescent HIV-1 TAR constructs. *Tetrahedron* **2007**, *63*, 3601–3607. [CrossRef] [PubMed]

17. Noé, M.S.; Ríos, A.C.; Tor, Y. Design, synthesis, and spectroscopic properties of extended and fused pyrrolo-dC and pyrrolo-C analogs. *Org. Lett.* **2012**, *14*, 3150–3153. [CrossRef] [PubMed]

18. Wicke, L.; Engels, J.W. Postsynthetic on column RNA labeling via Stille coupling. *Bioconjugate Chem.* **2012**, *23*, 627–642. [CrossRef]

19. Valis, L.; Mayer-Enthart, E.; Wagenknecht, H.A. 8-(Pyren-1-yl)-2′-deoxyguanosine as an optical probe for DNA hybridization and for charge transfer with small peptides. *Bioorg. Med. Chem. Lett.* **2006**, *16*, 3184–3187. [CrossRef] [PubMed]

20. Wanninger-Weiß, C.; Valis, L.; Wagenknecht, H.A. Pyrene-modified guanosine as fluorescent probe for DNA modulated by charge transfer. *Bioorg. Med. Chem.* **2008**, *16*, 100–106. [CrossRef] [PubMed]

21. Manfredini, S.; Baraldi, P.G.; Bazzanini, R.; Marangoni, M.; Simoni, D.; Balzarini, J.; De Clercq, E. Synthesis and cytotoxic activity of 6-vinyl- and 6-ethynyluridine and 8-vinyl- and 8-ethynyladenosine. *J. Med. Chem.* **1995**, *38*, 199–203. [CrossRef] [PubMed]

22. He, G.X.; Krawczyk, S.H.; Swaminathan, S.; Shea, R.G.; Dougherty, J.P.; Terhorst, T.; Law, V.S.; Griffin, L.C.; Coutré, S.; Bischofberger, N. *N*2- and C8-Substituted oligodeoxynucleotides with enhanced thrombin inhibitory activity *in vitro* and *in vivo*. *J. Med. Chem.* **1998**, *41*, 2234–2242. [CrossRef] [PubMed]

23. Cerna, I.; Pohl, R.; Hocek, M. The first direct C-H arylation of purine nucleosides. *Chem. Commun.* **2007**, 4729–4730. [CrossRef]

24. Čerňa, I.; Pohl, R.; Klepetářová, B.; Hocek, M. Direct C−H arylation of purines: Development of methodology and its use in regioselective synthesis of 2,6,8-trisubstituted purines. *Org. Lett.* **2006**, *8*, 5389–5392. [CrossRef] [PubMed]

25. Čerňa, I.; Pohl, R.; Klepetářová, B.; Hocek, M. Synthesis of 6,8,9-tri- and 2,6,8,9-tetrasubstituted purines by a combination of the Suzuki cross-coupling, *N*-arylation, and direct C−H arylation reactions. *J. Org. Chem.* **2008**, *73*, 9048–9054. [CrossRef] [PubMed]

26. Storr, T.E.; Baumann, C.G.; Thatcher, R.J.; De Ornellas, S.; Whitwood, A.C.; Fairlamb, I.J. S. Pd(0)/Cu(I)-mediated direct arylation of 2′-deoxyadenosines: Mechanistic role of Cu(I) and reactivity comparisons with related purine nucleosides. *J. Org. Chem.* **2009**, *74*, 5810–5821. [CrossRef] [PubMed]

27. Storr, T.E.; Firth, A.G.; Wilson, K.; Darley, K.; Baumann, C.G.; Fairlamb, I.J. S. Site-selective direct arylation of unprotected adenine nucleosides mediated by palladium and copper: Insights into the reaction mechanism. *Tetrahedron* **2008**, *64*, 6125–6137. [CrossRef]

28. Baumann, C.G.; De Ornellas, S.; Reeds, J.P.; Storr, T.E.; Williams, T.J.; Fairlamb, I.J.S. Formation and propagation of well-defined Pd nanoparticles (PdNPs) during C–H bond functionalization of heteroarenes: are nanoparticles a moribund form of Pd or an active catalytic species? *Tetrahedron* **2014**, *70*, 6174–6187. [CrossRef]

29. Storr, T.E.; Strohmeier, J.A.; Baumann, C.G.; Fairlamb, I.J. S. A sequential direct arylation/Suzuki-Miyaura cross-coupling transformation of unprotected 2'-deoxyadenosine affords a novel class of fluorescent analogues. *Chem. Commun.* **2010**, *46*, 6470–6472. [CrossRef]

30. Ornellas, S.D.; Storr, T.E.; Williams, T.J.; Baumann, C.G.; Fairlamb, I.J. S. Direct C-H/C-X coupling methodologies mediated by Pd/Cu or Cu: An examination of the synthetic applications and mechanistic findings. *Curr. Org. Synth.* **2011**, *8*, 79–101. [CrossRef]

31. Vaňková, B.; Krchňák, V.; Soural, M.; Hlaváč, J. Direct C–H arylation of purine on solid phase and its use for chemical libraries synthesis. *ACS Comb. Sci.* **2011**, *13*, 496–500. [CrossRef] [PubMed]

32. Sahnoun, S.; Messaoudi, S.; Peyrat, J.F.; Brion, J.D.; Alami, M. Microwave-assisted Pd(OH)$_2$-catalyzed direct C−H arylation of free-(NH$_2$) adenines with aryl halides. *Tetrahedron Lett.* **2008**, *49*, 7279–7283. [CrossRef]

33. Sahnoun, S.; Messaoudi, S.; Brion, J.D.; Alami, M. A site selective C-H arylation of free-(NH$_2$) adenines with aryl chlorides: Application to the synthesis of 6,8-disubstituted adenines. *Org. Biomol. Chem.* **2009**, *7*, 4271–4278. [CrossRef] [PubMed]

34. Ngassa, F.N.; DeKorver, K.A.; Melistas, T.S.; Yeh, E.A.H.; Lakshman, M.K. Pd−Xantphos-Catalyzed Direct Arylation of Nucleosides. *Org. Lett.* **2006**, *8*, 4613–4616. [CrossRef] [PubMed]

35. Bellina, F.; Cauteruccio, S.; Rossi, R. Palladium- and copper-mediated direct C-2 arylation of azoles—including free (NH)-imidazole, -benzimidazole and -indole—Under base-free and ligandless conditions. *Eur. J. Org. Chem.* **2006**, *2006*, 1379–1382. [CrossRef]

36. Western, E.C.; Shaughnessy, K.H. Inhibitory effects of the guanine moiety on Suzuki couplings of unprotected halonucleosides in aqueous media. *J. Org. Chem.* **2005**, *70*, 6378–6388. [CrossRef] [PubMed]

37. Gigante, A.; Priego, E.M.; Sánchez-Carrasco, P.; Ruiz-Pérez, L.M.; Vande Voorde, J.; Camarasa, M.J.; Balzarini, J.; González-Pacanowska, D.; Pérez-Pérez, M.J. Microwave-assisted synthesis of C-8 aryl and heteroaryl inosines and determination of their inhibitory activities against Plasmodium falciparum purine nucleoside phosphorylase. *Eur. J. Med. Chem.* **2014**, *82*, 459–465. [CrossRef] [PubMed]

38. Čerňa, I.; Pohl, R.; Klepetářová, B.; Hocek, M. Intramolecular direct C-H arylation approach to fused purines. Synthesis of purino[8,9-f]phenanthridines and 5,6-dihydropurino[8,9-a]isoquinolines. *J. Org. Chem.* **2010**, *75*, 2302–2308. [CrossRef] [PubMed]

39. Campeau, L.C.; Parisien, M.; Jean, A.; Fagnou, K. Catalytic direct arylation with aryl chlorides, bromides, and iodides: Intramolecular studies leading to new intermolecular reactions. *J. Am. Chem. Soc.* **2005**, *128*, 581–590. [CrossRef]

40. Iaroshenko, V.O.; Ostrovskyi, D.; Miliutina, M.; Maalik, A.; Villinger, A.; Tolmachev, A.; Volochnyuk, D.M.; Langer, P. Design and synthesis of polycyclic imidazole-containing N-heterocycles based on C-H activation/cyclization reactions. *Adv. Synth. Catal.* **2012**, *354*, 2495–2503. [CrossRef]

41. Barbero, N.; SanMartin, R.; Dominguez, E. Ligand-free copper(I)-catalysed intramolecular direct C-H functionalization of azoles. *Org. Biomol. Chem.* **2010**, *8*, 841–845. [CrossRef] [PubMed]

42. Meng, G.; Niu, H.Y.; Qu, G.R.; Fossey, J.S.; Li, J.P.; Guo, H.M. Synthesis of fused N-heterocycles via tandem C-H activation. *Chem. Commun.* **2012**, *48*, 9601–9603. [CrossRef]

43. Xi, P.; Yang, F.; Qin, S.; Zhao, D.; Lan, J.; Gao, G.; Hu, C.; You, J. Palladium(II)-catalyzed oxidative C−H/C−H cross-coupling of heteroarenes. *J. Am. Chem. Soc.* **2010**, *132*, 1822–1824. [CrossRef] [PubMed]

44. Stuart, D.R.; Fagnou, K. The catalytic cross-coupling of unactivated arenes. *Science* **2007**, *316*, 1172–1175. [CrossRef] [PubMed]

45. Vabre, R.; Chevot, F.; Legraverend, M.; Piguel, S. Microwave-Assisted Pd/Cu-Catalyzed C-8 Direct Alkenylation of Purines and Related Azoles: An Alternative Access to 6,8,9-Trisubstituted Purines. *J. Org. Chem.* **2011**, *76*, 9542–9547. [CrossRef] [PubMed]

46. Sahnoun, S.; Messaoudi, S.; Brion, J.D.; Alami, M. Pd/Cu-catalyzed direct alkenylation of azole heterocycles with alkenyl halides. *Eur. J. Org. Chem.* **2010**, *2010*, 6097–6102. [CrossRef]

47. Lagisetty, P.; Zhang, L.; Lakshman, M.K. Simple methodology for Heck arylation at C-8 of adenine nucleosides. *Adv. Synth. Catal.* **2008**, *350*, 602–608. [CrossRef]

48. Klecka, M.; Pohl, R.; Klepetarova, B.; Hocek, M. Direct C-H borylation and C-H arylation of pyrrolo[2,3-d]pyrimidines: synthesis of 6,8-disubstituted 7-deazapurines. *Org. Biomol. Chem.* **2009**, *7*, 866–868. [CrossRef] [PubMed]

49. Sabat, N.; Klecka, M.; Slavetinska, L.; Klepetarova, B.; Hocek, M. Direct C-H amination and C-H chloroamination of 7-deazapurines. *RSC Advances* **2014**, *4*, 62140–62143.

50. Sokolovs, I.; Lubriks, D.; Suna, E. Copper-catalyzed intermolecular C–H amination of (hetero)arenes via transient unsymmetrical λ3-iodanes. *J. Am. Chem. Soc.* **2014**, *136*, 6920–6928. [CrossRef] [PubMed]

51. Klecka, M.; Pohl, R.; Cejka, J.; Hocek, M. Direct C-H sulfenylation of purines and deazapurines. *Org. Biomol. Chem.* **2013**, *11*, 5189–5193. [CrossRef] [PubMed]

52. Qu, G.R.; Liang, L.; Niu, H.Y.; Rao, W.H.; Guo, H.M.; Fossey, J.S. Copper-catalyzed synthesis of purine-fused polycyclics. *Org. Lett.* **2012**, *14*, 4494–4497. [CrossRef] [PubMed]

53. Guo, H.M.; Jiang, L.L.; Niu, H.Y.; Rao, W.H.; Liang, L.; Mao, R.Z.; Li, D.Y.; Qu, G.R. Pd(II)-catalyzed *ortho* arylation of 6-arylpurines with aryl iodides via purine-directed C–H activation: A new strategy for modification of 6-arylpurine derivatives. *Org. Lett.* **2011**, *13*, 2008–2011. [CrossRef] [PubMed]

54. Lakshman, M.K.; Deb, A.C.; Chamala, R.R.; Pradhan, P.; Pratap, R. Direct arylation of 6-phenylpurine and 6-arylpurine nucleosides by Ruthenium-catalyzed C–H bond activation. *Angew. Chem. Int. Ed.* **2011**, *50*, 11400–11404. [CrossRef]

55. Chamala, R.R.; Parrish, D.; Pradhan, P.; Lakshman, M.K. Purinyl N1-directed aromatic C–H oxidation in 6-arylpurines and 6-arylpurine nucleosides. *J. Org. Chem.* **2013**, *78*, 7423–7435. [CrossRef] [PubMed]

56. Kim, J.Y.; Park, S.H.; Ryu, J.; Cho, S.H.; Kim, S.H.; Chang, S. Rhodium-catalyzed intermolecular amidation of arenes with sulfonyl azides via chelation-assisted C–H bond activation. *J. Am. Chem. Soc.* **2012**, *134*, 9110–9113. [CrossRef] [PubMed]

57. Čerňová, M.; Pohl, R.; Hocek, M. Switching the regioselectivity of direct C–H arylation of 1,3-dimethyluracil. *Eur. J. Org. Chem.* **2009**, *22*, 3698–3701. [CrossRef]

58. Čerňová, M.; Čerňá, I.; Pohl, R.; Hocek, M. Regioselective direct C–H arylations of protected uracils. Synthesis of 5- and 6-aryluracil bases. *J. Org. Chem.* **2011**, *76*, 5309–5319. [CrossRef] [PubMed]

59. Kim, K.H.; Lee, H.S.; Kim, J.N. Palladium-catalyzed direct 5-arylation of 1,3-dimethyluracil with aryl bromides: An electrophilic metalation–deprotonation with electrophilic arylpalladium intermediate. *Tetrahedron Lett.* **2011**, *52*, 6228–6233. [CrossRef]

60. Lee, H.S.; Kim, K.H.; Kim, S.H.; Kim, J.N. Palladium-catalyzed synthesis of benzo[c]pyrimido[1,6-a]azepine scaffold from Morita–Baylis–Hillman adducts: Intramolecular 6-arylation of uracil nucleus. *Tetrahedron Lett.* **2012**, *53*, 497–501. [CrossRef]

61. Cheng, C.; Shih, Y.C.; Chen, H.T.; Chien, T.C. Regioselective arylation of uracil and 4-pyridone derivatives via copper(I) bromide mediated C–H bond activation. *Tetrahedron* **2013**, *69*, 1387–1396. [CrossRef]

62. Gorelsky, S.I.; Lapointe, D.; Fagnou, K. Analysis of the palladium-catalyzed (aromatic) C–H bond metalation–deprotonation mechanism spanning the entire spectrum of arenes. *J. Org. Chem.* **2011**, *77*, 658–668. [CrossRef] [PubMed]

63. Bellina, F.; Cauteruccio, S.; Rossi, R. Efficient and practical synthesis of 4(5)-aryl-1H-imidazoles and 2,4(5)-diaryl-1H-imidazoles via highly selective palladium-catalyzed arylation reactions. *J. Org. Chem.* **2007**, *72*, 8543–8546. [CrossRef] [PubMed]

64. Sievers, A.; Wolfenden, R. Equilibrium of formation of the 6-carbanion of UMP, a potential intermediate in the action of OMP decarboxylase. *J. Am. Chem. Soc.* **2002**, *124*, 13986–13987. [CrossRef] [PubMed]

65. Yeoh, F.Y.; Cuasito, R.R.; Capule, C.C.; Wong, F.M.; Wu, W. Carbanions from decarboxylation of orotate analogs: Stability and mechanistic implications. *Bioorg. Chem.* **2007**, *35*, 338–343. [CrossRef] [PubMed]

66. Amyes, T.L.; Wood, B.M.; Chan, K.; Gerlt, J.A.; Richard, J.P. Formation and stability of a vinyl carbanion at the active site of orotidine 5'-monophosphate decarboxylase: pKa of the C-6 proton of enzyme-bound UMP. *J. Am. Chem. Soc.* **2008**, *130*, 1574–1575. [CrossRef] [PubMed]

67. Do, H.Q.; Daugulis, O. A general method for Copper-catalyzed arene cross-dimerization. *J. Am. Chem. Soc.* **2011**, *133*, 13577–13586. [CrossRef] [PubMed]

68. Mondal, B.; Hazra, S.; Roy, B. Pd(II)-catalyzed regioselective direct arylation of uracil via oxidative Heck reaction using arylboronic acids. *Tetrahedron Lett.* **2014**, *55*, 1077–1081. [CrossRef]

69. Kim, K.H.; Lee, H.S.; Kim, S.H.; Kim, J.N. Palladium(II)-catalyzed oxidative homo-coupling of 1,3-dimethyluracil derivatives. *Tetrahedron Lett.* **2012**, *53*, 1323–1327. [CrossRef]

23

70. Kianmehr, E.; Rezaeefard, M.; Rezazadeh Khalkhali, M.; Khan, K.M. Pd-catalyzed dehydrogenative cross-coupling of pyridine-*N*-oxides with uracils. *RSC Adv.* **2014**, *4*, 13764–13767. [CrossRef]

71. Jones, A.S.; Verhelst, G.; Walker, R.T. The synthesis of the potent anti-herpes virus agent, *E*-5-(2-bromovinyl)-2′-deoxyuridine and related compounds. *Tetrahedron Lett.* **1979**, *20*, 4415–4418. [CrossRef]

72. Ashwell, M.; Jones, A.S.; Kumar, A.; Sayers, J.R.; Walker, R.T.; Sakuma, T.; de Clercq, E. The synthesis and antiviral properties of (*E*)-5-(2-bromovinyl)-2′-deoxyuridine-related compounds. *Tetrahedron* **1987**, *43*, 4601–4608. [CrossRef]

73. Itahara, T. Oxidative coupling of uracil derivatives with maleimides by Palladium acetate. *Chem. Lett.* **1986**, *15*, 239–242. [CrossRef]

74. Hirota, K.; Isobe, Y.; Kitade, Y.; Maki, Y. A simple synthesis of 5-(1-alkenyl)uracil derivatives by Palladium-catalyzed oxidative coupling of uracils with olefins. *Synthesis* **1987**, *1987*, 495–496. [CrossRef]

75. Yu, Y.Y.; Georg, G.I. Dehydrogenative alkenylation of uracils via palladium-catalyzed regioselective C-H activation. *Chem. Commun.* **2013**, *49*, 3694–3696. [CrossRef]

76. Le Bras, J.; Muzart, J. Intermolecular dehydrogenative Heck reactions. *Chem. Rev.* **2011**, *111*, 1170–1214.

77. Lee, H.S.; Kim, S.H.; Kim, J.N. Pd(II)-Catalyzed acetoxylation of uracil via electrophilic palladation. *Bull. Korean Chem. Soc.* **2010**, *31*, 238–241. [CrossRef]

78. Roy, B.; Hazra, S.; Mondal, B.; Majumdar, K.C. Cu(OTf)$_2$-catalyzed dehydrogenative C–H activation under atmospheric oxygen: An expedient approach to pyrrolo[3,2-*d*]pyrimidine derivatives. *Eur. J. Org. Chem.* **2013**, *2013*, 4570–4577. [CrossRef]

79. Čerňová, M.; Pohl, R.; Klepetářová, B.; Hocek, M. C-H trifluoromethylations of 1,3-dimethyluracil and reactivity of the products in C-H arylations. *Heterocycles* **2014**, *89*, 1159–1171. [CrossRef]

80. Liang, Y.; Gloudeman, J.; Wnuk, S.F. Palladium-catalyzed direct arylation of 5-halouracils and 5-halouracil nucleosides with arenes and heteroarenes promoted by TBAF. *J. Org. Chem.* **2014**, *79*, 4094–4103. [CrossRef] [PubMed]

81. Lafrance, M.; Rowley, C.N.; Woo, T.K.; Fagnou, K. Catalytic intermolecular direct arylation of perfluorobenzenes. *J. Am. Chem. Soc.* **2006**, *128*, 8754–8756. [CrossRef] [PubMed]

molecules

MDPI

Review

Palladium-Catalyzed Modification of Unprotected Nucleosides, Nucleotides, and Oligonucleotides

Kevin H. Shaughnessy

Department of Chemistry, The University of Alabama, Box 870336, Tuscaloosa, AL 35487-0336, USA;
kshaughn@ua.edu; Tel.: +1-205-348-4435; Fax: +1-205-348-9104

Academic Editor: Mahesh Lakshman
Received: 24 March 2015; Accepted: 19 May 2015; Published: 22 May 2015

Abstract: Synthetic modification of nucleoside structures provides access to molecules of interest as pharmaceuticals, biochemical probes, and models to study diseases. Covalent modification of the purine and pyrimidine bases is an important strategy for the synthesis of these adducts. Palladium-catalyzed cross-coupling is a powerful method to attach groups to the base heterocycles through the formation of new carbon-carbon and carbon-heteroatom bonds. In this review, approaches to palladium-catalyzed modification of unprotected nucleosides, nucleotides, and oligonucleotides are reviewed. Polar reaction media, such as water or polar aprotic solvents, allow reactions to be performed directly on the hydrophilic nucleosides and nucleotides without the need to use protecting groups. Homogeneous aqueous-phase coupling reactions catalyzed by palladium complexes of water-soluble ligands provide a general approach to the synthesis of modified nucleosides, nucleotides, and oligonucleotides.

Keywords: nucleosides; nucleotides; oligonucleotides; palladium; cross-coupling; aqueous-phase catalysis

1. Introduction

Nucleosides are one of the fundamental building blocks in biochemistry. There has been long-standing interest in the synthesis of non-natural analogs of nucleosides. The modified nucleosides, or their nucleotide or oligonucleotide analogs, have been widely explored as pharmaceutically active compounds [1–4], in the study of carcinogenesis mechanisms [5–7], and to incorporate probe functionality in DNA and RNA [8–11]. Nucleoside derivatives can be made through manipulation of the sugar moiety, replacement of the purine or pyrimidine heterocycle, or through covalent modification of the natural heterocyclic base or their analogs. Functionalization of the canonical nucleoside scaffolds and their close derivatives (Figure 1) provides a general approach to the synthesis of base-modified nucleosides [12–14].

Metal-catalyzed cross-coupling reactions are powerful methods to form C-C and C-heteroatom bonds with aromatic and heteroaromatic structures [15]. These reactions can often be carried out under mild conditions and can be highly functional group tolerant. Palladium-catalyzed cross-coupling reactions have been widely used in the synthesis of base-modified nucleoside derivatives [16–22]. Halogenated nucleoside derivatives can be prepared by standard electrophilic aromatic halogenation reactions. The halonucleosides can be coupled with a wide variety of nucleophilic reagents to form carbon-carbon or carbon-heteroatom bonds (Figure 1). Amine-moieties on nucleoside rings can also be arylated. The rapid development of direct arylation methodologies has resulted in the reports of several methods for the direct arylation of C-H bonds on nucleoside bases.

Canonical nucleosides

Commonly used noncanonical analogs

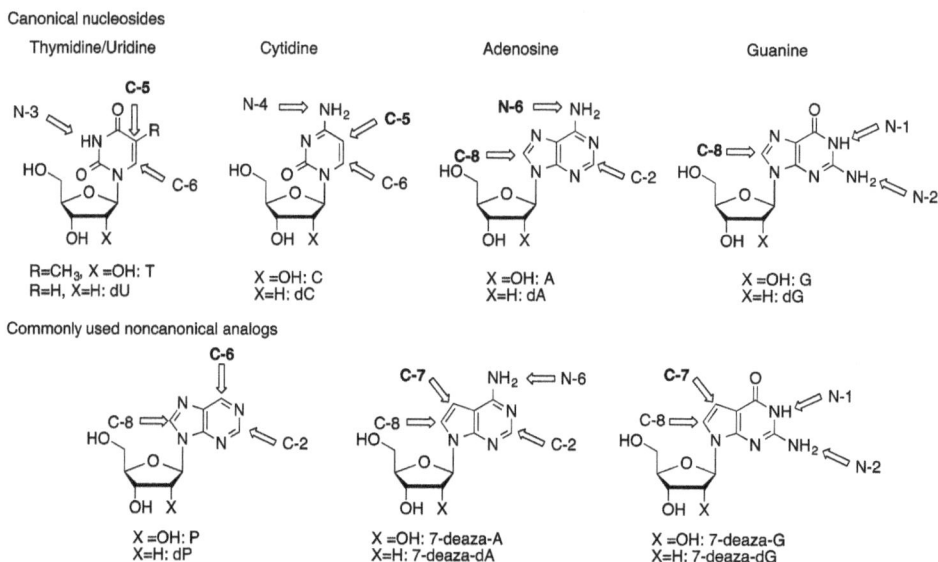

Figure 1. Typical functionalization sites in naturally occurring nucleosides and commonly used unnatural analogs. Most common functionalization sites are bolded.

Although palladium-catalyzed cross-coupling reactions are well precedented for a wide range of substrates, nucleosides present a number of challenges. Heterocycles are often challenging substrates in cross-coupling reactions because of their ability to coordinate to metal catalysts and deactivate them. In addition, heterocycles are often electronically very different from more typical aromatic substrates, which can hinder steps in the catalytic cycle. For example, electron-rich halogenated heterocycles, such as furans, thiophenes, and pyrroles, undergo slow oxidative addition to metal complexes. In addition to the challenges of cross-coupling highly functionalized heterocyclic substrates, the polar nature of nucleoside derivatives often results in them being poorly soluble in typical organic solvents. One common approach is to protect the ribose alcohols as esters or silyl ethers to provide a more hydrophobic substrate (Scheme 1, Path A). This approach introduces two additional synthetic steps, which results in decreased yields and poor atom economy. Protection strategies are typically not effective for the more hydrophilic nucleotides and oligonucleotides.

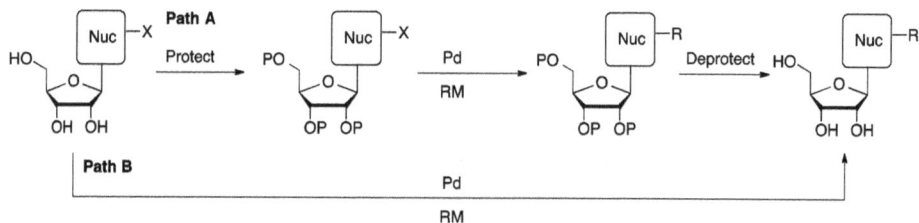

Scheme 1. Synthetic approaches to palladium-catalyzed nucleoside modification.

A more attractive approach would be to use unprotected nucleoside, nucleotide, or oligonucleotide derivatives directly in cross-coupling reactions (Scheme 1, Path B). This can be accomplished using polar organic solvents, such as DMF, in the case of nucleosides. Water represents a more attractive solvent for these types of reactions. Water alone or in combination with organic cosolvents effectively

Molecules **2015**, *20*, 9419–9454

dissolves nucleosides and nucleotide derivatives, which allows reactions to be carried out under homogeneous conditions without the need for protection strategies. In this review, development of methods for functionalization of unprotected nucleosides using palladium-catalyzed cross-coupling in water and polar organic solvents will be reviewed.

2. Cross-Coupling of Unprotected Nucleosides with Ligand-Free Palladium Catalysts

The earliest examples of cross-coupling of unprotected nucleosides were promoted by ligand-free palladium salts in polar organic solvents. Mertes reported a Heck-type coupling of 5-(HgCl)dU with styrene derivatives mediated by stoichiometric Li_2PdCl_4 in methanol to give 5-alkenyl-dU compounds (Scheme 2) [23]. The method was also applied to 5-(HgOAc)dUMP. Direct 5-alkenylation of uridine and 2'-deoxyuridine was achieved in modest yield (35%–57%) using catalytic $Pd(OAc)_2$ with *t*-butyl perbenzoate in acetonitrile (Scheme 3) [24].

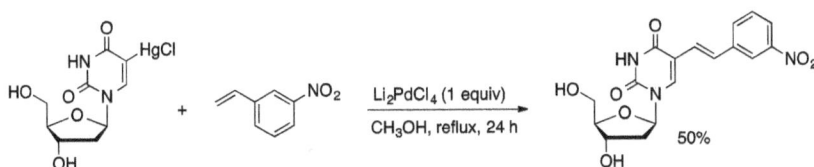

Scheme 2. Pd-mediated coupling of 5-HgCldU and styrenes.

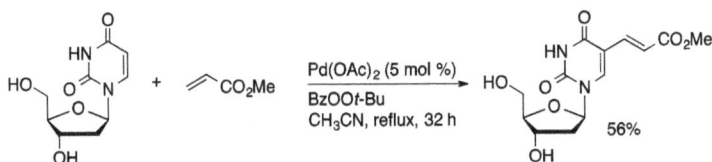

Scheme 3. Oxidative coupling of dU and methyl acrylate.

As palladium-catalyzed cross-coupling reactions developed into widely used synthetic methodologies, significant effort has been devoted to developing highly active and general catalyst systems. Much of this effort has focused on developing supporting ligands that increase the reactivity and stability of the palladium center compared to ligand-free palladium catalysts. Ligand free-systems have received renewed attention in recent years because they avoid the use of phosphines or related ligands, which can often be a larger cost in the reaction than the palladium source [25]. Colloidal or nanoparticle palladium catalysts can often provide good levels of activity with aryl iodides and bromides. Recently, ligand-free palladium catalysts have been applied to coupling reactions of unprotected nucleoside substrates.

2.1. Cross-Coupling with Organosilanes and Organoboranes

Hiyama reported the first examples of a modern cross-coupling of an unprotected halonucleoside with an organometallic reagent. The coupling of 5-IdU with 1-alkenyldifluoromethylsilanes catalyzed by [Pd(allyl)Cl]₂ was accomplished in THF [26]. A mixture of 1-alkenyl and 2-alkenyl products were obtained (Scheme 4).

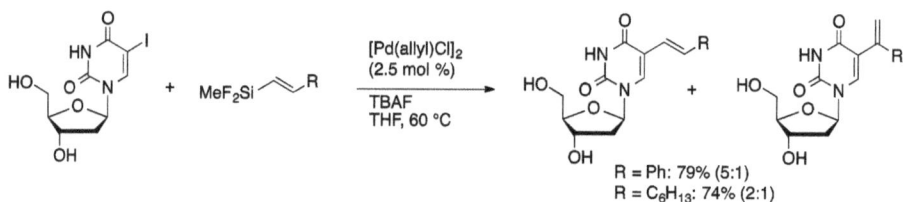

Scheme 4. Hiyama coupling of 5-IdU.

Suzuki coupling of 5-IdU with arylboronic acids has been successfully achieved using Na$_2$PdCl$_4$ without a supporting ligand in water with microwave heating at 100 °C [27,28]. Moderate to excellent yields (33%–85%) were achieved using 0.050–1 mol % palladium with reactions times of one hour or less. The more reactive 6-IU underwent coupling at room temperature with a variety of aryl boronic acids using 10 mol % Na$_2$PdCl$_4$ (Scheme 5) [29]. To date there are no examples of coupling of less reactive 8-bromopurine nucleosides using palladium without supporting ligands. Ligand-supported catalyst systems are generally required to activate electron-rich aryl bromides, such as 8-bromopurines.

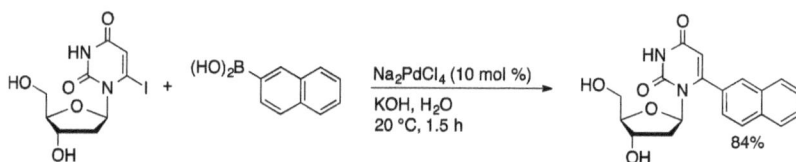

Scheme 5. Ligand-free Suzuki coupling of 6-IdU.

2.2. Heck Coupling

Heck couplings catalyzed by palladium without supporting ligands are well precedented [30]. The Heck coupling of 5-IdU and N-allyl trifluoroacetamide was achieved in 44% yield at room temperature in water using excess palladium (10 equiv) [31]. Use of commercially available 5-IdU was preferable to previous routes that relied on 5-mercurated uridine. Heck coupling of styrene derivatives and 5-IdU occur in good yield using stoichiometric K$_2$PdCl$_4$ in DMF at 80 °C [32]. The reaction can be achieved using catalytic amounts of palladium under similar conditions. Heck coupling of 5-IdU with acrylates was achieved in modest to high yield (35%–90%) using Pd(OAc)$_2$ (10 mol %) at 80 °C in water under microwave irradiation (Scheme 6) [33].

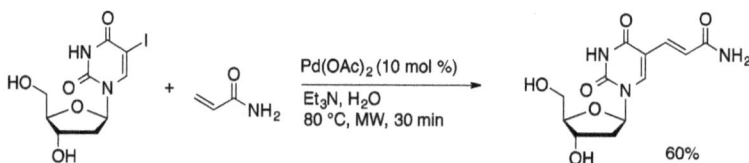

Scheme 6. Heck coupling of 5-IdU with a ligand-free palladium catalyst.

2.3. Direct Arylation via C-H Activation

The direct coupling of heterocycles with arenes has received significant interest in recent years [34, 35]. Electron-rich arenes can be coupled with aryl halides under oxidative conditions with high selectivity for the cross-coupled product. The purine nucleosides are effective coupling partners with aryl iodides. 6-(4-Methoxyphenyl)purine ribonucleoside was coupled with aryl iodides in modest yield

(27%–50% yield) catalyzed by Pd(OAc)$_2$ (5 mol %) using stoichiometric CuI in DMF at 125 °C [36]. The coupling reaction was specific for the electron-rich C8-position. The coupling protocol was also applied to adenosine to give a mixture of C8- and N6-arylated products in 4–5:1 ratios (Scheme 7). Switching the base from piperidine to Cs$_2$CO$_3$ improved the selectivity for C8-arylation of adenosine under otherwise similar conditions [37,38]. The less stable 2'-deoxyadenosine gave significant depurination at 125 °C. Lowering the reaction temperature 80 °C, allowed 8-ArdA derivatives to prepared in 84% yield.

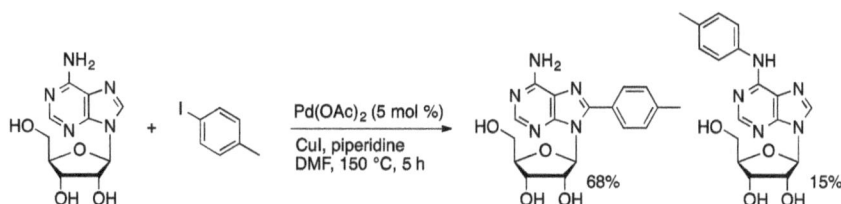

Scheme 7. Oxidative coupling of 4-iodotoluene and adenosine.

Electron-deficient pyrimidines are more difficult to C-H activate than purines. By reversing the nature of the reaction it is possible to coupling halogenated pyrimidine nucleosides with electron-rich heterocycles. Direct coupling of 5-IU and 5-IdU with furan, thiophene, and pyrrole was achieved using 5 mol % Pd(dba)$_2$ in the presence of TBAF in DMF (Scheme 8) [39]. Coupling occurred selectively to give 5-(2-heteroaryl)uridine or 2'-deoxyuridine derivatives.

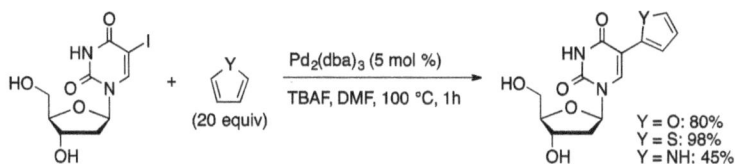

Scheme 8. Coupling of 5-IdU and thiophene.

3. Cross-Coupling of Unprotected Nucleosides with Palladium Complexes of Hydrophobic Ligands

Palladium catalysts without supporting ligands are effective catalysts for cross-coupling of iodopyrimidine nucleosides, but have not been effectively applied to coupling reaction of the less reactive 8-halopurine derivatives. Phosphine and N-heterocyclic carbene ligands provide more active catalyst systems that can effectively activate all classes of aryl halides. Typical supporting ligands, such as triphenylphosphine, are highly hydrophobic. Using these catalyst species for coupling reactions of unprotected nucleosides requires identifying solvent systems that can solubilize both the hydrophobic catalyst and the hydrophilic nucleoside. Aprotic dipolar solvents, such as DMF and NMP, can be effective solvents for these reactions. Alternatively, mixed aqueous-organic solvent systems can be used effectively. Hydrophobic catalysts can exhibit high activity in reactions run with water as the solvent, even when all reagents are hydrophobic [40]. These "on-water" reactions represent a potential future area of exploration for nucleoside coupling reactions.

3.1. Suzuki Couplings

Wagenknecht reported the first example of Suzuki coupling of an unprotected nucleoside [41,42]. 1-Pyrenylboronic acid was coupled with 5-IdU using Pd(PPh$_3$)$_4$ (10 mol %) in a solvent system composed of THF/MeOH/H$_2$O (2:1:2, Scheme 9). The coupled product was isolated in 70% yield. In comparison, coupling of acetyl-protected 5-IdU with 1-pyrenylboronic acid followed by deprotection with NaOMe gave a 55% overall yield. Anthraquinone-labeled uridine was prepared by coupling of 9,10-dimethylanthracen-2-yl pinacolatoborane with 5-IdU catalyzed by Pd(PPh$_3$)$_4$ (5 mol %) in THF/MeOH/H$_2$O followed by oxidation to afford 5-(2-anthraquinonyl)dU in 50% yield over two steps (Scheme 10) [43].

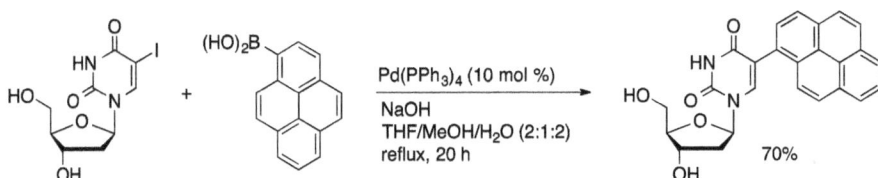

Scheme 9. Suzuki coupling of 5-IdU and 1-pyrenylboronic acid catalyzed by Pd(PPh$_3$)$_4$.

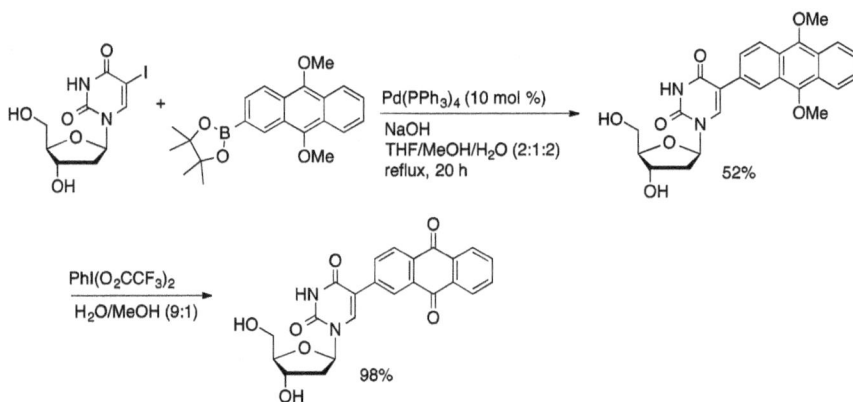

Scheme 10. Synthesis of anthraquinone-labeled uridine.

Other solvent systems have been used in the Pd/PPh$_3$-catalyzed Suzuki coupling of unprotected halonucleosides. The synthesis of a novel spin-labeled uridine derivative began with the coupling of 4-formylphenylboronic acid and 5-IdU catalyzed by Pd(PPh$_3$)$_4$ in methanol/water (Scheme 11) [44]. The 5-aryldU derivative was then converted to 5-(2'-phenyl-4',4',5',5',-tetramethylimidazoline-3'-oxy-1'-oxyl)dU by a sequence of steps. A protected version of the spin-labeled nucleoside was incorporated into oligonucleotides using the phosphoramidite method and used for EPR analysis of DNA structures. Aqueous DMF was used for the Pd(PPh$_3$)$_4$-catalyzed coupling of arylboronic acids and 5-IdU [45]. After monophosphorylation, the 5-aryldUMP derivatives were explored as potential anti-tuberculosis agents.

Despite the insolubility of triphenylphosphine in water, Suzuki coupling of arylboronic acids and 5-IdU have been successfully catalyzed by Pd(OAc)$_2$/PPh$_3$ with water as the only solvent at 120 °C using microwave irradiation [46]. Notably, the PPh$_3$-based catalyst provided comparable yields to those obtained using the water-soluble phosphine sodium tri(2,4-dimethyl-5-sulfonatophenyl)phosphine (TXPTS).

Triphenylphosphine-based catalysts have also been applied to Suzuki arylation of unprotected purine nucleosides. Coupling of polycyclic arylboronic acids with 8-BrdG catalyzed by Pd(PPh$_3$)$_4$ in THF/MeOH/H$_2$O (2:1:2) afforded 8-(1-pyrenyl)dG (65%), 8-(1-pyrenyl)dA (10%), and 8-(6-benzo[a]pyrenyl)dG (25%) [47,48]. A family of 8-arylA adducts were prepared using Pd(PPh$_3$)$_4$ in DME/H$_2$O (2:1). The products were then converted to 5-amino-1-β-D-ribofuranosylimidazole-4-carboxamide derivatives (AICAR, Scheme 12), which were explored as potential AMP-activated protein kinase activators [49]. Suzuki coupling of *trans*-β-styrylboronic acid with 8-BrdG catalyzed by Pd(PPh₃)₄ in DMF affords 8-(*trans*-styrenyl)dG in 64% yield as a single alkene isomer [50]. The product was used in the study of photochemical *E* to *Z* isomerization of the functionalized guanosine derivative.

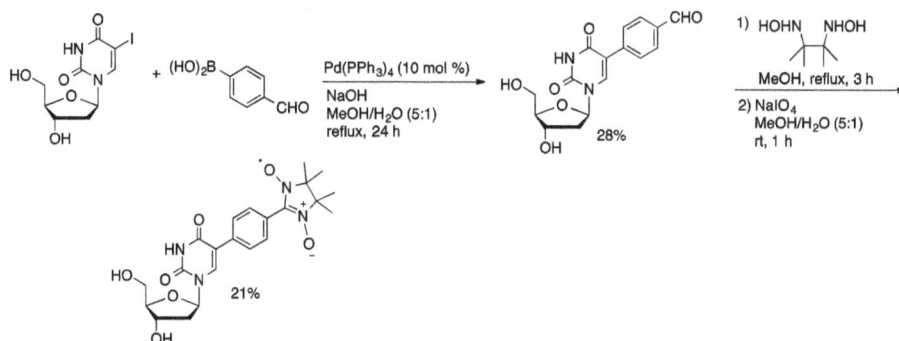

Scheme 11. Synthesis of a spin-labeled uridine derivative via a Suzuki coupling.

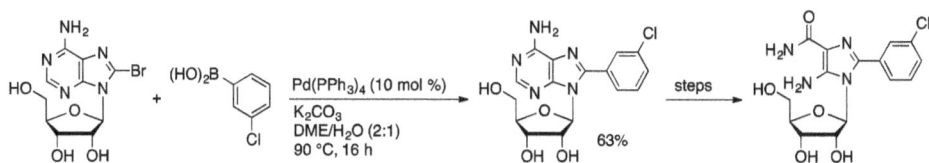

Scheme 12. Suzuki coupling of 8-BrdA as first step in AICAR synthesis.

1,1′-Bis(diphenylphosphino)ferrocene (dppf) has also been applied as a ligand in the Suzuki coupling of unprotected nucleosides. Wagenknecht used PdCl₂(dppf) (10 mol %) to catalyze the coupling of 10-methylphenothiazin-3-ylboronate ester **1** with 5-IdU in THF/MeOH/H₂O (2:1:2) to provide the coupled product in 34% yield (Scheme 13) [51]. This method afforded 5-(2-pyrenyl)dU in 62% yield from 5-IdU and pinacol pyrene-2-boronate ester [52]. A route to BODIPY-modified uridines starts with the PdCl₂(dppf)-catalyzed coupling of 4-formylboronic acid with 5-IdU in water/acetonitrile (2:1) to give the product in 76% yield (Scheme 14) [53]. The aldehyde was then condensed with 2,4-dimethylpyrrole, followed by complexation with BF₃ to afford the fluorescent uridine derivative. 2-Pyrenyl-dU was prepared in 65% by the Suzuki coupling of 2-pyrenylboronate pinacol ester with 5-IdU PdCl₂(dppf) (11 mol %) in THF/MeOH/H₂O (2:1:1) [47].

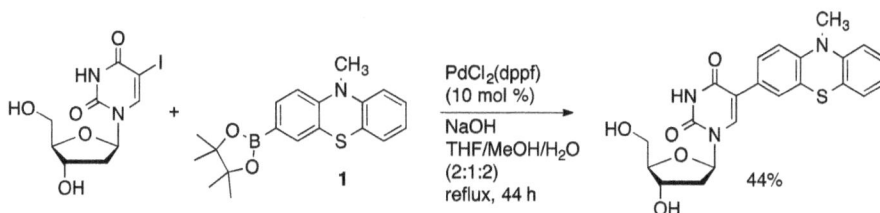

Scheme 13. Synthesis of 5-(10-methylphenothiazin-3-yl)dU.

3.2. Stille Coupling

The Stille coupling is an effective and mild method to introduce a wide range of carbon substituents. Because of the toxicity of stannanes, and the challenges associated with removing tin byproducts, the Suzuki coupling has largely supplanted the Stille coupling. The Stille coupling provides an early example of the introduction of heteroaryl, alkenyl, and allyl substituents to the 5-position of uridine (Scheme 15) [54]. Organostannanes were coupled with 5-IdU using $PdCl_2(PPh_3)_4$ as the catalyst in refluxing THF to give the coupled products in good yields (42%–72%). Vinylstannanes are more readily available than vinylboronic acid derivatives. A recent synthesis of 8-vinyldG relied on a $Pd(PPh_3)_4$-catalyzed Stille coupling of tributylvinylstannane and 8-BrdG in NMP at 110 °C [55].

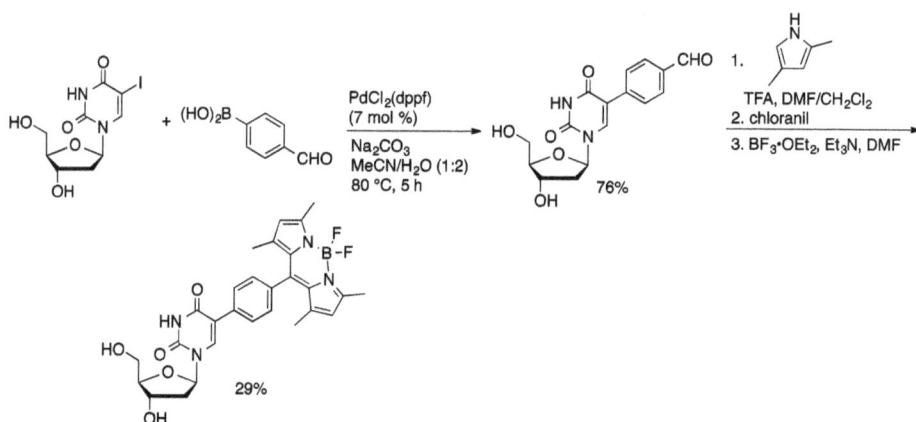

Scheme 14. Synthesis of BODIPY-functionalized uridine.

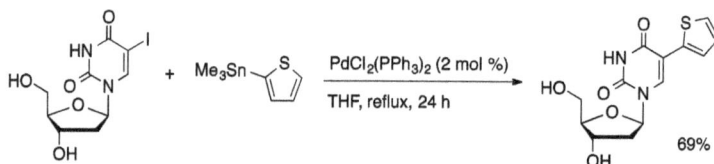

Scheme 15. Stille coupling of 2-thienylstannane and 5-IdU.

3.3. Sonogashira Coupling

Alkynylation of nucleosides is another important modification strategy. The alkynyl modification can be introduced with minimal effect on DNA structure. 5-Alkynyluridine can replace thymidine bases without significant structural change to the DNA conformation, for example. The Sonogashira coupling of aryl halides with alkynes is an effective method for preparing arylacetylene derivatives. This approach has been used widely with protected nucleosides in organic solvents. Because Sonogashira couplings are often performed in polar aprotic solvents, alkynylation of unprotected nucleosides using triphenylphosphine-based catalysts were some of the first examples of direct coupling of unprotected nucleosides.

The Sonogashira coupling was first demonstrated in the coupling of 2-IA with alkynes in DMF catalyzed by $PdCl_2(PPh_3)_2$ and CuI [56]. The 2-alkynyl-A products were obtained in excellent yields (84%–97%). The versatility of the methodology has been demonstrated by preparing alkynyl-substituted nucleosides bearing acidic, basic, and hydrophobic groups from 7-I-7-deaza-dA, 8-BrdA, 7-I-7-deaza-dG, 5-IdU, and 5-IdC (Scheme 16) [57]. The functionalized nucleotides were converted to triphosphates and incorporated into oligonucleotides using polymerase enzymes. Oligonucleotides with a high density of functionalized nucleoside residues could be prepared.

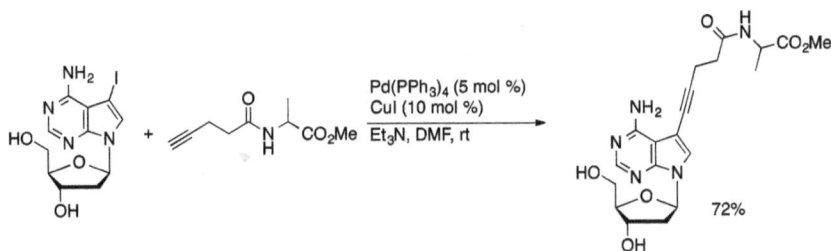

Scheme 16. Sonogashira coupling catalyzed by Pd(PPh₃)₄/CuI.

Couplings of alkynes and 8-BrdA catalyzed by PdCl₂(PPh₃)₂ and CuI gave a series of 8-alkynyldA derivatives [58]. The series was further diversified by reducing the alkynyl-dG compounds to 8-alkenyldG and 8-alkyldG derivatives. 8-EthynyldA, prepared from trimethylsilylacetylene followed by deprotection, was the most active of the compounds with micromolar inhibitory activity against a range of viral targets. Sonogashira coupling catalyzed by Pd(PPh₃)₄ in DMF at rt was used to prepare 5-alkynyl-dU derivatives containing lipophilic, amide, urea, and sulfonamide functionality [45]. Simple aliphatic alkynes, such as 1-dodecyne, were coupled directly with 5-IdU (Scheme 17). Alkynes with polar functionalities were prepared by coupling of acetoxy-protected 5-IdU with alkynes. The authors do not indicate why protected 5-IdU was used with the functionalized alkynes. The library of compounds was tested as inhibitors of mycobacterial thymidylate synthases (ThyX and ThyA) in *Mycobacterium tuberculosis*. 8-Alkynyl-A derivatives, prepared by the PdCl₂(PPh₃)₂/CuI-catalyzed coupling of 8-BrA and alkynes, gave selective antagonists of the A₃ adenosine receptor [59]. An attempt to prepare 8-(3-phenyl-3-hydroxypropyn-1-yl)A resulted in rearrangement of the alkyne to give 8-(3-phenyl-1-propyn-3-one)A instead (Scheme 18).

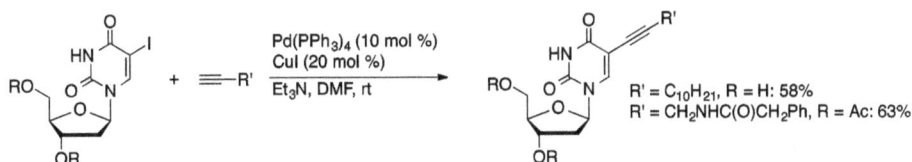

Scheme 17. Sonogashira coupling of protected and unprotected 5-IdU.

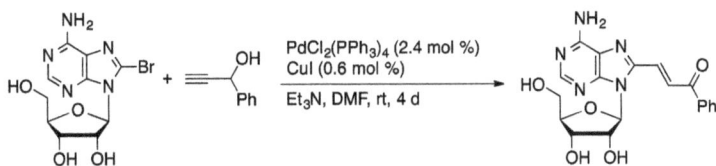

Scheme 18. Rearrangement of 1-phenyl-2-propyn-1-ol during Sonogashira coupling.

Boron-rich compounds are attractive pharmacophores of use in boron neutron capture therapy. Coupling of 8-BrdA with 2-ethynyl-*para*-carborane catalyzed by Pd(PPh₃)₄ and CuI in DMF afforded the 8-substituted adenosine in 80% yield (Scheme 19) [60]. Alternatively, 8-ethynyl-dA could be coupled with 2-iodo-*para*-carborane, but in only 27% yield. Coupling of tripropargyl amine (10 equivalents) and 7-I-7-deaza-dG catalyzed by Pd(PPh₃)₄ and CuI in DMF gave a 7-alkynylated product **2** with two free alkynyl units (Scheme 20) [61]. Guanosine derivative **2** was converted to a protected phosphoramidite and incorporated into oligonucleotides using solid phase synthetic techniques. The oligonucleotides were then reacted with 1-azidomethylpyrene to give the doubly functionalized structure. Single strand oligonucleotides containing residue **2** (**3**) do not show excimer fluorescence, nor do double strand (ds) DNA containing only one strand with residue **2**. In contrast, ds oligonucleotides containing two appropriately placed **2** residues show strong excimer excitation.

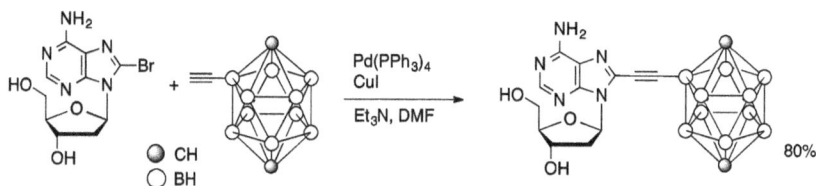

Scheme 19. Sonogashira coupling of 8-BrdA with 2-ethynyl-*para*-carborane.

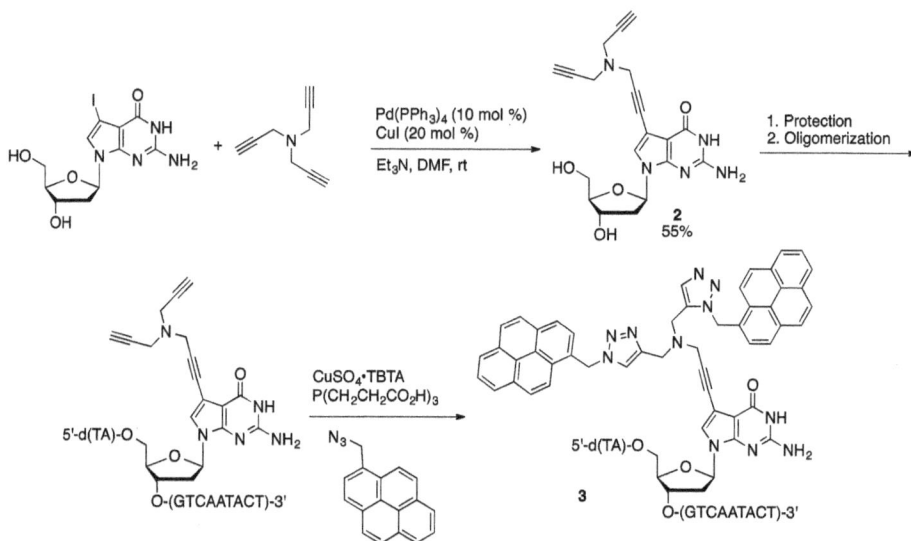

Scheme 20. Synthesis of doubly pyrene-substituted oligonucleotides.

3.4. Heck Couplings

In contrast to the other classic palladium-catalyzed C–C bond-forming reactions, there are limited examples of the Heck coupling of unprotected nucleosides. Baranger reported the coupling of 2-IA with allylbenzene mediated by a stoichiometric amount of Pd(OAc)$_2$/P(*o*-tolyl)$_3$ in acetonitrile (Scheme 21) [62]. 2-(3-Phenyl-1-propenyl)adenosine (**4**) was isolated in 53% yield. Compound **4** was then hydrogenated to give 2-(3-phenylpropyl)A. Palladium(PTA)$_2$(saccharinate)$_2$ (PTA = 1,3,5-triaza-7-phosphaadamantane, Figure 2) is an effective precatalyst for the Heck coupling of 5-IdU and alkenes in acetonitrile to give 5-alkenylated dU derivatives in high yield (Scheme 22) [63]. The saccharinate complex was more active than other imidate PTA complexes (phthalimidate, maleimidate, or succinimidate).

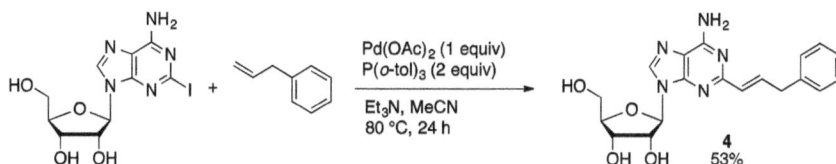

Scheme 21. Heck coupling of 2-IA mediated by Pd(OAc)$_2$/P(*o*-tolyl)$_3$.

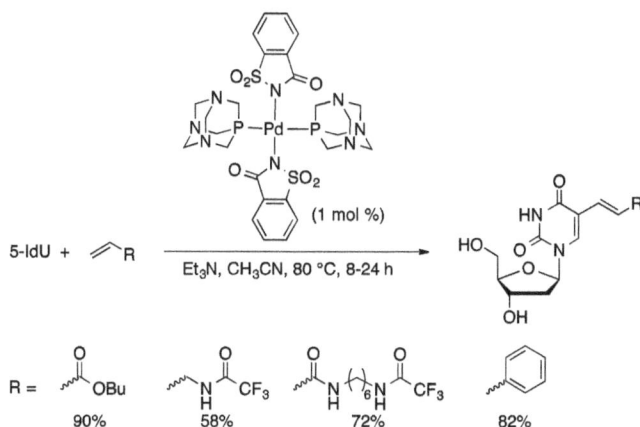

Scheme 22. Heck coupling of 5-IdU catalyzed by Pd(PTA)$_2$(saccharinate)$_2$.

4. Aqueous-Phase Cross-Coupling of Nucleosides, Nucleotides, and Oligonucleotides Using Hydrophilic Ligand-Supported Catalysts

Palladium-catalyzed cross-coupling reactions are typically performed in organic solvents using hydrophobic supporting ligands, such as triphenylphosphine. For typical hydrophobic substrates, these homogeneous conditions typically offer optimal catalyst activity. Aqueous-biphasic using water-soluble transition metal catalysts offers a number of potential advantages over traditional homogeneous organic-phase catalysis [64–67]. Water is an attractive solvent as it is non-toxic, non-flammable, and a renewable resource. Separation of homogeneous catalysts from organic products is a common challenge, particularly in pharmaceutical processes [68]. The potential to constrain the catalyst in the aqueous phase allows for easily separation from the organic products. The standard approach to design water-soluble ligands is to append hydrophilic functionality to commonly used ligand structures, such as triphenylphosphine (Figure 2). Hydrophilic catalysts have primarily been applied in coupling of hydrophobic substrates. They also provide the opportunity to perform homogeneous coupling of hydrophilic substrates, such as biomolecules.

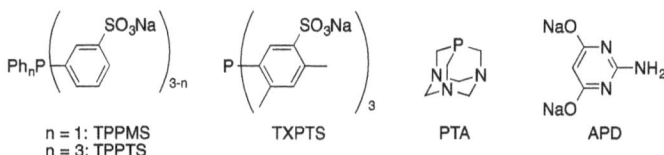

Figure 2. Hydrophilic ligands commonly applied in nucleoside cross-coupling reactions.

Casalnuovo [69] was the first to report the application of a water-soluble palladium/phosphine catalyst for the cross-coupling of aryl halides. He showed that Pd(TPPMS)$_3$ (TPPMS = sodium diphenyl(3-sulfonatophenyl)phosphine) provided an effective catalyst for Suzuki, Heck, and Sonogashira couplings of aryl and heteroaryl halides in aqueous acetonitrile. In addition to hydrophobic substrates, examples of Heck and Sonogashira couplings of 5-IdU and 5-IdCMP were reported. By using a water-soluble catalyst system, cross-coupling of these hydrophilic substrates was accomplished under homogeneous conditions. In the decade following Casalnuovo's report, significant effort was devoted to developing new water-soluble phosphine ligands and their application to cross-coupling of water-insoluble substrates. In contrast, no examples of the application of water-soluble catalyst systems to nucleoside modification were reported in the decade following Casalnuovo's seminal paper.

4.1. Nucleosides

4.1.1. Suzuki Coupling

Methodology Development. The Shaughnessy group revisited the aqueous-phase cross-coupling of nucleosides with the goal of making this a general methodology for modification of both purine and pyrimidine nucleosides. Casalnuovo's initial paper reported coupling of the more reactive iodopyrimidine nucleosides, whereas we had an interest in coupling the less reactive 8-bromopurines, such as 8-Br(d)G and 8-Br(d)A. Catalysts derived from a range of hydrophilic phosphine ligands were screened for the ability to couple phenylboronic acid and 8-BrdG in aqueous acetonitrile [70]. The sterically demanding, electron-rich phosphine *t*-Bu-Pip-phos (4-(di-*tert*-butylphosphino)-N,N-dimethylpiperidinium chloride), which provides high activity catalysts for simple aryl bromides [71], gave low conversion. Palladium in combination with TPPTS, provided an effective catalyst for this coupling despite being much less effective than *t*-Bu-Pip-phos in the aqueous-phase coupling of simple aryl halides.

The Pd(OAc)$_2$/TPPTS catalyst system is a general method for Suzuki coupling of halogenated purine and pyrimidine nucleosides and 2′-deoxynucleosides. Good to excellent yields are afforded with a range of aryl boronic acids and 5-IdU, 8-BrG, 8-BrdG, 8-BrA, and 8-BrdA (Scheme 23) [70]. The methodology has also been extended to heteroarylboronic acids [72]. The order of reactivity of halonucleosides is 5-IdU > 8-BrdA >> 8-BrdG. The low reactivity of guanosine is believed to be due to competitive coordination of the deprotonated form of guanosine to palladium through the anionic N-1 position [73].

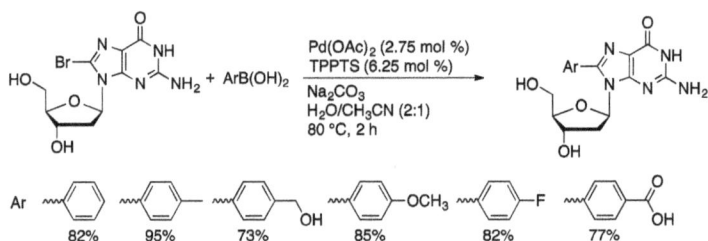

Scheme 23. Pd(OAc)$_2$/TPPTS-catalyzed Suzuki coupling of 8-BrdG.

The more sterically demanding TXPTS ligand provides a more active catalyst for Suzuki coupling of halonucleosides [70]. Using 10 mol % Pd/TXPTS, complete conversion of 8-BrdA to 8-PhdA was achieved at room temperature in 30 minutes. In comparison, the TPPTS-derived catalyst required 24 h to give 74% conversion to product. Even 8-BrdG gave 40% conversion to 8-PhdG after 18 h at room temperature. Although TXPTS provides a more active catalyst, the Pd(OAc)$_2$/TPPTS catalyst has become the standard system for these reactions. TPPTS is more widely commercially available and costs about half of TXPTS on a per mole basis [74]. For challenging cases, TXPTS may prove to be an attractive alternative.

Organic cosolvents are not required in these reactions. Good yields can be achieved using water as the only solvent [70,75,76]. In the case of the more reactive 5-IdU, the TPPTS ligand is not required [77]. Good yields could be achieved with electron-rich arylboronic acids, but much lower yields were obtained with electron-deficient boronic acids. Using TPPTS, good yields can be achieved with a broad range of arylboronic acids in water alone.

Other hydrophilic ligands have been used in the Suzuki coupling of halonucleosides with boronic acids in aqueous media. Pd(PTA)$_2$(phthalimidate)$_2$ (PTA = 1,3,5-triaza-7-phosphaadamantane, Figure 2) is an effective precatalyst for the synthesis of 5-aryl-2′-deoxypyridimine derivatives from 5-IdU and 5-IdC in water (Scheme 24) [78]. The phthalimidate complex gave higher yields than dihalide palladium-TPA complexes or the catalyst generated *in situ* from Pd(OAc)$_2$ and PTA. Suzuki coupling of aryl- and alkenylboronic acids with 5-IdU can also be accomplished using Pd(OAc)$_2$(2-aminopyrimidine-4,6-diolate)$_2$ as the precatalyst (Scheme 25) [79]. These catalyst systems have not been extended to the less reactive 8-halopurine nucleosides.

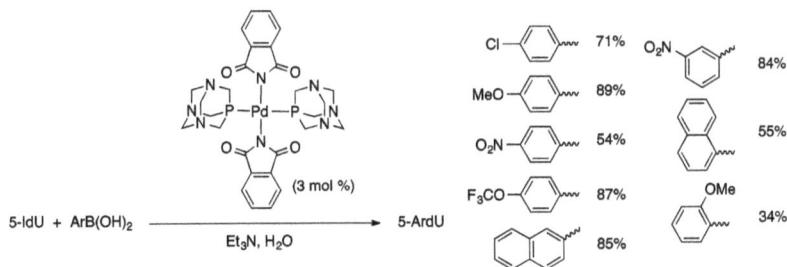

Scheme 24. Pd(PTA)$_2$(phthalimidate)$_2$-catalyzed Suzuki coupling of 5-IdU.

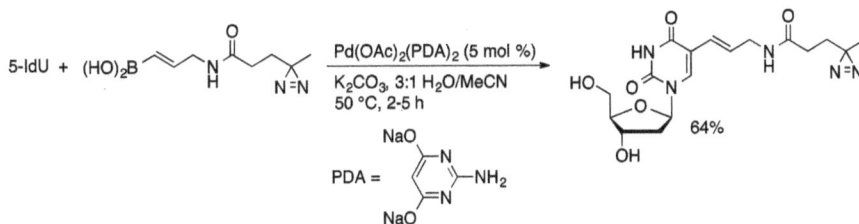

Scheme 25. Suzuki coupling of 5-IdU catalyzed by Pd-APD complexes.

Applications. The palladium/TPPTS methodology sparked interest in the direct functionalization of unprotected nucleosides in aqueous reaction media. The homogeneous conditions have proven to be more general than catalyst systems using hydrophobic ligands in water or polar organic solvents. This generality is particularly useful in the coupling of nucleotide and oligonucleotide substrates as discussed in Sections 4.2 and 4.3. The Pd/TPPTS catalyst system has been applied to the synthesis of wide variety of modified nucleosides that incorporated fluorescent or electrochemically active reporter groups, coordination sites, or have potential pharmaceutical activity.

An early demonstration of the utility of the Pd/TPPTS catalyst system was the synthesis of an amino acid-nucleoside adduct reported by Hocek [80]. The coupling of 8-BrA and 8-BrdA with phenylalanyl-4-boronic acid is achieved in 71 and 75% yield, respectively, in water-acetonitrile using the Pd/TPPTS catalyst system (Scheme 26). Notably, neither the nucleoside nor amino acid substrates contained protecting groups. In the coupling with 6-chloropurine nucleosides, improved yields were obtained using microwave irradiation compared to traditional thermal heating [81]. The Hocek group has also reported the attachment of polypyridyl ligands to 8-BrdA [82], 7-I-7-deaza-dA [83], and 5-IdU [84] via Suzuki couplings using Pd/TPPTS (Scheme 27). High yields were achieved using bipyridyl, phenanthryl, and terpyridyl boronic acids as either free ligands or preformed ruthenium complexes. The ruthenium complex-modified nucleosides are of interest as luminescent and electroactive probes in oligonucleotides.

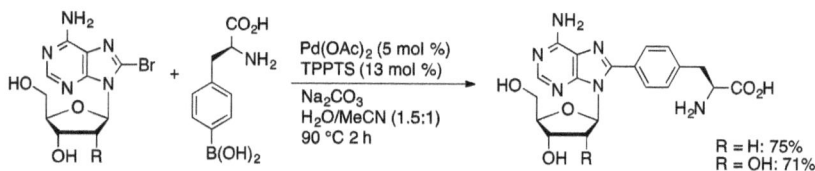

Scheme 26. Pd/TPPTS-catalyzed coupling of 8-bromoadenosines with phenylalanine.

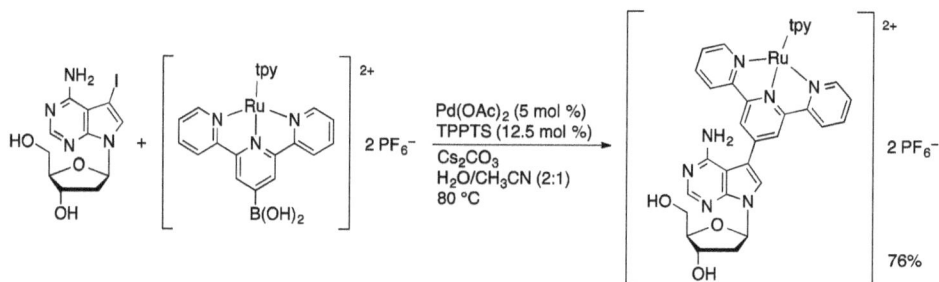

Scheme 27. Attachment of a ruthenium(II) terpyridine complex to 7-I-7-deaza-dA.

The Pd/TPPTS catalyzed Suzuki coupling has been used to introduce a range of fluorescent probes into nucleosides. This methodology has been used to attach five-membered ring heterocycles (2-pyrrolyl, 2-indolyl, 2-furyl, and 2-thiophenyl) to the 8 position of 2′-deoxyguanosine [85,86]. The fluorescence of the resulting heterocyclic adducts is sensitive to hydrogen bonding with other nucleobases as well as the nucleoside conformation. The 8-(2-benzo[*b*]thienyl)-dG group serves as a fluorescent reporter to probe for the preference for *syn*- or *anti*-conformations in duplex DNA [87].

Fluorescent pyrrolopyrimidopyrimidine [88] and pyrimidopyrimidoindole [89] nucleosides can be prepared by aqueous-phase coupling of 5-IdC with *N*-Boc-protected 2-pyrrolylboronic acid or *N*-Boc-protected 2-indolylboronic acid (Scheme 28). The coupling is followed by condensation of the 6-amino group of the cytidine ring with the Boc moiety to give the fluorescent tri- or tetracyclic nucleoside analogs. Fluorescent 5-substituted uridine and 2-deoxyuridine analogs can be prepared by Suzuki coupling of 5-I(d)U with aryl or styrylboronic acid derivatives [90]. An alternative approach to 5-styryluridine derivatives was achieved through the coupling of arylboronic acids with commercially available 5-(2-bromovinyl)uridine (BVDU, Scheme 29).

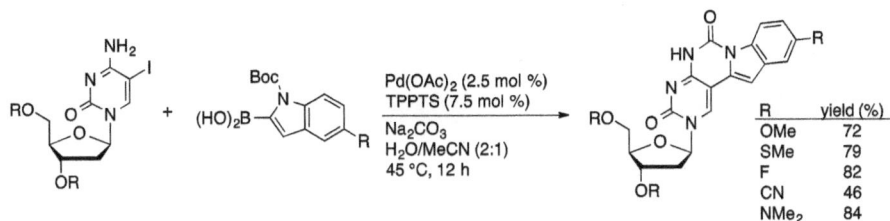

Scheme 28. Suzuki coupling/condensation sequence to prepare pyrimidopyrimidoindole nucleosides.

R	yield (%)
OMe	72
SMe	79
F	82
CN	46
NMe$_2$	84

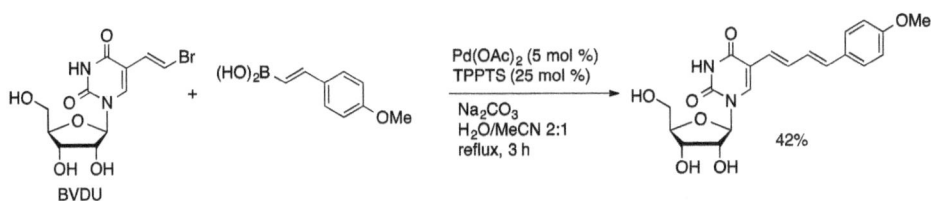

Scheme 29. Suzuki coupling of BVDU catalyzed by Pd(OAc)$_2$/TPPTS.

A library of C8-biaryl-modified dA derivatives were prepared by a sequence involving palladium-catalyzed oxidative coupling of iodobromobenzenes with dA (Scheme 30) [91]. The resulting 8-(bromoaryl)dA derivatives were then coupled with arylboronic acids using Pd(OAc)$_2$/TPPTS (1.25 mol % Pd) in aqueous acetonitrile. Suzuki coupling of the unnatural C-nucleoside 2′-deoxy-2′-(5-bromo-2-thiophenyl)ribose with arylboronic acids catalyzed by Pd(OAc)$_2$/TPPTS gave a series of 2-arylthiophenylribose nucleoside analogs (Scheme 31) [92].

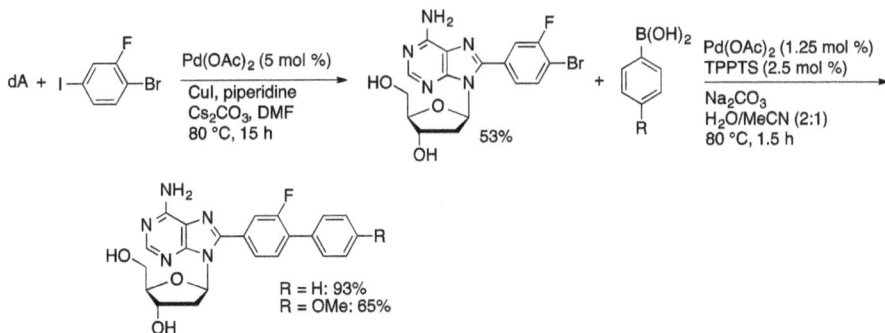

Scheme 30. Double arylation sequence to prepare 8-biaryl-dA derivatives.

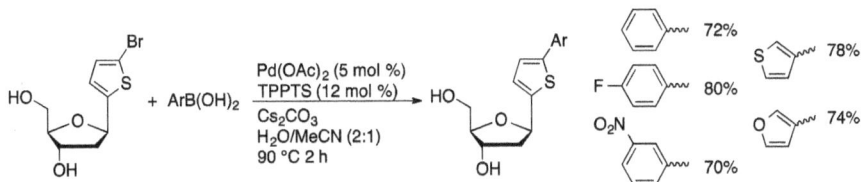

Scheme 31. Suzuki coupling of 2′-deoxy-2′-(5-bromo-2-thiophenyl)ribose.

Suzuki coupling with Pd/TPPTS was used to introduce a photoresponsive chemical switch based on a diarylethylene moiety in pyrimidine nucleosides. Suzuki coupling of boronate ester **5** with 5-IdU and 5-IdC afforded photoswitchable nucleosides that undergo reversible photochemical electrocyclic cyclization under UV irradiation (Scheme 32) [93]. The modified nucleoside reverts to the open form under visible light. The modified nucleosides could potentially be used to photochemically control the structure and function of oligonucleotides.

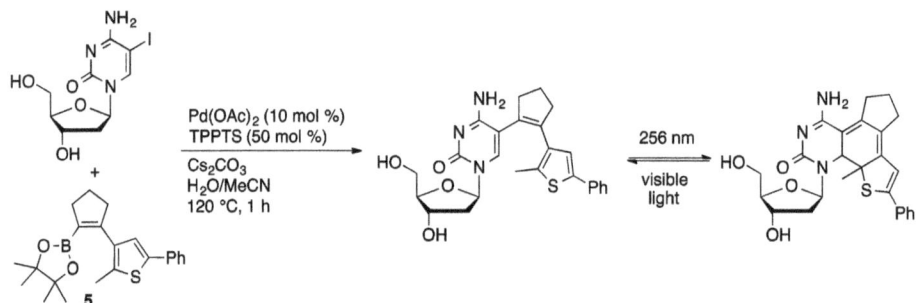

Scheme 32. Synthesis of a photoswitchable cytidine derivative.

Nucleosides containing arylthiol moieties were prepared by Suzuki coupling of protected 4-thiophenylboronic acid derivatives with 5-IdC and 7-I-7-deaza-dA using Pd(OAc)$_2$/TPPTS in aqueous acetonitrile (Scheme 33) [94]. No conversion occurred in the presence of the free thiol group, but high yields were achieved with boronic acids containing protected thiols. The thiol-functionalized nucleosides can be incorporated into oligonucleotides and used to attach the DNA to gold surfaces.

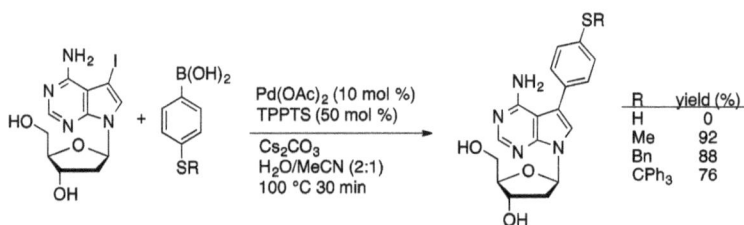

Scheme 33. Synthesis of phenyl sulfide-substituted 7-deaza-dA derivatives.

Guanosine forms self-assembled tetrameric structures (G-tetrads) through intermolecular hydrogen bonding. In the presence of cations, these can further self assemble into G-quadruplexes. Structural modifications can be used to enhance this self assembly. Aqueous-phase Suzuki coupling was used to prepare 8-(3- or 4-acetylphenyl)dG derivatives [95]. The acetylphenyl moiety enhances G-tetrad formation by providing additional hydrogen bonding opportunities, while also extending the aromatic surface to improve noncovalent interactions.

8-Aryl-substituted derivatives of purine nucleosides are of interest as models of adducts formed *in vivo* during the metabolism of aromatic hydrocarbons. The motivation of our original study of the aqueous-phase Suzuki coupling of 8-BrdG was to prepare adducts to study the effect of these modifications on DNA conformation [96]. The carcinogenesis of polyaromatic hydrocarbon is thought to involve covalent modification of guanosine during metabolism of these compounds *in vivo*. Suzuki coupling of 1-pyrenylboronic acid, 1-naphthylboronic acid, and 9-phenanthrenylboronic acid with 8-BrdA gave good yields of the adducts using Pd(OAc)$_2$/TPPTS (Scheme 34) [5]. The more hindered 8-anthracenylboronic acid and its benzannulated analogs could not be coupled under these conditions, however.

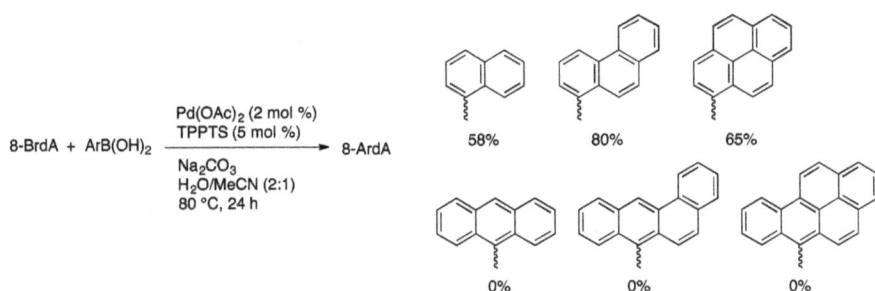

Scheme 34. Coupling of 8-BrdA with polyaromatic boronic acids.

Arylated nucleoside derivatives have been explored as potential pharmaceutically active compounds with antiviral and anticancer activity. The Hocek group has used the Pd(OAc)$_2$/TPPTS catalyst system to prepare a wide variety of natural and non-natural nucleoside derivatives through aqueous-phase Suzuki couplings. A library of 6-heteroarylpurine nucleosides were prepared from unprotected 6-cloropurine and 6-chloro-7-deazapurine nucleosides [97,98]. Good yields (40%–80%) were achieved with a variety of aryl and heteroaryl boronic acids. A similar library was prepared from 7-I-7-deaza-A [99]. A large library of 6-arylpurine nucleoside monophosphates were prepared by aqueous-phase Suzuki coupling of 6-chloropurine nucleoside with arylboronic acids followed by phosphorylation (Scheme 35) [100]. Further elaboration of 6-(3-bromophenyl)purine nucleoside with a subsequent Suzuki coupling afforded 6-biarylpurine nucleosides. The compounds were tested for their ability to inhibit 2'-deoxynucleoside 5'-phosphate *N*-hydrolase 1 (DNPH1), which is a potential anticancer target.

The Hocek group has also used aqueous-phase Suzuki couplings to prepare arylated nucleoside derivatives in which the ribose unit has been modified. Good yields are obtained in coupling of aryl and heteroarylboronic acids with 7-iodo-7-deazaadenine arabinoside (6) [101], 7-iodo-7-deaza-2'-C-methyladenosine (7) [101], 7-iodo-7-deazaadenine 2'-deoxy-2'-fluoroarabinoside (8) [101], 6-chloro-7-deazapurine 2'-deoxy-2'-fluororibinoside (9) [102], and 7-iodo-7-deazaadenine 2'-deoxy-2',2'-difluoro-β-D-*erythro*-pentofuranoside (10, Figure 3) [102]. The sugar modifications have little effect on the Suzuki coupling reaction. A family of 6-substituted-7-aryl-7-deazapurinenucleosides (13) was prepared starting from a protected 6-chloro-7-iodo-7-deazapurine nucleoside derivative (11, Scheme 36) [103]. Selective nucleophilic aromatic substitution at the 6-chloro position followed by deprotection provides the 7-iodo precursors (12). These were coupled with aryl or heteroarylboronic acids using Pd(OAc)$_2$/TPPTS.

4.1.2. Sonogashira Coupling

Palladium-catalyzed alkynylation of unprotected nucleosides with hydrophobic palladium/phosphine catalysts is well precedented. Recently, aqueous-phase Sonogashira couplings using hydrophilic ligands has received increasing attention. The Hocek group used Sonogashira couplings, in addition to Suzuki couplings, to introduce polypyridyl ligands to 8-BrdA (Scheme 37) [82], 7-I-7-deaza-dA [83], and 5-IdU [84] using the Pd/TPPTS catalyst system in combination with catalytic CuI in DMF. High yields were achieved with bipyridyl alkynes (82%–96%). In contrast to the Suzuki coupling, low yields were obtained in the coupling of ruthenium-coordinated analogs of the bipyridinyl alkynes (0%–57%). Decomposition of the alkyne-substituted ruthenium complexes competed with the desired cross-coupling under the Sonogashira reaction conditions.

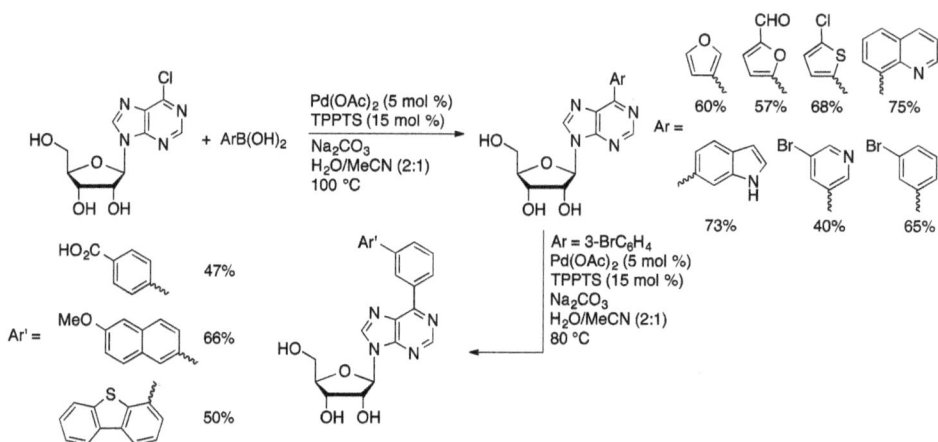

Scheme 35. Suzuki coupling of 6-chloropurine nucleoside catalyzed by Pd(OAc)$_2$/TPPTS.

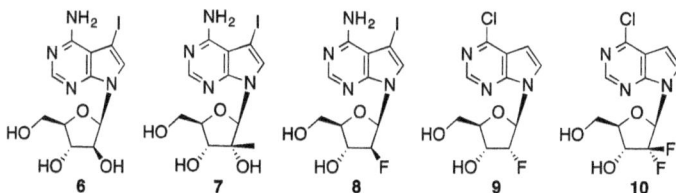

Figure 3. Sugar-modified purine halonucleosides.

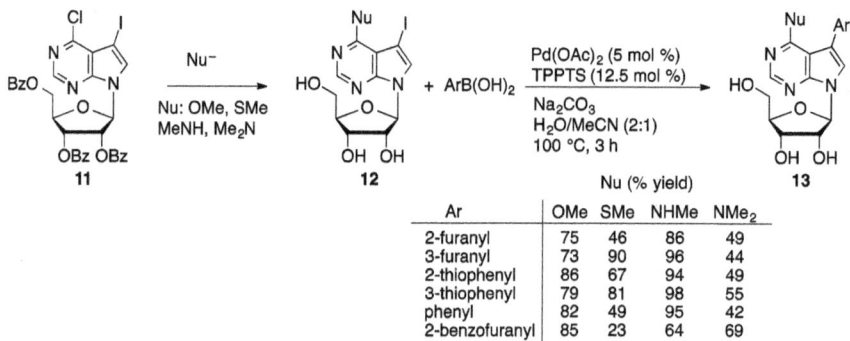

Ar	OMe	SMe	NHMe	NMe$_2$
2-furanyl	75	46	86	49
3-furanyl	73	90	96	44
2-thiophenyl	86	67	94	49
3-thiophenyl	79	81	98	55
phenyl	82	49	95	42
2-benzofuranyl	85	23	64	69

Scheme 36. S$_N$Ar/Suzuki sequence to 6-substituted-7-aryl-7-deazapurine nucleosides.

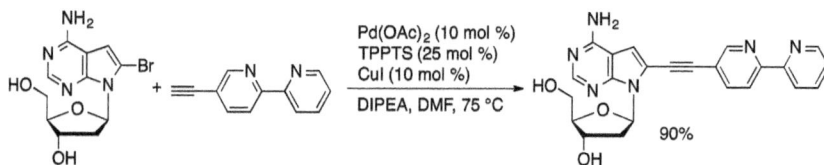

Scheme 37. Sonogashira coupling approach to bipyridine-dA adduct.

Sonogashira coupling with Pd(OAc)$_2$/TPPTS/CuI is effective for the coupling of 7-I-7-deaza-dA and 5-IdU with propargyl esters or amides of bile acids (Scheme 38) [104]. Yields of isolated products ranged from 31%–90%. The bile acid nucleoside adducts are of interest for oligonucleotide amphiphiles.

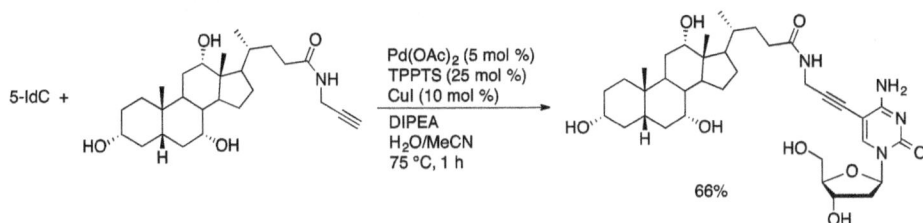

Scheme 38. Synthesis of bile acid-dC adducts through aqueous-phase Sonogashira coupling.

The more sterically demanding TXPTS ligand provides a more active catalyst for the Sonogashira coupling of 8-halopurines and 5-IdU than the catalyst derived from TPPTS [105,106]. In the coupling of 5-IdU with phenylacetylene catalyzed by TPPTS, Pd(OAc)$_2$ and CuI in aqueous acetonitrile reached 80% conversion after six hours at 50 °C. Under the same conditions, 90% conversion was achieved in one hour using TXPTS as the ligand. The Pd(OAc)$_2$/TXPTS system was effective for the coupling of aryl and alkyl acetylenes with 8-Br(d)A and 8-BrdG (Scheme 39). In the case of 5-IdU, the Sonogashira reaction was followed by nucleophilic attack of the C6-O on the alkyne resulting in formation of furano [2,3]-pyrimidin-2-one byproducts (**14**, Scheme 40).

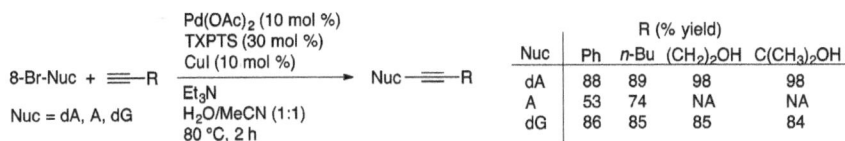

Nuc	R (% yield)			
	Ph	n-Bu	(CH$_2$)$_2$OH	C(CH$_3$)$_2$OH
dA	88	89	98	98
A	53	74	NA	NA
dG	86	85	85	84

Scheme 39. Pd/TXPTS-catalyzed Sonogashira coupling of 8-bromopurine nucleosides.

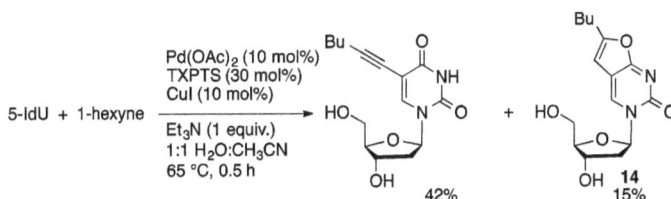

Scheme 40. Pd/TXPTS-catalyzed Sonogashira coupling of 5-IdU.

4.1.3. Heck Couplings

There are relatively few examples of Heck couplings of unprotected nucleosides reported with hydrophilic ligands. The Shaughnessy group reported the Heck coupling of 5-IdU with styrene and conjugated enones [107]. The Pd(OAc)$_2$/TPPTS gave good yields of 5-alkeynyl products, but similar yields were obtained when no ligand was used. More sterically demanding (TXPTS) or electron-rich ligands (*t*-Bu-Amphos) did not improve the catalyst performance. The Pd(OAc)$_2$/TPPTS catalyst was successfully used in the synthesis of 8-styryl-dG starting from 8-BrdG at 80 °C in 2:1 water/acetonitrile [55].

4.2. Nucleotides

Modified nucleosides are often prepared as precursors to nucleotides, which can be enzymatically incorporated into oligonucleotides. Selective 5′-phosphorylation of modified nucleosides represents one common approach to preparing base-modified nucleotides. Phosphorylation of lipophilic modified nucleosides can be difficult to achieve in high yield and good selectivity for the 5′-oxygen [108]. An alternative approach is to perform the cross-coupling reaction directly on the halogenated nucleotide. Water-soluble catalyst systems have proven to be effective in performing cross-coupling reactions with nucleoside mono-, di-, and triphosphate substrates. In contrast, there are no reported examples of the use of catalysts derived from hydrophobic ligands in the direct reaction of halogenated nucleotides.

4.2.1. Suzuki Coupling

The Pd/TPPTS catalyst system that is widely used for nucleoside Suzuki couplings is also effective in couplings with halogenated nucleotide derivatives. Early examples of Suzuki couplings of halonucleotides were reported by the Wagner [109] and Hocek groups [81,110]. Both groups used a catalyst derived from a palladium(II) salt and TPPTS (Scheme 41). Notably, the reaction could be carried out at a high pH (0.3 M CO_3^{2-}) at an elevated temperature (80–120 °C) without decomposition of the nucleotide.

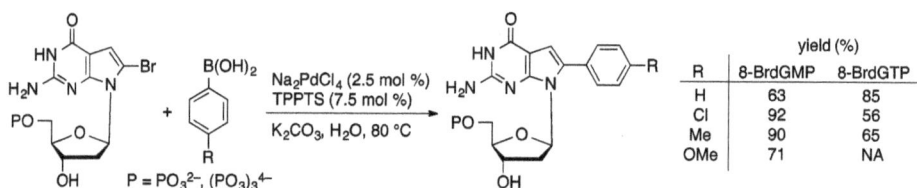

Scheme 41. Pd/TPPTS-catalyzed Suzuki coupling of 8-bromoguanosine phosphates.

Hocek reported similar conditions for the synthesis of 8-MedATP and 8-PhdATP from 8-BrdATP [111]. The methyl-substituted nucleoside was prepared using methylboronic acid, which is a rare example of the use of an alkylboron reagent in coupling reactions with nucleoside derivatives (Scheme 42). The resulting modified nucleosides could be incorporated into oligonucleotides using polymerase enzymes. This methodology was used to prepared 5-ArdUTP, 5-ArdCTP, and 7-Ar-7-deazadATP derivatives with 3-nitrophenyl and 3-aminophenyl aryl groups [112]. The modified nucleotides were incorporated into oligonucleotide sequences. The modifications serve as electrochemical labels in oligonucleotides that are sensitive to the local sequence.

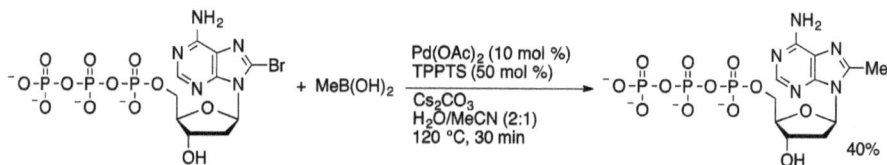

Scheme 42. Methylation of 8-BrATP.

These conditions have been applied to the synthesis of a variety of functionalized nucleotides that were then incorporated into oligonucleotides using polymerase enzymes. Alkylsulfanylphenyl-modified nucleotides were prepared modes yields (10%–50%) by coupling of the thioether-functionalized boronic acids with 7-I-7-deaza-dATP and 5-ICTP using the Pd/TPPTS catalyst

44

system [94]. The resulting functionalized nucleotides were enzymatically incorporated into oligonucleotides. The functionalized oligonucleotides were studied electrochemically and found to associate at the gold electrode surface.

Benzofurane has been proposed as a novel electrochemically active probe functionality for the study of oligonucleotides. The benzofurane moiety was attached to 7-deaza-dATP and dCTP via an aqueous-phase Suzuki coupling catalyzed by Pd/TPPTS [113]. Low yields (10%–22%) were obtained for the Suzuki coupling of the halonucleoside triphosphate substrates (Scheme 43). Suzuki coupling of the halonucleoside followed by phosphorylation gave a higher overall yield (17%–52%) of the benzofurane nucleotide adducts. Electrochemically active benzofurane moieties were attached to nucleotides in incorporated into oligonucleotides in parallel with nitrophenyl- and aminophenyl-modified nucleotides. The three electroactive groups could be addressed independently without apparent interference.

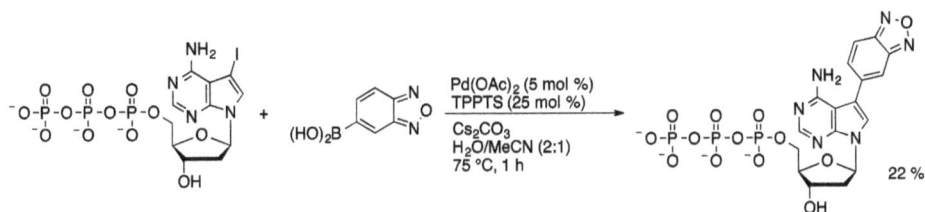

Scheme 43. Synthesis of benzofurane-dATP.

Solvatochromatic and pH-sensitive dual fluorescent ^{19}F-NMR probes were attached to 7-I-7-deaza-dATP and 5-IdUTP using an aqueous-phase Suzuki coupling (Scheme 44) [114]. The nucleotides were incorporated into oligonucleotides where they showed environment-dependent fluorescence properties. The Pd/TPPTS catalyst was used to prepare 5-formylthiophen-2-yl-modified nucleotides, which were then incorporated into oligonucleotides via primer extension or polymerase chain reaction protocols (Scheme 45) [115]. The aldehyde-modified oligonucleotides were then conjugated with amine derivatives through the formation of imine linkages. This methodology was used for selective staining of the aldehyde-containing oligonucleotides.

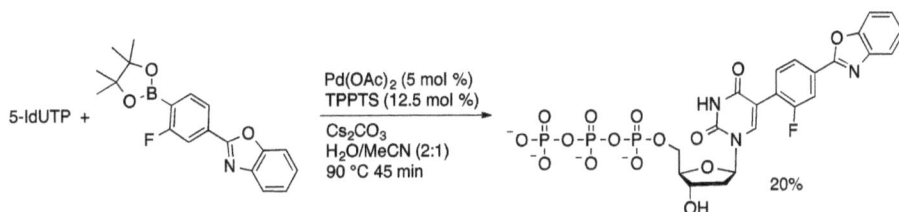

Scheme 44. Synthesis of pH-sensitive dual fluorescent ^{19}F-NMR probe nucleotides.

The aqueous-phase Suzuki coupling protocol has also been applied to the synthesis of dinucleotides and nucleotide-sugar conjugates. A series of arylboronic acids were coupled with 5-I-UDP glucoside (**15**) in water catalyzed by Na_2PdCl_4/TPPTS (1 mol %) to give 5-arylated products **16** in 40%–64% yield (Scheme 46) [116]. The corresponding 5-Br-UDP-glucose substrate gave no conversion under identical conditions. Analysis of the nucleotide conformation by NOE spectroscopy showed that 5-Br-UDP-glucose preferred an *anti*-configuration that places the bromide over the ribose ring, potentially making it less accessible. In contrast, 5-I-UDP-glucose preferred a *syn*-conformation in which the iodide is more readily accessible to the catalyst. It should also be noted that C–I bonds are generally found to be more reactive in cross-coupling reactions than C–Br bonds. A similar methodology was used to prepare 8-aryl-GDP-mannoside derivatives in 48%–82% yield for the Suzuki coupling step [117]. 8-Arylated nicotinamide adenine dinucleotides (**17**) were prepared by an alternate approach in which 8-BrAMP was arylated using Pd/TPPTS (Scheme 47) [118]. The arylated AMP was then coupled to nicotinamide ribose monophosphate through a phosphorylation reaction. The resulting compounds serve as fluorescent probes for NAD-consuming enzymes.

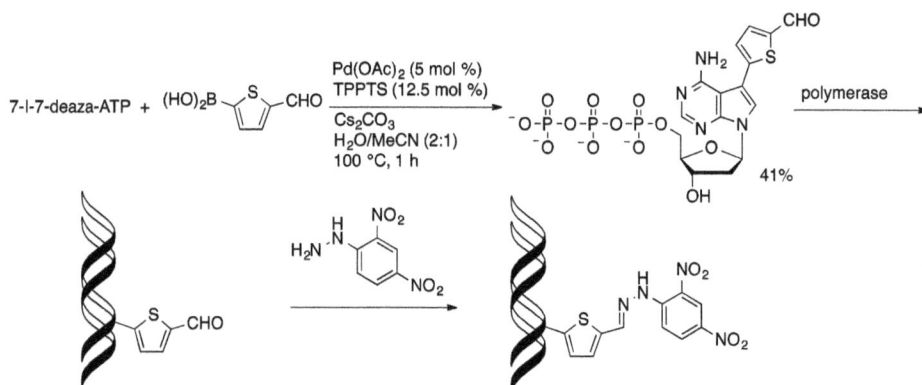

Scheme 45. Synthesis of aldehyde-functionalized nucleoside triphosphates.

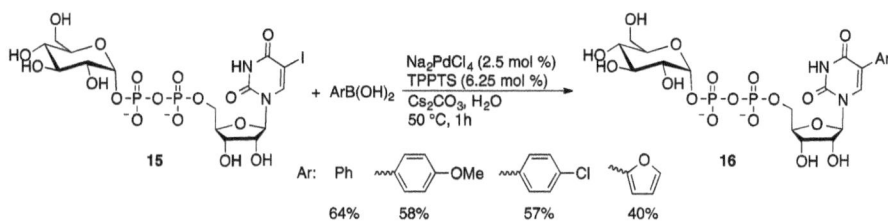

Scheme 46. Arylation of 5-I-UDP glucoside.

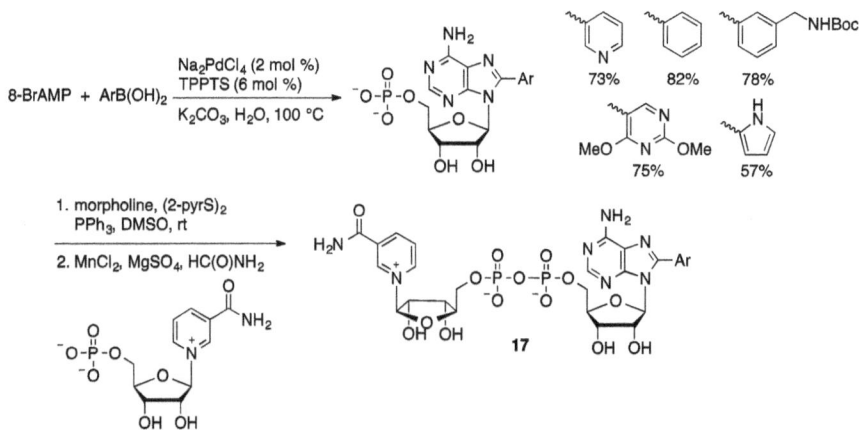

Scheme 47. Synthesis of 8-arylated nicotinamide adenine dinucleotides.

4.2.2. Sonogashira Coupling

Sonogashira coupling of alkynes with halogenated nucleotides is an effective method to introduce probe species. Burgess was the first to demonstrate this methodology in the synthesis of fluorescent dye-modified uridine triphosphates (**18**, Scheme 48) [108]. Fluorescein-based dyes with alkyne substituents were coupled with 5-I-dUTP using a preformed Pd/TPPTS catalyst and CuI at room temperature with triethylamine as the base in phosphate buffer. The preformed catalyst was generated by mixing Na_2PdCl_4, TPPTS, $NaBH_4$ and water at room temperature followed by removal of water and recovery of the resulting solid. Alkynylation of the nucleotide substrate was necessary to produce the desired compounds. In contrast, attempted phosphorylation of fluorescein-modified dU was unsuccessful.

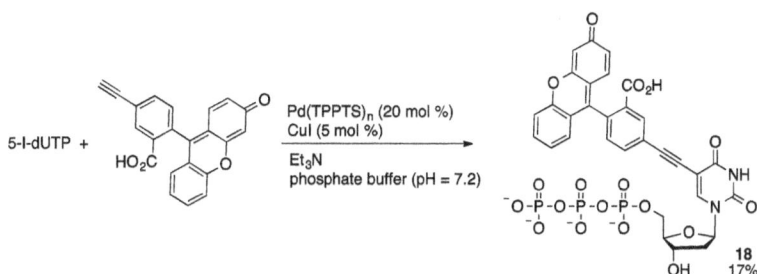

Scheme 48. Sonogashira coupling of 5-I-dUTP and ethynylfluorescein.

Phenylalanyl-modified nucleotide triphosphates can be prepared in 60%–67% yield by the Sonogashira coupling of 4-ethynylphenylalanine with halonucleotides using a catalyst derived from $Pd(OAc)_2$, TPPTS, and CuI in aqueous acetonitrile. This method was also used to prepare ethynylferrocene-modified nucleotide triphosphates (Scheme 49) [119]. Oligonucleotides prepared from the ferrocene-functionalized nucleotides are electrochemical probes of DNA binding. Bile acid conjugates of nucleotides were prepared in moderate yields (32%–57%) by coupling of propargyl bile acid amides with iodonucleotide triphosphates [104].

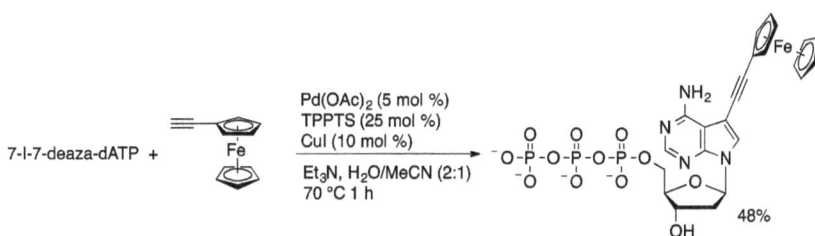

Scheme 49. Synthesis of ferrocenyl-modified 7-deaza-dATP.

4.2.3. Heck Coupling

To date there is only one example of a Heck coupling of a halonucleotide. Hocek reported the coupling of butyl acrylate with a range of 5-iodopurine and 7-iodo-7-deazapurine nucleoside mono- and triphosphates using $Pd(OAc)_2$ and TPPTS in water/acetonitrile at 80 °C (Scheme 50) [120]. The yields with nucleotides were modest (14%–55%) compared to the high yields obtained with the corresponding nucleosides (81%–98%). The monophosphates gave higher yields than triphosphates. Although modest yields were obtained in the Heck coupling of nucleotides, the yields were similar to those obtained when the Heck coupling was carried out on the nucleoside followed by phosphorylation.

Cytidine gave low yields in the Heck coupling and no conversion was obtained with cytidine mono- or triphosphate. The resulting alkene-modified nucleotides could be successfully incorporated into oligonucleotides by primer extension methods.

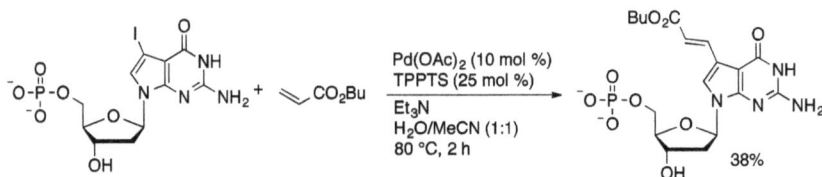

Scheme 50. Aqueous-phase Heck coupling of 7-iodo-7-deazaguanosine monophosphate.

4.3. Oligonucleotides

Palladium-catalyzed coupling reactions provide effective ways to attach a wide variety of moieties to nucleoside or nucleotide structures. The modified nucleosides can in many cases be effectively incorporated into oligonucleotides using primer extension, PCR, or solid-phase DNA synthesis methods. Although these approaches to modified oligonucleotides have been demonstrated, significant limitations to these approaches have been observed. For example, C8-arylated purine nucleosides are significantly more prone to acidic hydrolysis of the glycosidic bond and are more sensitive to oxidation than dA or dG [121]. As a result, they often are not compatible with solid-phase DNA synthesis techniques. A variety of modified nucleotide derivatives have been incorporated into oligonucleotides through enzymatic polymerase approaches, but steric limitations can limit the effectiveness of this approach [110,111].

A more general approach to preparing oligonucleotides with modified bases would be to build the nucleotide containing halogenated residues at the desired locations, followed by post-synthetic palladium-catalyzed coupling of the halonucleosides. The successful coupling of nucleotides suggests that the oligonucleotide backbone should be stable in the coupling reaction. An oligonucleotide represents a significantly more complex substrate than a simple nucleotide monomer, however. Successful development of this approach would allow a variety of modified oligonucleotides to be prepared from a common precursor containing a halogenated base residue at the desired position on the oligonucleotide.

Manderville reported the first successful example of the post-synthetic cross-coupling of a halogenated oligonucleotide [121]. Solution-phase coupling of oligonucleotides containing a single 8-bromoguanosine residue with arylboronic acids was performed with Pd(OAc)$_2$/TPPTS in water/acetonitrile with Na$_2$CO$_3$ as base (Scheme 51). Under optimized conditions, a guanine rich decanucleotide containing 8-BrdG (**19**) was arylated with 2-hydroxyphenylboronic acid to give **20** in 87% yield. In contrast, an attempt to prepare the same 8-arylguanine-containing decanucleotide by traditional solid-phase DNA synthetic methods resulted in the formation of a mixture of products. The major products were truncated oligomers that did not incorporate the 8-arylguanosine residue. Subsequent studies showed that a maximum of two bases could be incorporated after the 8-arylguanosine residue in the oligomerization process. The post-synthetic coupling approach was applied to the synthesis of oligonucleotides with up to 15 bases containing a single 8-BrdG residue.

The post-synthetic approach was applied to the synthesis of oligonucleotides containing diarylethylene photoswitches [93]. Oligonucleotides (15- and 19-mers) containing 5-IdC or 5-IdU residues were coupled with boronic acid **5** with Pd(OAc)$_2$/TPPTS in aqueous acetonitrile at 120 °C to give the coupled oligonucleotides in modest yields (16%–35%). The hindered boronic acid (**5**) is a much more challenging substrate than those used by Manderville. Coupling of oligonucleotides containing 5-IdU with vinylboronic acids occurs in high yields (49%–95%) using a preformed palladium complex of 2-aminopyrimidine-4,6-diolate (APD) as the precatalyst (Scheme 52) [79]. With the less-hindered boronic acids, the coupling could be performed under mild conditions (23–37 °C) in phosphate buffer. The reaction was highly selective with only small amounts of deiodination observed in some cases.

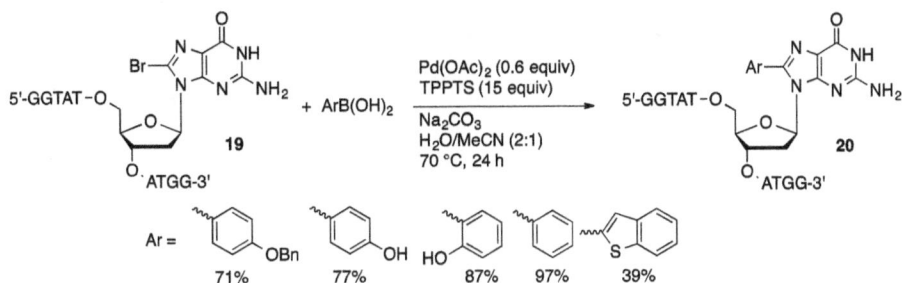

Scheme 51. Suzuki coupling of 8-BrdG-containing oligonucleotide.

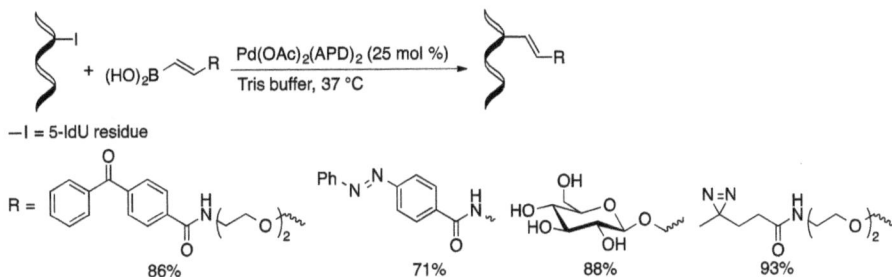

Scheme 52. Suzuki vinylation of 5-IdU-containing oligonucleotides.

The increased hydrolytic sensitivity of RNA makes it an even more challenging substrate than DNA for cross-coupling reactions, which often require strongly basic conditions. The Stille coupling can be carried out under mild conditions. Stille coupling of the dinucleotide 5-IUpG was catalyzed by Pd$_2$(dba)$_3$/AsPh$_3$ (50 mol % Pd) in DMF at 60 °C (Scheme 53) [122]. Excellent yields were obtained with electron-rich arylstannanes. This methodology was applied to the arylation of RNA oligomers containing a 5-IU residue supported on CPG solid support. Solid-supported 5-I-UAUAGGAGCU with stannane **20** gave the coupled product in 59% yield after removal from the solid support with ammonia.

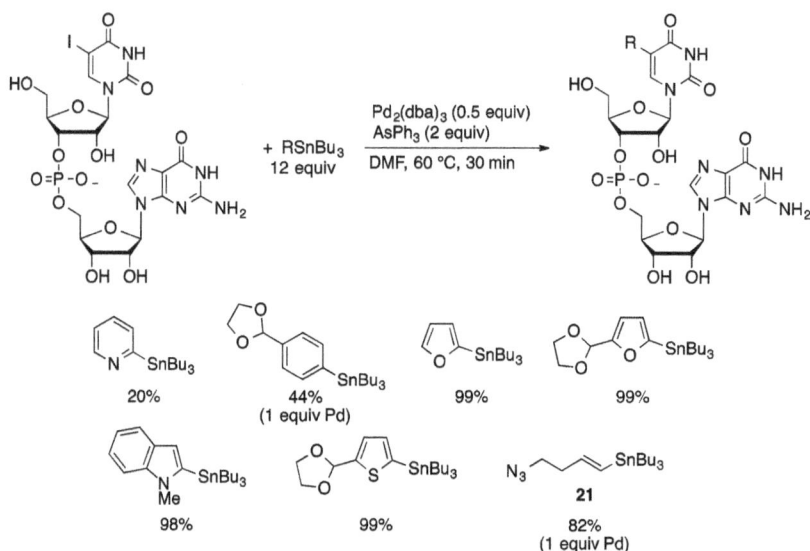

Scheme 53. Stille coupling of 5-I-UG.

Although only a few examples have been reported to date, the post-functionalization of oligonucleotides offers an exciting opportunity to prepare modified RNA and DNA structures. To date, only the Suzuki and Stille coupling of halogenated oligonucleotides has been demonstrated. Extension of this methodology to other important coupling reactions, such as the Sonogashira coupling, will further expand the usefulness of this transformation.

5. Conclusions

Palladium-catalyzed cross-coupling has developed into a highly effective method for the modification of unprotected nucleosides through carbon–carbon bond-forming reactions. Through the use of polar media, such as water, polar aprotic solvents, or combinations of the two, the hydrophilic nucleosides can be solubilized and converted to the desired adducts without the need to convert the nucleoside to a more lipophilic form. The ability to directly functionalize the nucleoside increases the overall yield and atom efficiency of the synthesis by avoiding the protection/deprotection sequence. In addition, protection strategies are generally not possible with the more hydrophilic nucleotide and oligonucleotide substrates. Catalysts supported with hydrophilic ligands provide the most general catalysts for modification of nucleoside derivatives. In aqueous solvents, hydrophilic catalysts are effective with nucleoside, nucleotides (mono- and triphosphates), and oligonucleotides. The ability to directly couple oligonucleotides is particularly noteworthy. Firstly, the ability to couple highly complex biomacromolecules is a true testament to the power and flexibility of palladium-catalyzed cross-coupling. Furthermore, traditional oligomerization strategies are often not compatible with modified nucleoside derivatives, so the ability to perform cross-coupling directly on oligonucleotides may be the only route to these materials.

Significant progress has been made in the area of palladium-catalyzed coupling of unprotected nucleosides, but there are still challenges left to be conquered. Although many of the classic C–C bond-forming coupling reactions have been demonstrated with unprotected nucleosides, examples of carbon-heteroatom bond formations remain unknown. In contrast, metal-catalyzed carbon-heteroatom coupling reactions of unprotected nucleosides are well precedented [17]. In general, Buchwald-Hartwig-type coupling reactions are less effective in aqueous solvent systems, although recent examples have been reported [123–125]. Developing these classes of coupling reactions

Molecules **2015**, *20*, 9419–9454

with unprotected nucleosides would provide access to new classes of nucleoside derivatives. The development of direct coupling reactions of arenes through C–H bond activation represents another attractive area of development. Heterocycles are common substrates for these types of reactions. In addition, the nucleobase heterocycles provide the opportunity for directed C–H functionalization reactions. Finally, further development of direct functionalization of oligonucleotides containing halogenated base residues will provide a route to prepare libraries of oligonucleotides containing modified bases. To date these reactions have been demonstrated with Suzuki and Stille couplings. Extension of these reactions to other classes of cross-coupling reactions would significantly increase the flexibility of these methodologies.

Acknowledgments: Support for our work in this area by the National Science Foundation (CHE-0124255) is gratefully acknowledged.

Conflicts of Interest: The authors declare no conflict of interest.

References and Notes

1. De Clercq, E. The unabated synthesis of new nucleoside analogs with antiviral potential: A tribute to Morris J. Robins. *Nucleos. Nucleot. Nucl. Acids* **2009**, *28*, 586–600. [CrossRef]
2. Robak, T. New Purine Nucleoside Analogs for Acute Lymphoblastic Leukemia. *Clin. Cancer Drugs* **2014**, *1*, 2–10. [CrossRef]
3. Jordheim, L.P.; Durantel, D.; Zoulim, F.; Dumontet, C. Advances in the development of nucleoside and nucleotide analogues for cancer and viral diseases. *Nat. Rev. Drug Discov.* **2013**, *12*, 447–464. [CrossRef] [PubMed]
4. Sofia, M.J.; Chang, W.; Furman, P.A.; Mosley, R.T.; Ross, B.S. Nucleoside, Nucleotide, and Non-Nucleoside Inhibitors of Hepatitis C Virus NS5B RNA-Dependent RNA-Polymerase. *J. Med. Chem.* **2012**, *55*, 2481–2531. [CrossRef] [PubMed]
5. Dai, Q.; Xu, D.; Lim, K.; Harvey, R.G. Efficient Syntheses of C8-Aryl Adducts of Adenine and Guanine Formed by Reaction of Radical Cation Metabolites of Carcinogenic Polycyclic Aromatic Hydrocarbons with DNA. *J. Org. Chem.* **2007**, *72*, 4856–4863. [CrossRef] [PubMed]
6. Champeil, E.; Pradhan, P.; Lakshman, M.K. Palladium-catalyzed synthesis of nucleoside adducts from bay- and fjord-region diol epoxides. *J. Org. Chem.* **2007**, *72*, 5035–5045. [CrossRef] [PubMed]
7. Lakshman, M.K.; Gunda, P. Palladium-catalyzed synthesis of carcinogenic polycyclic aromatic hydrocarbon epoxide-nucleoside adducts: The first amination of a chloro nucleoside. *Org. Lett.* **2003**, *5*, 39–42. [CrossRef] [PubMed]
8. Tanpure, A.A.; Pawar, M.G.; Srivatsan, S.G. Fluorescent Nucleoside Analogs: Probes for Investigating Nucleic Acid Structure and Function. *Isr. J. Chem.* **2013**, *53*, 366–378. [CrossRef]
9. Toseland, C.P.; Webb, M.R. Fluorescent nucleoside triphosphates for single-molecule enzymology. In *Single Molecule Enzymology: Methods and Protocols*; Mashanov, G.I., Batters, C., Eds.; Spring Science+Business Media, LLC: New York, NY, USA, 2011; Volume 778, pp. 161–174.
10. Matarazzo, A.; Hudson, R.H.E. Fluorescent adenosine analogs: A comprehensive survey. *Tetrahedron* **2015**, *71*, 1627–1657. [CrossRef]
11. Dodd, D.W.; Hudson, R.H.E. Intrinsically fluorescent base-discriminating nucleoside analogs. *Mini-Rev. Org. Chem.* **2009**, *6*, 378–391. [CrossRef]
12. Thomsen, N.M.; Vongsutilers, V.; Gannett, P.M. Synthesis of C8-Aryl Purines, Nucleosides, and Phosphoramidites. *Crit. Rev. Eukaryot. Gene Expr.* **2011**, *21*, 155–176. [CrossRef] [PubMed]
13. Kore, A.R.; Charles, I. Recent developments in the synthesis and applications of C5-substituted pyrimidine nucleosides and nucleotides. *Curr. Org. Chem.* **2012**, *16*, 1996–2013. [CrossRef]
14. Kore, A.R.; Yang, B.; Srinivasan, B. Recent Developments in the Synthesis of Substituted Purine Nucleosides and Nucleotides. *Curr. Org. Chem.* **2014**, *18*, 2072–2107. [CrossRef]
15. Colacot, T.J. *New Trends in Cross-Coupling: Theory and Applications*; Royal Society of Chemistry: London, UK, 2015; p. 864.
16. Agrofoglio, L.A.; Gillaizeau, I.; Saito, Y. Palladium-assisted routes to nucleosides. *Chem. Rev.* **2003**, *103*, 1875–1916. [CrossRef] [PubMed]

17. Lakshman, M.K. Synthesis of biologically important nucleoside analogs by palladium-catalyzed C–N bond formation. *Curr. Org. Synth.* **2005**, *2*, 83–112. [CrossRef]
18. Hocek, M.; Fojta, M. Cross-coupling reactions of nucleoside triphosphates followed by polymerase incorporation. Construction and applications of base-functionalized nucleic acids. *Org. Biomol. Chem.* **2008**, *6*, 2233–2241. [CrossRef] [PubMed]
19. De Ornellas, S.; Williams, T.J.; Baumann, C.G.; Fairlamb, I.J.S. Catalytic C-H/C-X bond functionalisation of nucleosides, nucleotides, nucleic acids, amino acids, peptides and proteins. In *C-H and C-X Bond Functionalization*; Ribas, X., Ed.; Royal Society of Chemistry: Cambridge, UK, 2013; Volume 11, pp. 409–447.
20. Hervé, G.; Sartori, G.; Enderlin, G.; MacKenzie, G.; Len, C. Palladium-catalyzed Suzuki reaction in aqueous solvents applied to unprotected nucleosides and nucleotides. *RSC Adv.* **2014**, *4*, 18558–18594. [CrossRef]
21. Hervé, G.; Len, C. Heck and Sonogashira couplings in aqueous media—Application to unprotected nucleosides and nucleotides. *Sust. Chem. Proc.* **2015**, *3*, 1–14. [CrossRef]
22. Hervé, G.; Len, C. Aqueous microwave-assisted cross-coupling reactions applied to unprotected nucleosides. *Front. Chem.* **2015**, *3*. [CrossRef]
23. Bigge, C.F.; Kalaritis, P.; Deck, J. R.; Mertes, M.P. Palladium-catalyzed coupling reactions of uracil nucleosides and nucleotides. *J. Am. Chem. Soc.* **1980**, *102*, 2033–2038. [CrossRef]
24. Hirota, K.; Isobe, Y.; Kitade, Y.; Maki, Y. A simple synthesis of 5-(1-alkenyl)uracil derivatives by palladium catalyzed oxidative coupling of uracils with olefins. *Synthesis* **1987**, *5*, 495–496. [CrossRef]
25. Deraedt, C.; Astruc, D. "Homeopathic" Palladium Nanoparticle Catalysis of Carbon–Carbon Coupling Reactions. *Acc. Chem. Res.* **2014**, *47*, 494–503. [CrossRef] [PubMed]
26. Matsuhashi, H.; Hatanaka, Y.; Kuroboshi, M.; Hiyama, T. Synthesis of 5-substituted pyrimidine nucleosides through a palladium-catalyzed cross-coupling of alkylhalosilanes. *Heterocycles* **1996**, *42*, 375–384. [CrossRef]
27. Gallagher-Duval, S.; Hervé, G.; Sartori, G.; Enderlin, G.; Len, C. Improved microwave-assisted ligand-free Suzuki-Miyaura cross-coupling of 5-iodo-2′-deoxyuridine in pure water. *New J. Chem.* **2013**, *37*, 1989–1995. [CrossRef]
28. Kumar, P.; Hornum, M.; Nielsen, L.J.; Enderlin, G.; Andersen, N.K.; Len, C.; Hervé, G.; Sartori, G.; Nielsen, P. High-Affinity RNA Targeting by Oligonucleotides Displaying Aromatic Stacking and Amino Groups in the Major Groove. Comparison of Triazoles and Phenyl Substituents. *J. Org. Chem.* **2014**, *79*, 2854–2863. [CrossRef] [PubMed]
29. Enderlin, G.; Sartori, G.; Hervé, G.; Len, C. Synthesis of 6-aryluridines via Suzuki-Miyaura cross-coupling reaction at room temperature under aerobic ligand-free conditions in neat water. *Tetrahedron Lett.* **2013**, *54*, 3374–3377. [CrossRef]
30. De Vries, J.G. A unifying mechanism for all high-temperature Heck reactions. The role of palladium colloids and anionic species. *Dalton Trans.* **2006**, 421–429.
31. Sakthivel, K.; Barbas, C.F., III. Expanding the Potential of DNA for Binding and Catalysis: Highly Functionalized dUTP Derivatives That Are Substrates for Thermostable DNA Polymerases. *Angew. Chem. Int. Ed.* **1998**, *37*, 2872–2875. [CrossRef]
32. Ding, H.; Greenberg, M.M. Hole Migration is the Major Pathway Involved in Alkali-Labile Lesion Formation in DNA by the Direct Effect of Ionizing Radiation. *J. Am. Chem. Soc.* **2007**, *129*, 772–773. [CrossRef] [PubMed]
33. Hervé, G.; Len, C. First ligand-free, microwave-assisted, Heck cross-coupling reaction in pure water on a nucleoside—Application to the synthesis of antiviral BVDU. *RSC Adv.* **2014**, *4*, 46926–46929. [CrossRef]
34. Hussain, I.; Singh, T. Synthesis of Biaryls through Aromatic C–H Bond Activation: A Review of Recent Developments. *Adv. Synth. Catal.* **2014**, *356*, 1661–1696. [CrossRef]
35. Rossi, R.; Bellina, F.; Lessi, M.; Manzini, C. Cross-Coupling of Heteroarenes by C–H Functionalization: Recent Progress towards Direct Arylation and Heteroarylation Reactions Involving Heteroarenes Containing One Heteroatom. *Adv. Synth. Catal.* **2014**, *356*, 17–117. [CrossRef]
36. Čerňa, I.; Pohl, R.; Hocek, M. The first direct C–H arylation of purine nucleosides. *Chem. Commun.* **2007**, *45*, 4729–4730. [CrossRef]
37. Storr, T.E.; Firth, A.G.; Wilson, K.; Darley, K.; Baumann, C.G.; Fairlamb, I.J.S. Site-selective direct arylation of unprotected adenine nucleosides mediated by palladium and copper: Insights into the reaction mechanism. *Tetrahedron* **2008**, *64*, 6125–6137. [CrossRef]

38. Storr, T.E.; Baumann, C.G.; Thatcher, R.J.; De Ornellas, S.; Whitwood, A.C.; Fairlamb, I.J.S. Pd(0)/Cu(I)-Mediated Direct Arylation of 2′-Deoxyadenosines: Mechanistic Role of Cu(I) and Reactivity Comparisons with Related Purine Nucleosides. *J. Org. Chem.* **2009**, *74*, 5810–5821. [CrossRef] [PubMed]

39. Liang, Y.; Gloudeman, J.; Wnuk, S.F. Palladium-Catalyzed Direct Arylation of 5-Halouracils and 5-Halouracil Nucleosides with Arenes and Heteroarenes Promoted by TBAF. *J. Org. Chem.* **2014**, *79*, 4094–4103. [CrossRef] [PubMed]

40. Chanda, A.; Fokin, V.V. Organic Synthesis "On Water". *Chem. Rev.* **2009**, *109*, 725–748. [CrossRef] [PubMed]

41. Amann, N.; Pandurski, E.; Fiebig, T.; Wagenknecht, H.A. Electron injection into DNA: Synthesis and spectrscopic properties of pyrenyl-modified oligonucleotides. *Chem. Eur. J.* **2002**, *8*, 4877–4883. [CrossRef] [PubMed]

42. Amann, N.; Wagenknecht, H.A. Preparation of pyrenyl-modified nucleosides via Suzuki-Miyaura cross-coupling reactions. *Synlett* **2002**, *5*, 687–691. [CrossRef]

43. Jacobsen, M.F.; Ferapontova, E.E.; Gothelf, K.V. Synthesis and electrochemical studies of an anthraquinone-conjugated nucleoside and derived oligonucleotides. *Org. Biomol. Chem.* **2009**, *7*, 905–908. [CrossRef] [PubMed]

44. Okamoto, A.; Inasaki, T.; Saito, I. Synthesis and ESR studies of nitronyl nitroxide-tethered oligodeoxynucleotides. *Tetrahedron Lett.* **2005**, *46*, 791–795. [CrossRef]

45. Kögler, M.; Vanderhoydonck, B.; De Jonghe, S.; Rozenski, J.; Van Belle, K.; Herman, J.; Louat, T.; Parchina, A.; Sibley, C.; Lescrinier, E.; *et al.* Synthesis and Evaluation of 5-Substituted 2′-deoxyuridine Monophosphate Analogs As Inhibitors of Flavin-Dependent Thymidylate Synthase in Mycobacterium tuberculosis. *J. Med. Chem.* **2011**, *54*, 4847–4862. [CrossRef] [PubMed]

46. Fresneau, N.; Hiebel, M.-A.; Agrofoglio, L.A.; Berteina-Raboin, S. Efficient synthesis of unprotected C-5-aryl/heteroaryl-2′-deoxyuridine via a Suzuki-Miyaura reaction in aqueous media. *Molecules* **2012**, *17*, 14409–14417. [CrossRef] [PubMed]

47. Mayer, E.; Valis, L.; Huber, R.; Amann, N.; Wagenknecht, H.A. Preparation of pyrene-modified purine and pyrimidine nucleosides via Suzuki-Miyaura cross-couplings and characterization of their fluorescent properties. *Synthesis* **2003**, *15*, 2335–2340.

48. Valis, L.; Wagenknecht, H.A. Synthesis and optical properties of the C-8 adduct of Benzo[*a*]pyrene and deoxyguanosine. *Synlett* **2005**, *15*, 2281–2284.

49. Kohyama, N.; Katashima, T.; Yamamoto, Y. Synthesis of novel 2-aryl AICAR derivatives. *Synthesis* **2004**, *17*, 2799–2804.

50. Ogasawara, S.; Saito, I.; Maeda, M. Synthesis and reversible photoisomerization of photo-switchable nucleoside, 8-styryl-2′-deoxyguanosine. *Tetrahedron Lett.* **2008**, *49*, 2479–2482. [CrossRef]

51. Wagner, C.; Wagenknecht, H.A. Reductive electron transfer in phenothiazine-modified DNA is dependent on the base sequence. *Chem. Eur. J.* **2005**, *11*, 1871–1876. [CrossRef] [PubMed]

52. Wanninger-Weiß, C.; Wagenknecht, H.A. Synthesis of 5-(2-pyrenyl)-2′-deoxyuridine as a DNA modification for electron-transfer studies: The critical role of the position of the chromophore attachment. *Eur. J. Org. Chem.* **2008**, *2008*, 64–71. [CrossRef]

53. Ehrenschwender, T.; Wagenknecht, H.A. Synthesis and spectroscopic characterization of BODIPY-modified uridines as potential fluorescent probes for nucleic acids. *Synthesis* **2008**, *2008*, 3657–3662. [CrossRef]

54. Hassan, M.E. Palladium-catalyzed cross-coupling reaction of organostannanes with nucleoside halides. *Coll. Czech. Chem. Commun.* **1991**, *56*, 1944–1947. [CrossRef]

55. Holzberger, B.; Strohmeier, J.; Siegmund, V.; Diederichsen, U.; Marx, A. Enzymatic synthesis of 8-vinyl- and 8-styryl-2′-deoxyguanosine modified DNA—Novel fluorescent molecular probes. *Bioorg. Med. Chem. Lett.* **2012**, *22*, 3136–3139. [CrossRef] [PubMed]

56. Matsuda, A.; Satoh, K.; Tanaka, H.; Miyasaka, T. Introduction of carbon substituents at C-2 position of purine nucleosides. *Nucleic Acids Symp. Ser.* **1983**, *12*, 5–8. [PubMed]

57. Jäger, S.; Rasched, G.; Komreich-Leshem, H.; Engesser, M.; Thum, O.; Famulok, M. A versatile toolbox for variable DNA functionalization at high density. *J. Am. Chem. Soc.* **2005**, *127*, 15071–15072. [CrossRef] [PubMed]

58. Sági, G.; Ötvös, L.; Ikeda, S.; Andrei, G.; Snoeck, R.; De Clercq, E. Synthesis and antiviral activities of 8-alkynyl-, 8-alkenyl-, and 8-alkyl-2′-deoxyadenosine analogs. *J. Med. Chem.* **1994**, *37*, 1307–1311. [CrossRef] [PubMed]

59. Volpini, R.; Costanzi, S.; Lambertucci, C.; Vittori, S.; Klotz, K.N.; Lorenzen, A.; Cristalli, G. Introduction of Alkynyl Chains on C-8 of Adenosine Led to Very Selective Antagonists of the A$_3$ Adenosine Receptor. *Bioorg. Med. Chem. Lett.* **2001**, *11*, 1931–1934. [CrossRef] [PubMed]

60. Olejniczak, A.; Wojtczak, B.; Lesnikowski, Z.J. 2′-Deoxyadenosine Bearing Hydrophobic Carborane Pharmacophore. *Nucleos. Nucleot. Nucl. Acids* **2007**, *26*, 1611–1613. [CrossRef]

61. Seela, F.; Ingale, S.A. "Double Click" Reaction on 7-Deazaguanine DNA: Synthesis and Excimer Fluorescence of Nucleosides and Oligonucleotides with Branched Side Chains Decorated with Proximal Pyrenes. *J. Org. Chem.* **2010**, *75*, 284–295. [CrossRef] [PubMed]

62. Zhao, Y.; Baranger, A.M. Design of an adenosine analogue that selectively improves the affinity of a mutant U1A protein for RNA. *J. Am. Chem. Soc.* **2003**, *125*, 2480–2488. [CrossRef] [PubMed]

63. Ardhapure, A.V.; Sanghvi, Y.S.; Kapdi, A.R.; García, J.; Sanchez, G.; Lozano, P.; Serrano, J.L. Pd-imidate complexes as recyclable catalysts for the synthesis of C5-alkenylated pyrimidine nucleosides via Heck cross-coupling reaction. *RSC Adv.* **2015**, *5*, 24558–24563. [CrossRef]

64. Li, C.J. Organic reactions in aqueous media with a focus on carbon-carbon bond formations: A decade update. *Chem. Rev.* **2005**, *105*, 3095–3165. [CrossRef] [PubMed]

65. Shaughnessy, K.H. Beyond TPPTS: New approaches to the development of efficient palladium-catalyzed aqueous-phase cross-coupling reactions. *Eur. J. Org. Chem.* **2006**, *2006*, 1827–1835. [CrossRef]

66. Polshettiwar, V.; Decottignies, A.; Len, C.; Fihri, A. Suzuki-Miyaura Cross-Coupling Reactions in Aqueous Media: Green and Sustainable Syntheses of Biaryls. *ChemSusChem* **2010**, *3*, 502–522. [CrossRef] [PubMed]

67. Shaughnessy, K.H. Greener approaches to cross-coupling. In *New Trends in Cross-Coupling: Theory and Application*; Colacot, T.J., Ed.; Royal Society of Chemistry: Cambridge, UK, 2015; pp. 645–696.

68. Garrett, C.E.; Prasad, K. The art of meeting palladium specifiations in active pharmaceutical ingredients produced by Pd-catalyzed reactions. *Adv. Synth. Catal.* **2004**, *346*, 889–900. [CrossRef]

69. Casalnuovo, A.L.; Calabrese, J.C. Palladium-catalyzed alkylation in aqueous media. *J. Am. Chem. Soc.* **1990**, *112*, 4324–4330. [CrossRef]

70. Western, E.C.; Daft, J.R.; Johnson, E.M., II; Gannett, P.M.; Shaughnessy, K.H. Efficient, one-step Suzuki arylation of unprotected halonucleosides using water-soluble palladium catalysts. *J. Org. Chem.* **2003**, *68*, 6767–6774. [CrossRef] [PubMed]

71. Shaughnessy, K.H.; Booth, R.S. Sterically demanding, water-soluble alkylphosphines as ligands for high activity Suzuki coupling of aryl bromides in aqueous solvents. *Org. Lett.* **2001**, *3*, 2757–2759. [CrossRef] [PubMed]

72. Hobley, G.; Gubala, V.; Rivera-Sánchez, M.D.C.; Rivera, J.M. Synthesis of 8-heteroaryl-2′-deoxyguanosine derivatives. *Synlett* **2008**, *2008*, 1510–1514. [CrossRef] [PubMed]

73. Western, E.C.; Shaughnessy, K.H. Inhibitory effects of the guanine moiety on the Suzuki couplings of unprotected halonucleosides in aqueous media. *J. Org. Chem.* **2005**, *70*, 6378–6388. [CrossRef] [PubMed]

74. Both ligands are available from STREM Chemical Co: TPPTS (15–8007, $79/mmol), TXPTS (15–7860, $149/mmol).

75. Collier, A.; Wagner, G.K. Suzuki-Miyaura cross-coupling of unprotected halopurine nucleosides in water-influence of catalyst and cosolvent. *Synth. Commun.* **2006**, *36*, 3713–3721. [CrossRef]

76. Sartori, G.; Enderlin, G.; Hervé, G.; Len, C. Highly effective synthesis of C-5-substituted 2′-deoxyuridine using Suzuki-Miyaura cross-coupling in water. *Synthesis* **2012**, *44*, 767–772. [CrossRef]

77. Sartori, G.; Hervé, G.; Enderlin, G.; Len, C. New, efficient approach for the ligand-free Suzuki-Miyaura reaction of 5-iodo-2′-deoxyuridine in water. *Synthesis* **2013**, *45*, 330–333. [CrossRef]

78. Kapdi, A.; Gayakhe, V.; Sanghvi, Y.S.; García, J.; Lozano, P.; da Silva, I.; Pérez, J.; Serrano, J.L. New water soluble Pd-imidate complexes as highly efficient catalysts for the synthesis of C5-arylated pyrimidine nucleosides. *RSC Adv.* **2014**, *4*, 17567–17572. [CrossRef]

79. Lercher, L.; McGouran, J.F.; Kessler, B.M.; Schofield, C.J.; Davis, B.G. DNA Modification under Mild Conditions by Suzuki-Miyaura Cross-Coupling for the Generation of Functional Probes. *Angew. Chem. Int. Ed.* **2013**, *52*, 10553–10558. [CrossRef]

80. Čapek, P.; Hocek, M. Efficient one-step synthesis of optically pure (adenin-8-yl)phenylalanine nucleosides. *Synlett* **2005**, *19*, 3005–3007.

81. Čapek, P.; Pohl, R.; Hocek, M. Cross-coupling reactions of unprotected halopurine bases, nucleosides, and nucleoside triphosphates with 4-boronophenylalanine in water. Synthesis of (purin-8-yl)- and (purin-6-yl)phenylalanines. *Org. Biomol. Chem.* **2006**, *4*, 2278–2284. [CrossRef] [PubMed]

82. Vrábel, M.; Pohl, R.; Klepetářová, B.; Votruba, I.; Hocek, M. Synthesis of 2'-deoxyadenosine nucleosides bearing bipyridine-type ligands and their Ru-complexes in position 8 through cross-coupling reactions. *Org. Biomol. Chem.* **2007**, *5*, 2849–2857. [CrossRef] [PubMed]

83. Vrábel, M.; Pohl, R.; Votruba, I.; Sajadi, M.; Kovalenko, S.A.; Ernsting, N.P.; Hocek, M. Synthesis and photophysical properties of 7-deaza-2'-deoxyadenosines bearing bipyridine ligands and their Ru(II)-complexes in position 7. *Org. Biomol. Chem.* **2008**, *6*, 2852–2860. [CrossRef] [PubMed]

84. Kalachova, L.; Pohl, R.; Hocek, M. Synthesis of 2'-deoxyuridine and 2'-deoxycytidine nucleosides bearing bipyridine and terpyridine ligands at position 5. *Synthesis* **2009**, *2009*, 105–112. [CrossRef]

85. Rankin, K.M.; Sproviero, M.; Rankin, K.; Sharma, P.; Wetmore, S.D.; Manderville, R.A. C8-Heteroaryl-2'-deoxyguanosine Adducts as Conformational Fluorescent Probes in the NarI Recognition Sequence. *J. Org. Chem.* **2012**, *77*, 10498–10508. [CrossRef] [PubMed]

86. Schlitt, K.M.; Millen, A.L.; Wetmore, S.D.; Manderville, R.A. An indole-linked C8-deoxyguanosine nucleoside acts as a fluorescent reporter of Watson-Crick *versus* Hoogsteen base pairing. *Org. Biomol. Chem.* **2011**, *9*, 1565–1571. [CrossRef] [PubMed]

87. Manderville, R.A.; Omumi, A.; Rankin, K.M.; Wilson, K.A.; Millen, A.L.; Wetmore, S.D. Fluorescent C-linked C8-aryl-guanine probe for distinguishing syn from anti structures in duplex DNA. *Chem. Res. Toxicol.* **2012**, *25*, 1271–1282. [CrossRef] [PubMed]

88. Miyata, K.; Mineo, R.; Tamamushi, R.; Mizuta, M.; Ohkubo, A.; Taguchi, H.; Seio, K.; Santa, T.; Sekine, M. Synthesis and Fluorescent Properties of Bi- and Tricyclic 4-N-Carbamoyldeoxycytidine Derivatives. *J. Org. Chem.* **2007**, *77*, 102–108. [CrossRef]

89. Mizuta, M.; Seio, K.; Miyata, K.; Sekine, M. Fluorescent Pyrimidopyrimidoindole Nucleosides: Control of Photophysical Characterizations by Substituent Effects. *J. Org. Chem.* **2007**, *72*, 5046–5055. [CrossRef] [PubMed]

90. Segal, M.; Fischer, B. Analogs of uracil nucleosides with intrinsic fluorescence (NIF-analogs): Synthesis and photophysical properties. *Org. Biomol. Chem.* **2012**, *10*, 1571–1580. [CrossRef] [PubMed]

91. Storr, T.E.; Strohmeier, J.A.; Baumann, C.G.; Fairlamb, I.J.S. A sequential direct arylation/Suzuki-Miyaura cross-coupling transformation of unprotected 2'-deoxyadenosine affords a novel class of fluorescent analogs. *Chem. Commun.* **2010**, *46*, 6470–6472. [CrossRef]

92. Bárta, J.; Pohl, R.; Klepetářová, B.; Ernsting, N.P.; Hocek, M. Modular Synthesis of 5-Substituted Thiophen-2-yl C-2'-Deoxyribonucleosides. *J. Org. Chem.* **2008**, *73*, 3798–3806. [CrossRef] [PubMed]

93. Cahová, H.; Jäschke, A. Nucleoside-Based Diarylethene Photoswitches and Their Facile Incorporation into Photoswitchable DNA. *Angew. Chem. Int. Ed.* **2013**, *52*, 3186–3190. [CrossRef]

94. Macíčková-Cahová, H.; Pohl, R.; Horáková, P.; Havran, L.; Špaček, J.; Fojta, M.; Hocek, M. Alkylsulfanylphenyl Derivatives of Cytosine and 7-Deazaadenine Nucleosides, Nucleotides and Nucleoside Triphosphates: Synthesis, Polymerase Incorporation to DNA and Electrochemical Study. *Chem. Eur. J.* **2011**, *17*, 5833–5841. [CrossRef] [PubMed]

95. Gubala, V.; Betancourt, J.E.; Rivera, J.M. Expanding the Hoogsteen Edge of 2'-Deoxyguanosine: Consequences for G-Quadruplex Formation. *Org. Lett.* **2004**, *6*, 4735–4738. [CrossRef] [PubMed]

96. Vongsutilers, V.; Phillips, D.J.; Train, B.C.; McKelvey, G.R.; Thomsen, N.M.; Shaughnessy, K.H.; Lewis, J.P.; Gannett, P.M. The conformational effect of para-substituted C8-arylguanine adducts on the B/Z-DNA equilibrium. *Biophys. Chem.* **2011**, *154*, 41–48. [CrossRef] [PubMed]

97. Nauš, P.; Kuchař, M.; Hocek, M. Cytostatic and antiviral 6-arylpurine ribonucleosides IX. synthesis and evaluation of 6-substituted 3-deazapurine ribonucleosides. *Coll. Czech. Chem. Comm.* **2008**, *73*, 665–678. [CrossRef]

98. Nauš, P.; Pohl, R.; Votruba, I.; Džubák, P.; Hajdúch, M.; Ameral, R.; Birkuš, G.; Wang, T.; Ray, A.S.; Mackman, R.; *et al.* 6-(Het)aryl-7-Deazapurine Ribonucleosides as Novel Potent Cytostatic Agents. *J. Med. Chem.* **2010**, *53*, 460–470. [CrossRef] [PubMed]

99. Bourderioux, A.; Nauš, P.; Perlíková, P.; Pohl, R.; Pichová, I.; Votruba, I.; Džubák, P.; Konečný, P.; Hajdúch, M.; Stray, K.M.; *et al.* Synthesis and Significant Cytostatic Activity of 7-Hetaryl-7-deazaadenosines. *J. Med. Chem.* **2011**, *54*, 5498–5507. [CrossRef] [PubMed]

100. Amiable, C.; Paoletti, J.; Haouz, A.; Padilla, A.; Labesse, G.; Kaminski, P.A.; Pochet, S. 6-(Hetero)Arylpurine nucleotides as inhibitors of the oncogenic target DNPH1: Synthesis, structural studies and cytotoxic activities. *Eur. J. Med. Chem.* **2014**, *85*, 418–437. [CrossRef] [PubMed]

101. Nauš, P.; Perlíková, P.; Bourderioux, A.; Pohl, R.; Slavětínská, L.; Votruba, I.; Bahador, G.; Birkuš, G.; Cihlář, T.; Hocek, M. Sugar-modified derivatives of cytostatic 7-(het)aryl-7-deazaadenosines: 2'-C-methylribonucleosides, 2'-deoxy-2'-fluoroarabinonucleosides, arabinonucleosides and 2'-deoxyribonucleosides. *Bioorg. Med. Chem.* **2012**, *20*, 5202–5214. [CrossRef] [PubMed]

102. Perlíková, P.; Jornet Martínez, N.; Slavětínská, L.; Hocek, M. Synthesis of 2'-deoxy-2'-fluororibo- and 2'-deoxy-2',2'-difluororibonucleosides derived from 6-(het)aryl-7-deazapurines. *Tetrahedron* **2012**, *68*, 8300–8310. [CrossRef]

103. Nauš, P.; Caletková, O.; Konečny, P.; Džubak, P.; Bogdanova, K.; Kolář, M.; Vrbková, J.; Slavětínská, L.; Tloušťová, E.; Perlíková, P.; *et al.* Synthesis, Cytostatic, Antimicrobial, and Anti-HCV Activity of 6-Substituted 7-(Het)aryl-7-deazapurine Ribonucleosides. *J. Med. Chem.* **2014**, *57*, 1097–1110. [CrossRef] [PubMed]

104. Ikonen, S.; Maćičková-Cahová, H.; Pohl, R.; Šanda, M.; Hocek, M. Synthesis of nucleoside and nucleotide conjugates of bile acids, and polymerase construction of bile acid-functionalized DNA. *Org. Biomol. Chem.* **2010**, *8*, 1194–1201. [CrossRef] [PubMed]

105. Cho, J.H.; Prickett, C.D.; Shaughnessy, K.H. Efficient Sonogashira Coupling of Unprotected Halonucleosides in Aqueous Solvents Using Water-Soluble Palladium Catalysts. *Eur. J. Org. Chem.* **2010**, 3678–3683. [CrossRef]

106. Cho, J.H.; Shaughnessy, K.H. Aqueous-phase Sonogashira alkynylation to synthesize 5-substituted pyrimidine and 8-substituted purine nucleosides. *Curr. Prot. Nucleic Acid Chem.* **2012**, *49*. [CrossRef]

107. Cho, J.H.; Shaughnessy, K.H. Aqueous-Phase Heck Coupling of 5-Iodouridine and Alkenes under Phosphine-Free Conditions. *Synlett* **2011**, *20*, 2963–2966.

108. Thoresen, L.H.; Jiao, G.S.; Haaland, W.C.; Metzker, M.L.; Burgess, K. Rigid, conjugated, fluoresceinated thymidine triphosphates: Syntheses and polymerase mediated incorporation into DNA analogs. *Chem. Eur. J.* **2003**, *9*, 4603–4610. [CrossRef] [PubMed]

109. Collier, A.; Wagner, G. A facile two-step synthesis of 8-arylated guanosine mono- and triphosphates (8-aryl GXPs). *Org. Biomol. Chem.* **2006**, *4*, 4526–4532. [CrossRef] [PubMed]

110. Čapek, P.; Cahová, H.; Pohl, R.; Hocek, M.; Gloeckner, C.; Marx, A. An efficient method for the construction of functionalized DNA bearing amino acid groups through cross-coupling reactions of nucleoside triphosphates followed by primer extension or PCR. *Chem. Eur. J.* **2007**, *13*, 6196–6203. [CrossRef] [PubMed]

111. Cahová, H.; Pohl, R.; Bednárová, L.; Nováková, K.; Cvačka, J.; Hocek, M. Synthesis of 8-bromo, 8-methyl- and 8-phenyl-dATP and their polymerase incorporation into DNA. *Org. Biomol. Chem.* **2008**, *6*, 3657–3660. [CrossRef] [PubMed]

112. Cahová, H.; Havran, L.; Brázdilová, P.; Pivoňková, H.; Pohl, R.; Fojta, M.; Hocek, M. Aminophenyl- and nitrophenyl-labeled nucleoside triphosphates: Synthesis, enzymatic incorporation, and electrochemical detection. *Angew. Chem. Int. Ed.* **2008**, *47*, 2059–2062. [CrossRef]

113. Balintová, J.; Plucnara, M.; Vidláková, P.; Pohl, R.; Havran, L.; Fojta, M.; Hocek, M. Benzofurazane as a New Redox Label for Electrochemical Detection of DNA: Towards Multipotential Redox Coding of DNA Bases. *Chem. Eur. J.* **2013**, *19*, 12720–12731. [CrossRef] [PubMed]

114. Riedl, J.; Pohl, R.; Rulíšek, L.; Hocek, M. Synthesis and Photophysical Properties of Biaryl-Substituted Nucleosides and Nucleotides. Polymerase Synthesis of DNA Probes Bearing Solvatochromic and pH-Sensitive Dual Fluorescent and ^{19}F-NMR Labels. *J. Org. Chem.* **2012**, *77*, 1026–1044. [CrossRef] [PubMed]

115. Raindlová, V.; Pohl, R.; Šanda, M.; Hocek, M. Direct Polymerase Synthesis of Reactive Aldehyde-Functionalized DNA and Its Conjugation and Staining with Hydrazines. *Angew. Chem. Int. Ed.* **2010**, *49*, 1064–1066. [CrossRef]

116. Pesnot, T.; Wagner, G.K. Novel derivatives of UDP-glucose: Concise synthesis and fluorescent properties. *Org. Biomol. Chem.* **2008**, *6*, 2884–2891. [CrossRef] [PubMed]

117. Collier, A.; Wagner, G.K. A fast synthetic route to GDP-sugars modified at the nucleobase. *Chem. Commun.* **2008**, *2*, 178–180. [CrossRef]

118. Pergolizzi, G.; Butt, J.N.; Bowater, R.P.; Wagner, G.K. A novel fluorescent probe for NAD-consuming enzymes. *Chem. Commun.* **2011**, *47*, 12655–12657. [CrossRef]

119. Brázdilová, P.; Vrábel, M.; Pohl, R.; Pivoňková, H.; Havran, L.; Hocek, M.; Fojta, M. Ferrocenylethynyl derivatives of nucleoside triphosphates: Synthesis, incorporation, electrochemistry, and bioanalytical applications. *Chem. Eur. J.* **2007**, *13*, 9527–9533. [CrossRef] [PubMed]

120. Dadová, J.; Vidláková, P.; Pohl, R.; Havran, L.; Fojta, M.; Hocek, M. Aqueous Heck Cross-Coupling Preparation of Acrylate-Modified Nucleotides and Nucleoside Triphosphates for Polymerase Synthesis of Acrylate-Labeled DNA. *J. Org. Chem.* **2013**, *78*, 9627–9637. [CrossRef] [PubMed]

121. Omumi, A.; Beach, D.G.; Baker, M.; Gabryelski, W.; Manderville, R.A. Postsynthetic Guanine Arylation of DNA by Suzuki-Miyaura Cross-Coupling. *J. Am. Chem. Soc.* **2011**, *133*, 42–50. [CrossRef] [PubMed]

122. Krause, A.; Hertl, A.; Muttach, F.; Jäschke, A. Phosphine-Free Stille-Migita Chemistry for the Mild and Orthogonal Modification of DNA and RNA. *Chem. Eur. J.* **2014**, *20*, 16613–16619. [CrossRef] [PubMed]

123. Hirai, Y.; Uozumi, Y. C-N and C-S bond forming cross coupling in water with amphiphilic resin-supported palladium complexes. *Chem. Lett.* **2011**, *40*, 934–935. [CrossRef]

124. Lipshutz, B.H.; Ghorai, S.; Abela, A.R.; Moser, R.; Nishikata, T.; Duplais, C.; Krasovskiy, A.; Gaston, R.D.; Gadwood, R.C. TPGS-750-M: A Second-Generation Amphiphile for Metal-Catalyzed Cross-Couplings in Water at Room Temperature. *J. Org. Chem.* **2011**, *76*, 4379–4391. [CrossRef] [PubMed]

125. Tardiff, B.J.; Stradiotto, M. Buchwald-Hartwig Amination of (Hetero)aryl Chlorides by Employing Mor-DalPhos under Aqueous and Solvent-Free Conditions. *Eur. J. Org. Chem.* **2012**, 3972–3977. [CrossRef]

![molecules logo] *molecules*

MDPI

Article

Formation of Mixed-Ligand Complexes of Pd²⁺ with Nucleoside 5'-Monophosphates and Some Metal-Ion-Binding Nucleoside Surrogates

Oleg Golubev, Tuomas Lönnberg and Harri Lönnberg *

Department of Chemistry, University of Turku, FIN-20014 Turku, Finland; oleg.golubev@utu.fi (O.G.); tuanlo@utu.fi (T.L.)

* Author to whom correspondence should be addressed; harlon@utu.fi; Tel.: +358-2-333-6770; Fax: +358-2-333-6700.

External Editors: Mahesh K. Lakshman and Fumi Nagatsugi

Received: 16 September 2014; in revised form: 8 October 2014; Accepted: 17 October 2014; Published: 22 October 2014

Abstract: Formation of mixed-ligand Pd²⁺ complexes between canonical nucleoside 5'-monophosphates and five metal-ion-binding nucleoside analogs has been studied by ^{1}H-NMR spectroscopy to test the ability of these nucleoside surrogates to discriminate between unmodified nucleobases by Pd²⁺-mediated base pairing. The nucleoside analogs studied included 2,6-bis(3,5-dimethylpyrazol-1-yl)-, 2,6-bis(1-methylhydrazinyl)- and 6-(3,5-dimethylpyrazol-1-yl)-substituted 9-(β-D-ribofuranosyl)purines **1–3**, and 2,4-bis(3,5-dimethylpyrazol-1-yl)- and 2,4-bis(1-methylhydrazinyl)-substituted 5-(β-D-ribofuranosyl)-pyrimidines **4–5**. Among these, the purine derivatives **1-3** bound Pd²⁺ much more tightly than the pyrimidine derivatives **4, 5** despite apparently similar structures of the potential coordination sites. Compounds **1** and **2** formed markedly stable mixed-ligand Pd²⁺ complexes with UMP and GMP, UMP binding favored by **1** and GMP by **2**. With **3**, formation of mixed-ligand complexes was retarded by binding of two molecules of **3** to Pd²⁺.

Keywords: Pd²⁺ complexes; nucleosides; NMR; mixed-ligand complexes

---

1. Introduction

It has been well established that linear-coordinating Hg²⁺ and Ag⁺ ions may stabilize TT and CC mismatches within oligonucleotide duplexes while square-planar-coordinating Cu²⁺ ion is able to bridge various modified metal-ion-binding bases on opposite strands [1–3]. Much less is known about discrimination between canonical nucleobases by oligonucleotide probes incorporating metal-ion-binding surrogate bases. Cu²⁺ and Zn²⁺ ions have been shown to enhance hybridization of 2'-O-methyl oligoribonucleotides containing a 2,6-bis(3,5-dimethylpyrazol-1-yl)purine base with complementary unmodified 2'-O-methyl oligoribonucleotides [4]. The magnitude of duplex stabilization does not, however, depend only on the identity of the opposite base, but also on the flanking sequences.

Square-planar-coordinating Pd²⁺ ion is known to exhibit exceptionally high affinity to nucleic acid bases [5–7]. Accordingly, Pd²⁺ complexes are interesting candidates for base moiety discrimination, although so far no convincing examples of Pd²⁺-mediated base-pairing at oligonucleotide level are available. Only some indications of recognition of thymine within an oligodeoxyribonucleotide by 2,6-bis(3,5-dimethylpyrazol-1-yl)purine in the presence of Pd²⁺ have been observed [8]. We have previously tried to evaluate the formation of mixed-ligand Pd²⁺ complexes between some metal ion binding nucleoside analogs and pyrimidine nucleosides [9]. Owing to severe solubility problems, the results obtained remain scanty. To learn more about the discrimination power of Pd²⁺ complexes, we now report on NMR studies concerning formation of mixed-ligand Pd^{2+} complexes between metal-ion-binding nucleosides **1–5** (Figure 1) and six nucleoside 5'-monophosphates (NMPs, Figure 2).

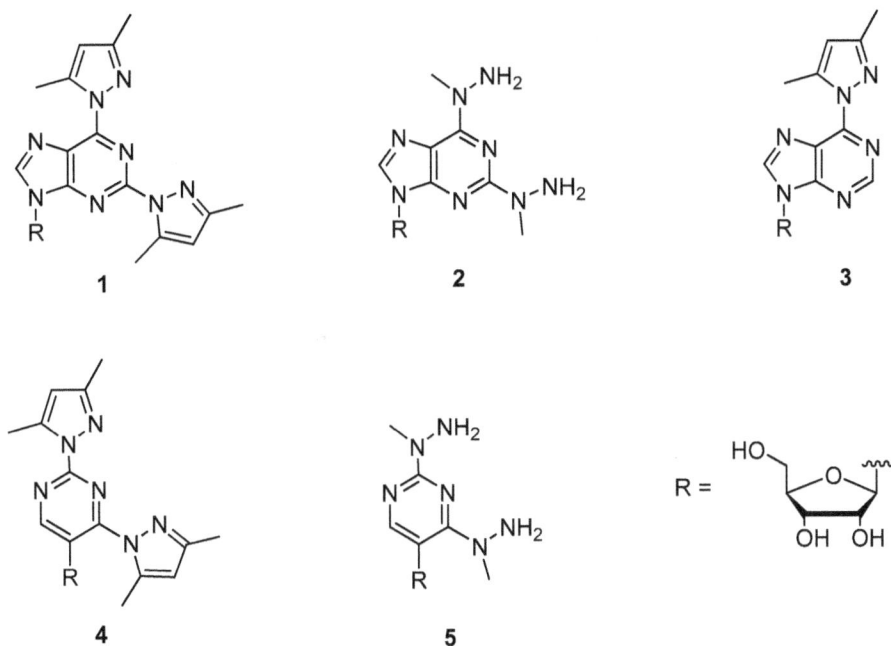

Figure 1. Structures of the metal-ion-binding nucleosides **1–5** used in the present study.

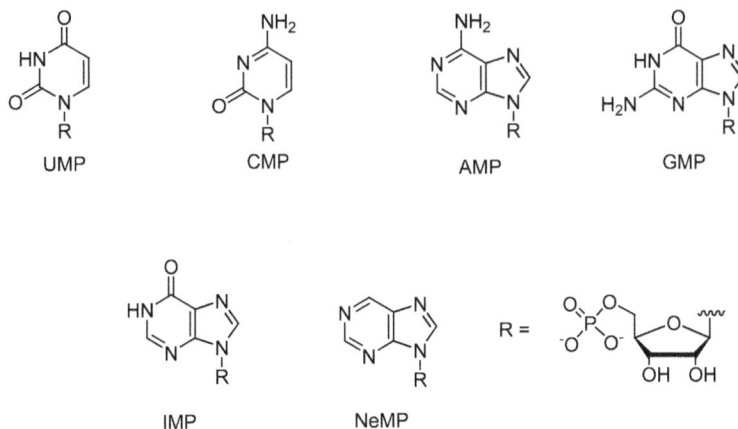

Figure 2. Structures of nucleoside 5'-monophosphates used in the present study.

2. Results and Discussion

2.1. Compounds Employed

The preparation of metal-ion-binding nucleosides **1–5** has been described previously [9,10]. Among the NMPs employed, UMP, CMP, AMP, GMP and IMP were commercial products and the preparation of the 5'-monophosphate of nebularine (NeMP), *i.e.*, unsubstituted 9-(β-D-ribofuranosyl) purine, has been described previously [11].

2.2. Pd^{2+} Complexes of Modified Nucleosides (**1–5**)

Interaction of the modified nucleosides **1–5** with Pd^{2+} was studied first. For this purpose, K$_2$PdCl$_4$ was added portionwise into a 5.0 mmol·L^{-1} solution of the nucleoside in phosphate buffered D$_2$O (0.12 mol·L^{-1}, pD 7.6, 25 °C) keeping the nucleoside concentration constant. After each addition a ^{1}H-NMR spectrum was recorded. Table 1 records the chemical shifts of the signals referring to the Pd^{2+} complexes formed.

Upon addition of K$_2$PdCl$_4$ to 2,6-bis(3,5-dimethylpyrazol-1-yl)purine riboside (**1**), the intensity of the H8 singlet of **1** at 8.59 ppm gradually decreased and two new pairs of singlets (8.85 and 8.83 and 8.63 and 8.61) appeared (Figure S1 in Supplementary Files). When 0.5 equiv. of K$_2$PdCl$_4$ had been added, the H8 singlet at 8.59 had almost entirely disappeared and a new singlet at 8.71 appeared. On approaching 1.0 equiv. addition of K$_2$PdCl$_4$, the two pairs of singlets weakened while the singlet at 8.71 became more intense (Figure S2 in Supplementary Files). Corresponding changes occurred in the anomeric proton region. Formation of the two pairs of singlets at 8.85 and 8.83 and 8.63 and 8.61 was accompanied by the appearance of three anomeric proton doublets at 5.98 (*J* 7.6), 5.89 (*J* 4.4 Hz) and 5.84 (*J* 4.8 Hz), the first one being twice as intense as the latter ones. A doublet at 6.19 (*J* 4.2 Hz), in turn, appeared parallel to the singlet at 8.71. Accordingly, at low concentration of K$_2$PdCl$_4$, a 2:1 (1:Pd) complex is formed and on increasing the concentration of K$_2$PdCl$_4$, the 1:1 complex predominates.

Table 1. Chemical shifts for the aromatic and anomeric protons of the modified nucleosides **1–5** and their Pd^{2+} complexes in D$_2$O at pD 7.6 (0.12 M phosphate buffer, 25 °C).

Compd.	Aromatic Proton Shifts	Anomeric Proton Shifts
1	s 8.59(H8), s 6.19 and 6.11(H4'')	d 6.15 (*J* 5.2)
(**1**)Pd	s 8.71(H8) [a]	d 6.19 (*J* 4.2)
(**1**)$_2$Pd	s 8.85 and 8.83 and 8.63 and 8.61(H8) [b]	d5.98(*J* 7.6), d 5.89 (*J* 4.4), d 5.84(*J* 4.8)
2	s 7.86(H8)	d 5.84 (*J* 5.3)
(**2**)Pd	s 8.06(H8)	d 5.86 [c]
3	s 8.81(H2), s 8.63(H8), s 6.23(H4'')	d 6.16 (*J* 5.6)
(**3**)Pd	s 8.92(H2/8), s 8.70(H2/8), s 6.40(H4'')	d 6.19 (*J* 3.8)
	s 8.84(H2/8), s 8.42(H2/8), s 6.38(H4'')	d 6.19 (*J* 3.8)
4	s 9.03(H6), s 6.12 and 6.07(H4'')	d 5.06 (*J* 5.4)
(**4**)Pd	d	e
5	s 7.92(H6)	s 5.60

[a] H4'' signals at 6.20–6.45 overlap with the corresponding signals of (**1**)$_2$Pd. [b] Several H4'' signals at 6.20–6.64.
[c] Overlaps with H1' of **2**. [d] The disappearance of H6 singlet of uncomplexed **4** was accompanied by appearance of 6 new singlets at 9.28, 9.22, 9.20, 8.93, 8.63 and 8.60. [e] The disappearance of the H1' doublet of **4** was accompanied by appearance of 3 new signals at 5.73, 5.39 and 5.30.

The 1:1 complex, exhibiting only one set of ^{1}H-NMR signals, most likely is a (**1**)PdCl$^+$ complex, the metal ion being coordinated to N1 of the purine base and N2 atoms of the pyrazolyl substituents. When the concentration of **1** is high compared to that of K$_2$PdCl$_4$, the chlorido ligand is replaced with another molecule of **1** which undergoes either N1 or N7 binding. N7 binding appears more likely, since this site is sterically less hindered than the N1 site flanked by the 3,5-dimethylpyrazol-1-yl groups, and since a reasonably large (0.26 and 0.24 ppm) downfield shift of the H8 signal is observed [7,12–15]. The H8 resonances of both modified purine bases engaged in the complex appear as two singlets, most likely due to the fact that two mutual orientations of the ligands are possible: the sugar moieties may be situated on the same or opposite sides of the plane of Pd^{2+} and the purine bases.

6-(3,5-Dimethylpyrazol-1-yl)purine riboside (**3**) bound Pd^{2+} much more weakly. Only half of **3** was complexed at an equimolar 5 mmol·L^{-1} concentration (Figure S3 in Supplementary Files). Two complexes were formed in parallel, evidently due to almost as efficient binding to N1 and N7 in addition to binding to the pyrazolyl N2 atom. The markedly lower affinity compared to **1** lends substantial additional evidence for the assumption that both pyrazolyl groups of **1** participate in binding of Pd^{2+}.

Another important observation is that replacement of aromatic 3,5-dimethylpyrazol-1-yl groups with aliphatic 1-methylhydrazinyl groups markedly weakens the binding of Pd^{2+}. Only half of 2,6-bis(1-methylhydrazinyl)purine riboside (**2**) was engaged in complex formation at 5.0 mmol·L^{-1} concentration of K_2PdCl_4 and **2**, although a tridentate coordination, as with **1**, apparently is possible (Figure S4 in Supplementary Files). However, binding to the terminal amino groups of the hydrazinyl substituents is evidently impeded by the fact that the lone electron pair of the nitrogen atoms participates in the π-electron resonance of the purine ring, which lowers the electron density at the potential donor atoms. With **1** the situation is different, since the lone electron pair of the N2 atoms of the pyrazolyl substituents is not delocalized but the N2 atoms are pyridine type nitrogens. Additionally, hydrogen bonding of the NH_2 group to N1 gives an expectedly moderately stable five membered structure, which may still retard the complexing ability of **2**. A marked broadening of the signals took place at high concentrations of K_2PdCl_4 and unidentified broad signals appeared, which may well refer to formation of polymeric complexes.

Quite unexpectedly, 2,4-bis(3,5-dimethylpyrazol-1-yl)-5-(β-D-ribofuranosyl)pyrimidine (**4**) also turned out to bind Pd^{2+} very weakly in spite of the fact that the expected binding site, *viz.* the N2 atoms of the two pyrazolyl groups and the intervening N3 of the pyrimidine ring, appears very similar to the binding site in **1**. At 5.0 mmol·L^{-1} concentration of both K_2PdCl_4 and **4**, more than 50% of **4** was complexed, but instead of a single clearly recognizable tridentate complex, several species in comparable amounts were formed (Figure S5 in Supplementary Files). Presumably, steric repulsion between the ribosyl group and the 5-methyl substituent of the pyrazolyl group at C4 prevents this group to adopt a coplanar orientation with the pyrimidine and the other pyrazolyl ring required for tridentate binding of Pd^{2+} (Figure 3). In other words, owing to this repulsion, the N2 side of the prazolyl group is turned away from the vicinity of the N3 binding site.

2,4-Bis(1-methylhydrazinyl)-5-(β-D-ribofuranosyl)pyrimidine (**5**) binds Pd^{2+} even more weakly than **4**. In fact, no signals referring to complex formation could be detected upon addition of K_2PdCl_4 into a 5 mmol·L^{-1} solution of **5**. As with **2**, involvement of the lone electron pair of the N2 atom of the 1-methylhydrazinyl groups in the π-electron resonance of the heteroaromatic ring makes the hydrazinyl amino groups poor donor atoms, but does not explain why binding of Pd^{2+} to **5** is even weaker than binding to **2**. Tentatively, the presence of the bulky ribofuranosyl group next to one of the hydrazinyl groups still sterically retards the binding of Pd^{2+}.

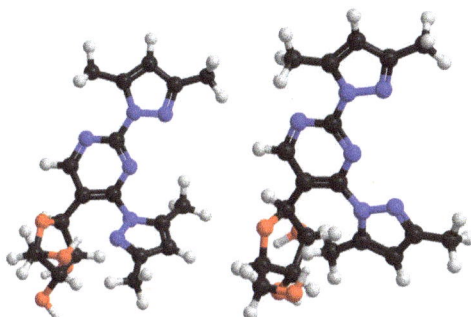

Figure 3. Semi-empirical (PM6) minimized structure for 2,4-bis(3,5-dimethylpyrazol-1-yl)-5-(β-D-ribofuranosyl)pyrimidine (**4**, **left**) and the structure allowing tridentate binding to the N2 atoms of the 3,5-dimethylpyrazolyl groups and the intervening N3 of the pyrimidine ring (**right**). In the latter structure the steric repulsion is much more pronounced than in the former.

2.3. Mixed-Ligand Pd^{2+} Complexes of Modified Nucleosides **1–5** with Nucleoside 5'-Monophosphates

Since 2,6-bis(3,5-dimethylpyrazol-1-yl)purine riboside (**1**) formed by far the most stable Pd^{2+} complexes among the modified nucleosides studied, the formation of mixed-ligand Pd^{2+} complexes between this nucleoside and various NMPs was then studied. For this purpose, equimolar amounts of **1** and K$_2$PdCl$_4$ were stepwise added into a 5.0 mmol·L^{-1} solution of NMP in phosphate buffered D$_2$O (0.12 mol·L^{-1}, pD 7.6, 25 °C), keeping the concentration of NMP constant. Upon addition of **1** and K$_2$PdCl$_4$ into the solution of UMP, the ^{1}H-NMR signals of UMP gradually disappeared and a set a signals referring to a mixed-ligand Pd^{2+} complex of **1** and UMP appeared. Figure 4 shows as an illustrative example the spectrum obtained when the total concentration of **1** and K$_2$PdCl$_4$ was 3.4 mmol·L^{-1}. The chemical shifts of the aromatic and anomeric proton resonances are given in Table 2. No signals referring to the complex (**1**)Pd^{2+} or (**1**)$_2$Pd^{2+} appeared. Most likely, deprotonated N3 of UMP occupies the fourth coordination site of Pd^{2+} bound tridentately to **1**. When the total concentration of **1**, K$_2$PdCl$_4$ and UMP was 4.0, 4.0 and 5.0 mmol·L^{-1}, respectively, 78% of UMP was engaged in the mixed ligand complex, the theoretical maximum being 80% (Table 3). These conditions were selected as reference conditions in the present study, since at equimolar 5.0 mmol concentration considerable broadening of NMR signals in some cases occurred, which may be taken as an indication of polymeric complex formation or precipitation.

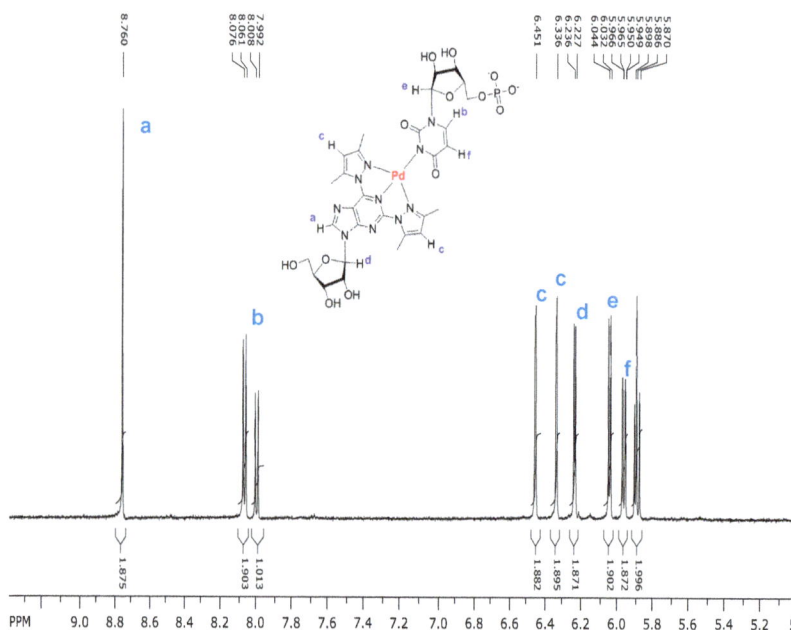

Figure 4. Partial ^{1}H-NMR spectrum of a mixture of 2,6-bis(3,5-dimethylpyrazol-1-yl)purine riboside (**1**; 3.4 mmol·L^{-1}), K$_2$PdCl$_4$ (3.4 mmo·L^{-1}) and UMP (5.0 mmol·L^{-1}) in D$_2$O (phosphate buffer 0.12 mol·L^{-1}, pD 7.6, 25 °C). The doublet at 8.00 and overlapping doublets at 5.89 refer to uncomplexed UMP.

Table 2. Chemical shifts for the aromatic and anomeric protons of the mixed ligand Pd^{2+} complexes of 2,6-bis(3,5-dimethylpyrazol-1-yl)purine riboside (**1**) with nucleoside 5'-monophosphates in D$_2$O at pD 7.6 (0.12 M phosphate buffer, 25 °C).

Compd.	Aromatic Proton Shifts	Anomeric Proton Shifts
UMP	d 8.00(H6) ^{a}, d 5.88(H5) ^{a}	d 5.88 ^{b}
(**1**)Pd(UMP)	s 8.76(H8 of **1**), s 6.45 and s 6.33(H4" of **1**), d 8.07(H6 of UMP) ^{c}, d 5.96(H5 of UMP) ^{c}	d 6.23(J 4.6), d 6.04(J 5.6)
CMP	d 7.98(H6) ^{d}, d 6.03(H5) ^{d}	d 5.89(J 5.2)
(CMP)Pd	d 7.93(H6) ^{d}, d 5.92(H5) ^{d}	d 5.82(J 5.5)
(**1**)Pd(CMP)	s 8.59(H8 of **1**), s 6.48 and 6.37(H4" of **1**), d 7.86(H6 of CMP) ^{d}, d 5.94(H5 of CMP) ^{d}	d 5.83(J 3) d 5.83(J 3.9), d 5.88 ^{e}
GMP	s 8.09(H8)	d 5.82(J 6.1)
(**1**)Pd(GMP)	s 8.59(H8 of **1**), s 6.43 and 6.32 (H4" of **1**), s 8.11(H8 of GMP)	d 6.14(J 5.3), d 5.85(J 6.3)
IMP	s 8.46(H8), s 8.12(H2)	d 6.03(J 5.8)
(**1**)Pd(IMP)	s 8.75 (H8 of **1**), s 6.40 and 6.29(H4" of **1**), s 8.66(H8 of IMP), s 8.44(H2 of IMP)	d 6.02 ^{e}, d 6.14(J 5.3)
AMP ^{f}	s 8.48(H8), s 8.14(H2)	d 6.02 (J 5.9)

^{a} J 8.2 Hz. ^{b} Overlaps with H5. ^{c} J 7.8 Hz. ^{d} J 7.6 Hz. ^{e} Overlaps with the H1' resonance of the uncomplexed NMP. ^{f} (AMP)Pd and (**1**)Pd(AMP) precipitated.

Table 3. Mole fraction of NMPs engaged in a mixed-ligand Pd^{2+} complex with modified nucleosides **1–5**, when the total concentration of **1**, K$_2$PdCl$_4$ and NMP is 4.0, 4.0 and 5.0 mmol·L^{-1}, respectively.

NMP	1	2	3	4	5
UMP	0.78	0.41	d	d	b
CMP	≈0.2	0.26	d	b	d
GMP	0.45	0.61	d	b	b
IMP	0.57	c	c	c	c
AMP	a	d	b	b	b
NeMP	b	c	c	c	c

^{a} Precipitation occurred. ^{b} No mixed-ligand complex formed. ^{c} Not studied. ^{d} Traces of several species formed in parallel.

Mixed ligand complex formation of **1** with the other NMPs is considerably weaker: 45% of GMP and 57% of IMP was engaged in the mixed ligand complex under the reference conditions ([**1**] = [K$_2$PdCl$_4$] = 4.0 mmol·L^{-1} and [NMP] = 5.0 mmol·L^{-1}) (Figures S6 and S7 in Supplementary Files). The binding site cannot be definitely assigned. The H8 signal of GMP and the H2 and H8 signals of IMP all undergo a modest downfield shift, the shift of H2 of IMP being the largest. This suggests that the binding site is deprotonated N1, since N7 binding to a purine base usually shifts the H8 signal downfield by approximately 0.5 ppm, leaving the H2 shift almost unchanged [7,11–14]. Now the H2 signal of IMP is shifted more than the H8 signal.

In the case of CMP, formation of binary (CMP)Pd^{2+} complex competed with formation of the mixed-ligand complex. Only around 20% of CMP was engaged in the mixed ligand complex under the reference conditions (Figure S8 in Supplementary Files). Similarly, the interaction with purine riboside 5'-monophosphate turned out to be weak; no assignable mixed ligand complex was formed. The binary and mixed ligand Pd^{2+} complexes of AMP precipitated. Accordingly, only NMPs having a displaceable proton at N1 seem to form reasonably stable mixed ligand Pd^{2+} complexes with **1**, UMP being bound considerably more firmly than IMP or GMP.

Among the other modified nucleosides studied (**2–5**), only 2,6-bis(1-methylhydrazinyl)purine riboside (**2**) formed mixed ligand complexes stable enough to be reliably detected (Table 4). The complex with UMP was less stable than the corresponding complex of **1**, consistent with the lower affinity of **3** for Pd^{2+}. Only 41% of UMP was engaged in the mixed-ligand complex under the reference conditions ($[K_2PdCl_4]$ = [**3**] = 4.0 mmol·L^{-1}, [NMP] = 5.0 mmol·L^{-1}] (Figure S9 in Supplementary Files). For comparison, the observed 78% engagement of UMP in the mixed ligand complex with **1** was close to the theoretical maximum, 80%. Ternary complexes (**2**)Pd^{2+}(CMP) and (**2**)Pd^{2+}(GMP) were, in turn, formed even slightly more readily than the corresponding complexes of **1**; 26% of CMP and 61% of GMP were engaged in a mixed ligand complex under the reference conditions (Figures S10 and S11 in Supplementary Files). In fact, GMP was now bound slightly more firmly than UMP. The marked downfield shift of the H8 resonance of GMP (0.66 ppm) suggests N7 coordination. Interaction with AMP appeared rather weak, and several species were formed in parallel. Upon mixing of 6-bis(3,5-dimethylpyrazol-1-yl)purine riboside (**3**) and K_2PdCl_4 with UMP, CMP or GMP, so complicated mixtures were formed that assignment of any single mixed-ligand complex was impossible. With AMP, no complexes were formed. As discussed above, the pyrimidine derivatives **4** and **5** did not form stable Pd^{2+} complexes. Expectedly, they did not form an assignable mixed-ligand complex with any of the NMPs studied. The only species that could be assigned referred to binary Pd^{2+} complexes of NMPs.

Table 4. Chemical shifts for the aromatic and anomeric protons of the mixed ligand Pd^{2+} complexes of 2,6-bis(1-methylhydrazinyl)purine riboside (**2**) with nucleoside 5'-monophosphates in D$_2$O at pD 7.6 (0.12 M phosphate buffer).

Compd.	Aromatic Proton Shifts	Anomeric Proton Shifts
(**2**)Pd(UMP)	s 8.13(H8 of **2**), d 7.81(H6 of UMP) ^{b}, d 5.77(H5 of UMP) ^{b}	m 5.91–5.95 ^{a}
(**2**)Pd(CMP)	s 8.14(H8 of **2**), d 8.11(H6 of CMP) ^{c}, d 6.13(H5 of CMP) ^{c}	m 5.90–5.95 ^{a}
(**2**)Pd(GMP)	s 8.04(H8 of **2**), s 8.75(H8 of GMP)	br s 5.81 ^{d}
(**2**)Pd(AMP)	^{e}	^{e}

^{a} The H1' resonances of both ligands overlap. ^{b} J 7.7 Hz. ^{c} J 7.6 Hz. ^{d} Overlaps with the H1' resonance of GMP. ^{e} Could not be reliably assigned.

3. Experimental Section

3.1. General Information

The ^{1}H-NMR spectra were recorded on Bruker Avance 500- or 400-MHz NMR spectrometers using Me$_4$Si as an external standard. The chemical shifts, δ, are given in ppm and the coupling constants, J, in Hz. HR-ESI-MS spectra were obtained by a Bruker Daltonics MicrOTOF-Q instrument.

3.2. ^{1}H-NMR Spectroscopic Studies of the Interaction of K_2PdCl_4 with Nucleosides **1–5**

To a solution of nucleoside **1–5** (5.0 mmol·L^{-1}) in a phosphate buffer in D$_2$O (0.12 mol·L^{-1}, pH 7.2), K_2PdCl_4 was added portionwise keeping the concentration of nucleoside constant. After each addition, a ^{1}H-NMR spectrum was recorded at 25 °C and the signals appearing in the region of aromatic and anomeric protons (chemical shift > 5 ppm) were carefully integrated. The species distribution at different concentrations was calculated on the basis of the relative intensities of the resonances of their aromatic and anomeric protons.

3.3. ^{1}H-NMR Spectroscopic Studies of the Formation of Mixed-Ligand Pd^{2+} Complexes of Nucleosides **1–5** with NMPs

To a solution of NMP (5.0 mmol·L^{-1}) in a phosphate buffer in D$_2$O (0.12 mol·L^{-1}, pD 7.6), a 1:1 mixture of K_2PdCl_4 and one of nucleosides **1–5** in the same buffer was portionwise added. The concentration of K_2PdCl_4 and the nucleoside (**1–5**) was in this manner varied from zero to 5 mmol·L^{-1}, while the concentration of NMP was kept constant (5.0 mmol·L^{-1}). After each addition, a ^{1}H-NMR spectrum was recorded at 25 °C and the signals appearing in the region of aromatic and anomeric

Molecules **2014**, *19*, 16976–16986

protons (chemical shift > 5 ppm) were carefully integrated. The species distribution at different concentrations was calculated on the basis of the relative intensities of the resonances of their aromatic and anomeric protons.

4. Conclusions

In spite of the apparent similarity of tridentate coordination sites in purine ribosides **1** and **2**, on the one hand, and in pyrimidine *C*-ribosides **4** and **5**, on the other hand, only the purine derivatives turned out to be able to form stable Pd^{2+} complexes. Evidently the bulky ribosyl group at C5 of the pyrimidine ring sterically prevents the donor atom of the C4-substituent from adopting an orientation allowing tridentate binding. Among the purine derivatives, 2,6-bis(3,5-dimethylpyrazol-1-yl)purine riboside (**1**) forms mixed-ligand Pd^{2+} complexes with NMPs much more readily than its 2,6-bis(1-methylhydrazinyl) counterpart, probably due to participation of the lone electron pair of the terminal amino groups of the latter in the purine π-electron resonance. UMP is recognized most efficient, followed by IMP, GMP and CMP, in this order. Interestingly, interaction with unsubstituted purine riboside 3'-monophosphate is very weak. Mixed-ligand complexes with AMP precipitate. Bidentately coordinating 6-(3,5-dimethylpyrazol-1-yl)purine riboside give a complicated product mixture upon mixing with K_2PdCl_4 and any of the canonical NMPs.

Supplementary Materials: Supplementary materials can be accessed at: http://www.mdpi.com/1420-3049/19/10/16976/s1.

Acknowledgments: Financial aid from the Foundation of Turku University is gratefully acknowledged.

Author Contributions: O.G., T.L. and H.L. designed the research together. O.G. performed the experimental measurements. O.G., T.L. and H.L. participated in analyzing the data and writing the paper. H.L. prepared the final version. O.G., T.L. and H.L. read and approved the final manuscript.

Conflicts of Interest: The authors declare no conflict of interest

References

1. Takezawa, Y.; Shinoya, M. Metal-mediated DNA base pairing: Alternatives to hydrogen-bonded Watson-Crick base pairs. *Acc. Chem. Res.* **2012**, *45*, 2066–2076. [CrossRef]
2. Müller, J. Metal-ion-mediated base pairs in nucleic acids. *Eur. J. Inorg. Chem.* **2008**, 3749–3763.
3. Clever, G.H.; Kaul, C.; Carell, T. DNA-metal base pairs. *Angew. Chem. Int. Ed.* **2007**, *46*, 6226–6236. [CrossRef]
4. Taherpour, S.; Lönnberg, H.; Lönnberg, T. 2,6-Bis(functionalized) purines as metal-ion-binding surrogate nucleobases that enhance hybridization with unmodified 2'-*O*-methyl oligoribonucleotides. *Org. Biomol. Chem.* **2013**, *11*, 991–1000. [CrossRef]
5. Shehata, M.R.; Shoukry, M.M.; Ali, S. Mono- and binuclear complexes involving [Pd(*N*,*N*-dimethylethylenediamine)(H$_2$O)$_2$]$^{2+}$, 4,4'-bipiperidine and DNA constituents. *J. Coord. Chem.* **2012**, *65*, 1311–1323. [CrossRef]
6. Kiss, A.; Farkas, E.; Sovago, I.; Thormann, B.; Lippert, B. Solution equilibria of the ternary complexes of [Pd(dien)Cl]$^{+}$ and [Pd(terpy)Cl]$^{+}$ with nucleobases and *N*-acetyl amino acids. *J. Inorg. Biochem.* **1997**, *68*, 85–92. [CrossRef]
7. Kim, S.H.; Martin, R.B. Stabilities and ^{1}H-NMR studies of (diethylenetriamine)Pd(II) and (1,1,4,7,7-pentamethyldien)Pd(II) with nucleosides and related ligands. *Inorg. Chim. Acta* **1984**, *91*, 11–18. [CrossRef]
8. Taherpour, S.; Lönnberg, T. Metal ion chelates as surrogates of nucleopbases for the recognition of nucleic acid sequences: The Pd^{2+} complex of bis(3,5-dimethylpyrazol-1-yl)purine riboside. *J. Nucleic Acids* **2012**. [CrossRef]
9. Golubev, O.; Lönnberg, T.; Lönnberg, H. Metal-ion-binding analogs of ribobucleosides: Preparation and formation of ternary Pd^{2+} and Hg^{2+} complexes with natural pyrimidine nucleosides. *Helv. Chim. Acta* **2013**, *96*, 1658–1669. [CrossRef]
10. Taherpour, S.; Golubev, O.; Lönnberg, T. Metal-ion-mediated base pairing between natural nucleobases and bidentate 3,5-dimethylpyrazolyl-substituted purine ligands. *J. Org. Chem.* **2014**, *79*, 8990–8999. [CrossRef]

Molecules **2014**, *19*, 16976–16986

11. Golubev, O.; Lönnberg, T.; Lönnberg, H. Interaction of Pd^{2+} complexes of 2,6-disubstituted pyridines with nucleoside 5'-monophosphates. *J. Inorg. Biochem.* **2014**, *139*, 21–29. [CrossRef]

12. Scheller, K.H.; Scheller-Krattiger, V.; Martin, R.B. Equilibria in solutions of nucleosides, 5-nucleotides and dienPd^{2+}. *J. Am. Chem. Soc.* **1981**, *103*, 6833–6839. [CrossRef]

13. Häring, U.K.; Martin, R.B. Complexes of (ethylenediamine)Pd(II) with inosine, guanosine, adenosine and their phosphates. *J. Inorg. Nucl. Chem.* **1976**, *38*, 1915–1918. [CrossRef]

14. Matczak-Jon, E.; Jezowska-Trzebiatowska, B.; Kozlowski, H. Interaction of Pd(II) glycyl-L-histidine complex with cytidine and GMP. Proton and carbon-13 nmr studies. *J. Inorg. Biochem.* **1980**, *12*, 143–156. [CrossRef]

15. Jezowska-Trzebiatowska, B.; Kozlowski, H.; Wolowiec, S. Coordination of Gly-Typ·Pd(II) complex to GMP nucleotide. *Acta Biochim. Pol.* **1980**, *27*, 99–109.

Sample Availability: *Sample Availability*: Samples of the compounds are not available.

Communication

Design and Synthesis of a Series of Truncated Neplanocin Fleximers

Sarah C. Zimmermann [1], Elizaveta O'Neill [1], Godwin U. Ebiloma [2], Lynsey J. M. Wallace [2], Harry P. De Koning [2] and Katherine L. Seley-Radtke [1,*]

[1] Department of Chemistry & Biochemistry, University of Maryland, Baltimore County, 1000 Hilltop Circle, Baltimore, MD 21250, USA

[2] Institute of Infection, Immunity and Inflammation, College of Medical, Veterinary and Life Sciences, University of Glasgow, 120 University Place, Glasgow G12 8TA, UK

* Correspondence: kseley@umbc.edu; Tel.: +1-410-455-8684; Fax: +1-410-455-2608

External Editor: Mahesh K. Lakshman

Received: 23 October 2014; in revised form: 8 December 2014; Accepted: 9 December 2014; Published: 16 December 2014

Abstract: In an effort to study the effects of flexibility on enzyme recognition and activity, we have developed several different series of flexible nucleoside analogues in which the purine base is split into its respective imidazole and pyrimidine components. The focus of this particular study was to synthesize the truncated neplanocin A fleximers to investigate their potential anti-protozoan activities by inhibition of S-adenosylhomocysteine hydrolase (SAHase). The three fleximers tested displayed poor anti-trypanocidal activities, with EC_{50} values around 200 μM. Further studies of the corresponding ribose fleximers, most closely related to the natural nucleoside substrates, revealed low affinity for the known *T. brucei* nucleoside transporters P1 and P2, which may be the reason for the lack of trypanocidal activity observed.

Keywords: fleximer; carbocyclic nucleosides; 3-deazaneplanocin A; SAHase; trypanomiasis

1. Introduction

Modified nucleosides, in particular carbocyclic nucleosides, are potent inhibitors of S-adenosyl homocysteine hydrolase (SAHase) [1]. SAHase is a critical enzyme that hydrolyzes S-adenosyl homocysteine, the byproduct of biomethylations that utilize S-adenosylmethionine (SAM) [2,3]. By inhibiting SAHase, an excess of SAH is produced, which in turn exhibits potent inhibitory effects on methyltransferases [4]. Thus, inhibition of SAHase leads to incomplete methylation of nucleic acids, phospholipids, proteins, and other small molecules, disrupting various biochemical pathways [5]. As a result, carbocyclic nucleosides have proven useful in a number of chemotherapeutic applications [6–8].

Neplanocin A (NpcA, Figure 1) and aristeromycin (Ari) are both naturally occurring carbocyclic adenosine analogues that have shown significant antiviral, antiparasitic and anticancer properties [4,7,9,10]. Unfortunately, NpcA and Ari both exhibit deleterious cytotoxicity due to intracellular conversion to their triphosphate forms by adenosine kinase as well as their recognition and metabolism by adenosine deaminase [11–13]. Removal of the 4'-CH$_2$OH from Ari and NpcA, as shown in the truncated analogues shown in Figure 1 (R = H), significantly lowers the cytotoxicity [14].

Interestingly, nucleosides with base modifications such as 3-deazaadenosine have also been found to act as substrates, with similar K_m's found for adenosine and 3-deazaadenosine [7,10,15]. To date, the truncated 3-deaza analogues of Ari and NpcA (truncated DZNepA, Figure 1) lacking the 4'-hydroxy-methyl group have both exhibited greater levels of inhibition than their parent counterparts [5,6,13,16].

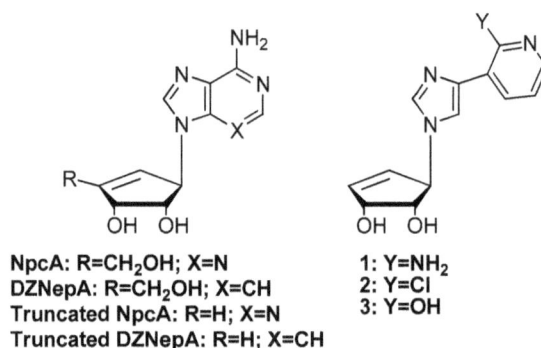

NpcA: R=CH₂OH; X=N
DZNepA: R=CH₂OH; X=CH
Truncated NpcA: R=H; X=N
Truncated DZNepA: R=H; X=CH

1: Y=NH₂
2: Y=Cl
3: Y=OH

Figure 1. Neplanocin A (NpcA) and analogues and the target NcpA fleximers (**1–3**).

More importantly, these compounds have also shown potent inhibition against chloroquine-resistant and chloroquine-susceptible strains of *P. falciparum* [5]. In protozoan parasites, methylation of the four nucleosides present in the "cap-four" terminal end of mRNA requires SAM as the methyl donor. This cap structure is important for RNA recognition and stability, is highly conserved across almost all protozoan species, and is critical for replication [17–19]. Thus, inhibition of SAHase results in an accumulation of SAH, causing methylations to cease, which then disrupts the methylation of the cap structure, thereby providing an important target for the development of potential antiparasitic chemotherapeutics [5].

The Seley-Radtke group has long been interested studying the effects of flexibility on the nucleobase. This flexibility is achieved by "splitting" the purine base into its respective imidazole and pyrimidine (or pyridine) components, which remain connected by a single carbon-carbon bond between the two heteroaromatic moieties. This connectivity allows for free rotation, while still retaining the elements essential for base pairing and molecular recognition [20–28]. This modification has led to enhanced enzyme binding and recognition, as well as the ability to overcome point mutations in enzyme binding sites [29,30]. These analogues have also been studied for their potential therapeutic properties [20–28]. Interestingly, when the fleximer analogues of adenosine (Flex-A), inosine (Flex-I) and guanosine (Flex-G) were studied in SAHase, which is a flexible enzyme, Flex-A and Flex-I acted as substrates, whereas Flex-G proved to be an inhibitor [22]. This is significant because it is, to our knowledge, the only report of a G-nucleoside inhibiting an adenosine metabolizing enzyme. It has been postulated that this is due to an intramolecular hydrogen bond between the pyrimidine and the 5'-OH of the sugar, which then positions the amino group into the binding site where the amino group on adenosine would normally reside, thus essentially creating an adenosine mimic [22].

Historically, a number of nucleoside analogues have been evaluated for trypanocidal activity [31–34]. For example, Cai *et al.* showed that the antiviral drug ribavirin was an inhibitor of *Trypanosoma cruzi* SAHase [33]. Additionally, 7-deaza-5'-noraristeromycin was shown to be a potent inhibitor of four strains of *Trypanosoma brucei* [34]. To further explore the potential of base flexibility and antiparasitic activity, we combined the fleximer base with the carbocyclic nucleoside scaffold, to determine whether the flexible base motif would enhance the biological results previously observed with carbocyclic analogues such as NpcA and Ari. Thus, a series of 3-deaza fleximers (compounds **1–3**, Figure 1) were designed and synthesized to evaluate their anti-parasitic properties.

2. Results and Discussion

2.1. Chemistry

As shown in Scheme 1, cyclopentenol **5** was available from known literature procedures starting from D-cyclopentenone **4** [35], which can be obtained following stereospecific reduction to the "down" hydroxyl using Luche reduction conditions [36]. Alcohol **5** was then coupled to 4,5-diiodoimidazole [29] using standard Mitsunobu [37] conditions to give **6**. Initially the Mitsunobu reaction was attempted with diisopropylazodicarboxylate (DIAD) and triphenylphosphine (TPP) in dichloromethane at room temperature to yield **5**, however only in a 12% yield. Attempts at heating the reaction only served to give additional side products, as well as to lower the yield even further. Changing the solvent to THF increased the solubility of the diiodoimidazole and subsequently resulted in an improved yield of 40%. Unfortunately, contaminates from the byproduct, triphenylphospine oxide (TPPO), still proved to be problematic during purification. Altering the phosphine reagent to DPPE (1,2-bis(diphenylphosphino)ethane) drastically improved the ease in purification. Other coupling methods were also tried, such as using Hendrickson's "POP" reagent, bis(triphenyl)oxodiphosphonium trifluoromethanesulfonate [38], or using bases such as NaH or K_2CO_3 [39] to form the imidazole nucleophile, proved unsuccessful when compared to the Mitsunobu coupling using DPPE.

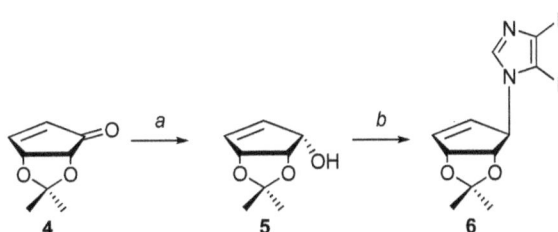

Scheme 1. Synthesis of compound **6**. *Reagents and Conditions: a.* $CeCl_3 \cdot 7H_2O$, MeOH, $NaBH_4$; *b.* DPPE, DIAD, 4,5-diiodoimidazole, THF, rt.

Next, as shown below in Scheme 2, removal of the 5-iodo group of **6** to give compound **7** was achieved via selective deiodination using ethyl magnesium bromide (EtMgBr) followed by quenching with water. Coupling to the pyridine ring was then accomplished using Stille [40] coupling.

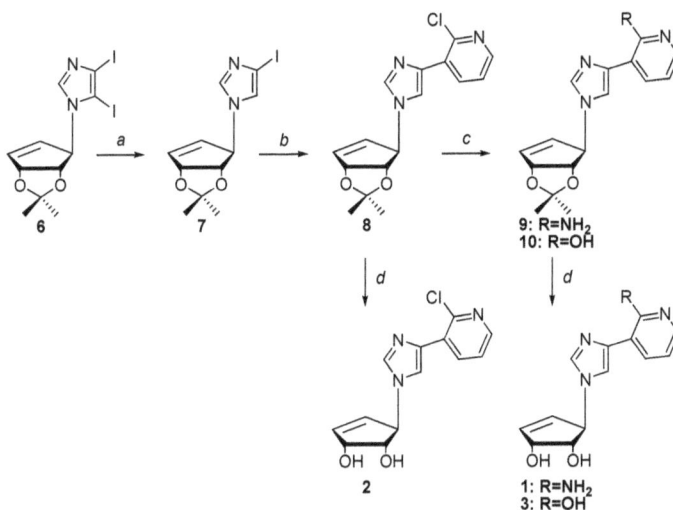

Scheme 2. Synthesis of compounds **1–3**. *Reagents and Conditions: a.* EtMgBr, THF, 0 °C; *b.* 3-tributyltin-2-chloropyridine, Pd(PPh₃)₄, Cu(I) Br, 1,4-dioxane, reflux; *c.* for compound **9**: (i) hydrazine neat, 80 °C; (ii) TiCl₃; for compound **10**: concentrated acetic acid neat, 110 °C; *d.* TFA/H₂O (1/1) in THF.

The 3-tributyltin-2-chloropyridine was prepared from the commercially available 3-bromo-2-chloropyridine. Stille coupling of **7** with the 3-tributyltin-2-chloropyridine provided **8** in a 23% yield, however when copper (I) bromide was used, the yield improved to 71%. Following Stille coupling, transformation of the chloro group into the exocyclic amine group was necessary. Standard procedures using $MeOH/NH_3$ or converting the chloro to an azide using sodium or lithium azide proved unsuccessful. A literature search revealed a palladium-assisted method developed by Hartwig using sodium *t*-butoxide in ammonia saturated 1,4-dioxane [41]. Unfortunately this method also proved unsuccessful.

Related to this latter route, Buchwald developed a similar method, where the catalyst is made *in situ* using a more common phosphine ligand [42]. This method seemed promising since one of the examples utilized 2-chloropyridine, which was successfully converted in a 96% yield [28], but it too proved to be unsuccessful. Use of $NaNH_2$ in ammonia was also tried but the conditions proved to be too harsh and decomposition ensued [43]. Another approach involved converting the chloro group using hydrazine followed by reduction. Initial attempts at reducing the hydrazine employed zinc in acetic acid, but this resulted in a complex mixture that could not be purified. Using titanium chloride $(TiCl_3)$ [44] proved to be successful, although there was evidence of some isopropylidene deprotected product(s) as well as protected products, thus treatment of the mixture with dilute TFA in THF gave the desired final product **1**.

Next, deaminated compound **10** was obtained from **8** using concentrated acetic acid at high temperature. Although this conversion also led to partial deprotection of the isopropylidene on the 2'- and 3'-hydroxyls, the protected pyridine **10** was the major product. Subsequent deprotection of the isopropylidene of **10** led to the fleximer inosine **3**.

2.2. Trypanosomiasis Screening

The three NpcA fleximers (**1–3**) were tested for trypanocidal activity against the laboratory *Trypanosoma brucei brucei* strain Lister 427, using a standard protocol based on the fluorescent format of 23 doubling dilutions, starting at 500 μM, in 96-well plates. All three fleximers tested displayed very similar activities against this strain, with EC_{50} values around 200 μM; in contrast, the control drug pentamidine displayed activity in the low nM range (Table 1), consistent with previous results [45,46].

Table 1. Trypanosomiasis results.

Compound	Average EC_{50} (μM)
1	216 ± 21
2	212 ± 31
3	287 ± 24
pentamidine	0.0044 ± 0.0001

EC_{50} = concentration of drug required to give a 50% response. Data are the average of three independent experiments and SEM.

We considered that the relatively low activity might be related to a lack of recognition of these molecules by the *T. brucei* nucleoside transporters. We therefore investigated whether fleximers in general display reduced uptake kinetics in these parasites, compared to their fixed-ring counterparts (Figure 2). Using the fleximers [21] most closely related to the original nucleoside substrates, it is clear from Table 2 that fleximers indeed show low affinity for the known *T. brucei* nucleoside transporters P1 and P2 [47].

Molecules **2014**, *19*, 21200–21214

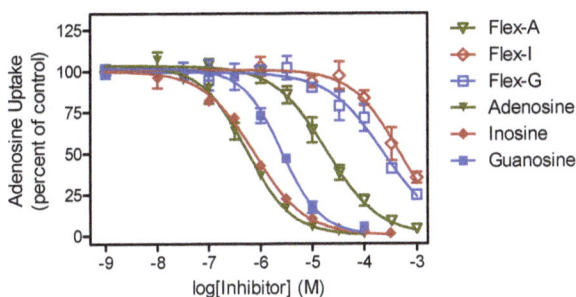

Figure 2. Transport of 0.1 μM [³H]-adenosine by *Trypanosoma brucei brucei* bloodstream form parasites.

Table 2. Comparison of affinity of purine nucleosides and corresponding fleximers for *T. brucei* transporters.

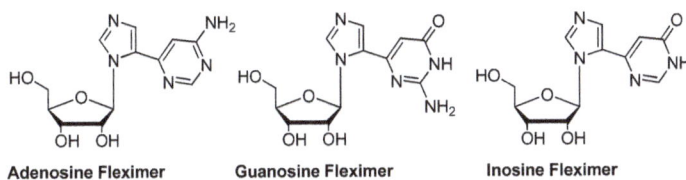

	Adenosine Fleximer		**Guanosine Fleximer**		**Inosine Fleximer**	
	P1 Kᵢ (μM)			**P2 Kᵢ (μM)**		
	Nucleoside [1]	Fleximer	δ(ΔG⁰)	Nucleoside [1]	Fleximer	δ(ΔG⁰)
Adenosine	0.36 ± 0.05	35 ± 11	11.4	0.91 ± 0.29	37 ± 3	9.2
Guanosine	1.8 ± 0.3	251 ± 75	12.2	>500	>500	
Inosine	0.44 ± 0.10	387 ± 30	16.8	>500	>500	

Data are the average inhibition constants (K_i) and SEM of at least three independent experiments; Values for adenosine are Michaelis-Menten constants (K_m). [1] values were taken from previous findings of De Koning [47] and included here for comparison; δ(ΔG⁰) is the difference in Gibbs free energy of interaction of the nucleoside and the fleximer with the transporter, given in kJ/mol.

Transport, mediated by the P1 nucleoside transporter, was measured in the presence or absence of various concentrations of nucleosides (filled symbols) or their corresponding fleximers (open symbols), in the presence of 100 μM adenine to block potential adenosine transport through the P2 transporter. Data shown are the average and SEM of triplicate determinations in a single experiment, representative of three independent experiments with essentially identical outcomes.

It is thus clear that the fleximers generally display about two orders of magnitude less affinity for the *T. brucei* nucleoside transporters than the corresponding nucleosides, limiting cellular uptake as there are no other nucleoside uptake mechanisms in these parasites than the P1 and P2 systems, although P1 consists of a cluster of multiple genes with slightly divergent sequences [48,49]. In addition, the truncated NpcA fleximers lack a 4'-hydroxymethyl group and an equivalent of the purine N3 residue, and both required for high affinity for P1 [50]. Moreover, the P2 transporter does not recognize any oxopurine nucleoside analogues [47]. The loss of approximately 10 kJ/mol in Gibbs free energy for the fleximer-transporter interaction may in part be due to the increased entropy in the orientation of the fleximer orientation in the binding pocket, as well as the slightly larger volume of the base. We thus conclude that the low effectiveness of the Npc fleximers is at least partially due to unfavorable interactions with the parasite's nucleoside transporters. As important differences exist between nucleoside transporters of even closely related pathogenic parasites including *Trypanosoma congolense* [51] and *Leishmania* species [48], it would be worthwhile to follow this study with a wider screening of anti-parasite activity for a diverse panel of protozoa.

3. Experimental Section

3.1. General Information

All chemicals were obtained from commercial sources and used without further purification unless otherwise noted. Anhydrous DMF, MeOH, DMSO and toluene were purchased from Fisher Scientific (Pittsburgh, PA, USA). Anhydrous THF, acetone, CH_2Cl_2, CH_3CN and ether were obtained using a solvent purification system (mBraun Labmaster 130, MBRAUN, Stratham, NH, USA). 3-Bromo-2-chloropyridine was obtained from Sigma-Aldrich (St. Louis, MO, USA). Melting points are uncorrected. NMR solvents were purchased from Cambridge Isotope Laboratories (Andover, MA, USA). All ^{1}H- and ^{13}C-NMR spectra were obtained on a JEOL ECX 400 MHz NMR, operated at 400 and 100 MHz respectively, and referenced to internal tetramethylsilane (TMS) at 0.0 ppm. The spin multiplicities are indicated by the symbols s (singlet), d (doublet), dd (doublet of doublets), t (triplet), q (quartet), m (multiplet), and b (broad). Reactions were monitored by thin-layer chromatography (TLC) using 0.25 mm Whatman Diamond silica gel 60-F_{254} precoated plates. Column chromatography was performed using silica gel (63–200 μm) from Dynamic Adsorptions Inc. (Norcross, GA, USA), and eluted with the indicated solvent system. Yields refer to chromatographically and spectroscopically (^{1}H- and ^{13}C-NMR) homogeneous materials. High resolution mass spectra were recorded at the Johns Hopkins Mass Spectrometry Facility (Baltimore, MD, USA) using fast atom bombardment for ionization.

3.2. Synthesis

Preparation of (4R,5R)-4,5-O-isopropylidene-2-cyclopenten-1-ol (**5**): (4R,5R)-4,5-O-isopropylidene-2-cyclopentenone **4** [35] (4.63 g, 0.03 mol) was dissolved in dry methanol (20 mL) at room temperature. $CeCl_3 \cdot 7H_2O$ was added to the reaction followed by the portionwise addition of $NaBH_4$ (1.36 g, 0.04 mol). Once TLC analysis showed the complete disappearance **4**, the reaction was extracted into ethyl acetate (50 mL) and washed with water (10 mL). The organic layer was dried over $MgSO_4$ and the solvent was removed under rotary evaporation. The crude oil was used in the following reaction without further purification.

Preparation of (1'R,2'S,3'R)-1-[(2',3'-O-isopropylidene)-4'-cyclopenten-1'-yl]-4,5-diiodoimidazole (**6**): Thoroughly dried **5** (0.78 g, 0.05 mol) was dissolved in dry THF (100 mL) under N_2. Ethylenebis (diphenylphosphine) (2.0 g, 0.05 mol) and 4,5-diiodoimidazole (3.2 g, 0.01 mol) added to the reaction, followed by the dropwise addition of diisopropyl azodicarboxylate. The reaction was allowed to stir for 48 h and then the solvent was removed using reduced pressure. The crude material was purified by silica gel column chromatography hexanes–ethyl acetate (2:1) to yield a yellow waxy solid (0.95 g, 0.02 mol, 41% yield). ^{1}H-NMR ($CDCl_3$): δ 1.32 (s, 3H), 1.45 (s, 3H), 4.44 (d, 1H, $J = 5.5$ Hz), 5.24 (d, 1H, $J = 1.4$ Hz), 5.31 (m, 1H), 5.90 (dt, 1H, $J = 4.6, 5.9$ Hz), 6.34 (dt, 1H, $J = 1.8$ Hz, 5.9 Hz), 7.38 (s, 1H). ^{13}C-NMR ($CDCl_3$): δ 25.9, 27.4, 70.8, 82.7, 84.2, 84.3, 97.1, 112.6, 128.9, 139.0, 139.4. HRMS calculated for $C_{11}H_{12}I_2N_2O_2$ $[M+H]^+$ 458.9066, found 458.9073.

Preparation of (1'R,2'S,3'R)-1-[(2',3'-O-isopropylidene)-4'-cyclopenten-1'-yl]-4-iodoimidazole (**7**): Dried **6** (1.83 g, 0.004 mol) was dissolved in anhydrous THF under N_2. The reaction was dropped to 0 °C, and ethyl magnesium bromide (3.0 M, 1.3 mL, 0.004 mol) was added dropwise to the reaction. After 1 h the reaction was quenched with saturated NH_4Cl (5 mL) and then the solvent was removed. The crude mixture was dissolved in ethyl acetate (50 mL) and washed with water (20 mL) and then dried over $MgSO_4$. The solvent was removed under reduced pressure and purified using silica gel chromatography petroleum ether–ethyl acetate (2:1) to yield a yellow oil (1.12 g, 0.003 mol, 85% yield). ^{1}H-NMR ($CDCl_3$): δ 1.28 (s, 3H), 1.40 (s, 3H), 4.44 (d, 1H, $J = 5.5$ Hz), 5.15 (d, 1H, $J = 1.4$ Hz), 5.30 (dq, 1H, $J = 1.8, 5.5$ Hz), 5.87 (dt, 1H, $J = 0.9, 5.5$ Hz), 6.22 (dt, 1H, $J = 1.8, 5.9$ Hz), 6.89 (d, 1H, $J = 1.4$ Hz), 7.34 (d, 1H, $J = 1.4$ Hz). ^{13}C-NMR ($CDCl_3$): δ 25.7, 27.3, 68.3, 82.7, 84.2, 84.9, 112.5, 123.2, 129.8, 137.5, 138.1. HRMS calculated for $C_{11}H_{13}IN_2O_2$ $[M+H]^+$ 333.0100, found 333.0010.

Preparation of 3-tributylstannyl-2-chloropyridine: Commercially available 3-bromo-2-chloropyridine (0.30 g, 0.002 mol) was dissolved in anhydrous THF (20 mL) under N_2. Ethyl magnesium bromide (3.0 M, 0.5 mL, 0.002 mol) was added dropwise at room temperature. The reaction was allowed to stir for 2 h, and then tributyltin chloride (0.42 mL, 0.002 mol) was added and the reaction was left to stir overnight. The reaction was concentrated *in vacuo* and then purified on silica gel chromatography hexanes-ethyl acetate (15:1) to yield a colorless oil (0.40 g, 0.001 mol, 64% yield). ^{1}H-NMR (CDCl$_3$): δ 1.13 (m, 5 H), 1.30 (m, 13H), 1.56 (m, 6 H), 1.65 (m, 3 H), 7.13 (dd, 1H, *J* = 4.6, 7.8 Hz), 7.67 (dd, 1H, *J* = 1.8, 4.6 Hz), 8.27 (dd, 1H, *J* = 1.8, 7.8 Hz). ^{13}C-NMR (CDCl$_3$): δ 26.9, 27.0, 27.3, 27.9, 28.9, 29.0, 29.1, 122.2, 139.5, 146.9, 147.1, 147.2, 149.6, 159.2.

Preparation of (1'R,2'S,3'R)-3-[((2',3'-O-isopropylidene)-4'-cyclopenten-1'-yl)-(imidazol-4-yl)]-2-chloropyridine (**8**): Intermediate **7** (0.34 g, 0.001 mol) and 2-chloro-3-(tributylstannyl)pyridine (3.50 g, 0.007 mol) were dissolved in 1,4-dioxane under N_2. Pd(PPh$_3$)$_4$ (0.05 g, 0.04 mmol) and CuBr (0.08 g, 0.5 mmol) were added to the reaction and the reaction and was refluxed at 120 °C for 12 h. The reaction was cooled and filtered through a pad of Celite. The filtrate was diluted in ethyl acetate (20 mL) and washed with a saturated solution of NH$_4$Cl (20 mL), water (20 mL), brine (20 mL) and then dried over MgSO$_4$. The organic solvent was removed under reduced pressure and the crude material was purified using 5% MeOH in CH$_2$Cl$_2$ to yield a yellow oil (0.23 g, 0.7 mmol, 71% yield). ^{1}H-NMR (CDCl$_3$): δ 1.36 (s, 3H), 1.48 (s, 3H), 4.58 (d, 1H, *J* = 5.5 Hz), 5.28 (d, 1H, *J* = 1.4 Hz), 5.40 (dt, 1H, *J* = 0.9, 4.6 Hz), 6.00 (dd, 1H, *J* = 1.2, 5.5 Hz), 6.32 (dt, 1H, *J* = 1.8, 5.5 Hz), 7.30 (dd, 1H, *J* = 4.6, 7.7 Hz), 7.55 (d, 1H, *J* = 1.4 Hz), 7.67 (d, 1H, *J* = 0.9 Hz), 8.26 (dd, 1H, *J* = 1.8, 4.6 Hz), 8.50 (dd, 1H, *J* = 1.8, 7.8 Hz). ^{13}C-NMR (CDCl$_3$): δ 25.7, 27.3, 68.5, 84.4, 85.1, 112.6, 118.5, 122.8, 129.4, 130.1, 132.1, 135.9, 136.9, 137.7, 137.9, 147.2. HRMS calculated for C$_{16}$H$_{16}$ClN$_3$O$_2$ [M+H ^{35}Cl]$^+$ 318.1009, [M+H ^{37}Cl]$^+$ 320.0980, found, 318.1001, 320.0976.

Preparation of (1'R,2'S,3'R)-3-[((2',3'-O-isopropylidene)-4'-cyclopenten-1'-yl)-(imidazol-4-yl)]-2-pyrimidone (**10**): Analogue **9** (0.1 g, 0.3 mmol) was dissolved in concentrated acetic acid in a sealed glass tube and heated to 120 °C overnight. The acetic acid was evaporated and the crude material was extracted into ethyl acetate. Silica gel chromatography using ethyl acetate-acetone-methanol (6:1:1) yielded an off white solid (0.05 g, 0.2 mmol, 56% yield). ^{1}H-NMR (DMSO-d_6): δ 1.24 (s, 3H), 1.32 (s, 3H), 4.41 (d, 1H, *J* = 6.9 Hz), 5.05 (d, 1H, *J* = 4.6 Hz), 5.98 (d, 1H, *J* = 5.9 Hz), 6.12 (dt, 1H, *J* = 2.3, 5.9 Hz), 6.26 (t, 1H, *J* = 6.4 Hz), 7.24 (d, 1H, *J* = 4.6 Hz), 7.65 (d, 1H, *J* = 0.9 Hz), 7.79 (d, 1H, *J* = 1.4 Hz), 8.07 (dd, 1H, *J* = 1.8, 7.3 Hz), 11.78 (bs, 1H). ^{13}C-NMR (DMSO-d_6): δ 25.5, 27.3, 67.4, 73.0, 78.9, 105.8, 117.6, 125.1, 131.0, 131.6, 132.9, 133.9, 136.7, 136.8, 136.9, 160.7. HRMS calculated for C$_{16}$H$_{17}$N$_3$O$_3$ [M+H]$^+$ 300.1348, found 300.1342.

Preparation of (1'R,2'S,3'R)-3-[(2',3'-Dihydroxy)-4'-cyclopenten-1'-yl]-(imidazol-4-yl)-2-amino-pyridine (**1**): Compound **8** (80 mg, 0.25 mmol) was refluxed in hydrazine (2 mL) for 1 h. The solvent was removed under reduced pressure and the residue dissolved in THF (5 mL). Titanium (III) chloride (1.7 mmol, 0.5 mL, 20% in 3% HCl) was neutralized using NaOH (0.5 mL, 20%), and 0.6 mL of the solution was added dropwise to the reaction. The mixture was refluxed at 70 °C for 4 h, cooled to room temperature, and brought to pH > 10 using NaOH (20%) while cooling in an ice-bath. The solvents were removed under vacuum, and the product was extracted with CH$_2$Cl$_2$ (10 mL × 5). The organic layer was dried over MgSO$_4$. The solvents were evaporated, and the crude material, 3-[((2', 3'-O-isopropylidene)-4'-cyclopenten-1'-yl)-imidazol-4-yl]-2-aminopyridine (**9**), was used directly in preparation of **1**. Crude **9** was dissolved in THF (5 mL) and TFA:H$_2$O (1 mL:1 mL) was added dropwise. This was allowed to stir overnight at room temperature. The solvent was evaporated and co-evaporated with ethanol (3 × 5 mL) to yield an off-white solid (0.03 g, 0.1 mmol, 46% yield over 2 steps). ^{1}H-NMR (DMSO-d_6): δ 3.93 (m, 1H), 4.45 (m, 1H), 4.96 (m, 1H), 5.08 (m, 1H), 5.14 (d, 1H, *J* = 6.9 Hz), 5.96 (dd, 1H, *J* = 1.4, 6.4 Hz), 6.11 (dt, 1H, *J* = 2.3, 6.4 Hz), 6.54 (dd, 1H, *J* = 5.0, 7.3 Hz), 6.95 (bs, 2H), 7.59 (d, 1H, *J* = 0.9 Hz), 7.70 (dd, 1H, *J* = 1.8, 7.4 Hz), 7.78 (d, 1H, *J* = 1.0 Hz), 7.81 (m, 1H). ^{13}C-NMR (DMSO-d_6): δ 67.6, 73.1, 78.6, 112.2, 112.5, 115.1, 132.9, 133.7, 136.2, 136.9, 140.1, 146.5, 156.5 HRMS calculated for C$_{13}$H$_{14}$N$_4$O$_2$ [M+H]$^+$ 259.1195, found 259.1193.

Preparation of (1'R,2'S,3'R)-3-[(2',3'-Dihydroxy)-4'-cyclopenten-1'-yl]-(imidazol-4-yl)-2-chloro-pyridine (**2**): Intermediate **8** (0.16 g, 0.5 mmol) was dissolved in THF (5 mL) and TFA:H_2O (1 mL:1 mL) was added dropwise. This was allowed to stir overnight at room temperature. The solvent was evaporated and co-evaporated with ethanol (3 × 5 mL). Column chromatography in 10% MeOH in CH_3CN returned an off-white solid (0.12 g, 0.4 mmol, 86% yield). ^{1}H-NMR (DMSO-d_6): δ 3.90 (m, 1H), 4.42 (m, 1H), 4.98 (m, 1H), 5.08 (m, 1H), 5.17 (d, 1H, J = 6.4 Hz), 6.00 (dd, 1H, J = 1.4, 5.9 Hz), 6.11 (dt, 1H, J = 2.3, 5.9 Hz), 7.43 (dd, 1H, J = 4.6, 7.8 Hz), 7.77 (s, 1H), 7.80 (s, 1H), 8.22 (dd, 1H, J = 2.3, 4.6 Hz), 7.81 (dd, 1H, J = 1.8, 7.8 Hz). ^{13}C-NMR (DMSO-d_6): δ 67.6, 73.0, 78.9, 119.0, 123.8, 129.9, 132.7, 135.4, 137.1, 137.6, 137.9, 146.5, 147.5. HRMS calculated for $C_{13}H_{12}ClN_3O_2$ [M+H ^{35}Cl]$^+$ 278.0696, [M+H ^{37}Cl]$^+$ 280.0667, found, 278.0689, 280.0668.

Preparation of (1'R,2'S,3'R)-3-[(2',3'-Dihydroxy)-4'-cyclopenten-1'-yl]-imidazol-4-yl)-2-hydroxypyridine (**3**): Intermediate **10** (0.08 g, 0.3 mmol) was dissolved in THF (5 mL) and TFA:H_2O (1 mL:1 mL) was added dropwise. This was allowed to stir overnight at room temperature. The solvent was evaporated and co-evaporated with ethanol (3 × 5 mL). Column chromatography in ethyl acetate–acetone–methanol–water (6:1:1:0.5) produced an off-white solid (0.03 g, 0.1 mmol, 43% yield). ^{1}H-NMR (DMSO-d_6): δ 3.82 (d, 1H, J = 4.6 Hz), 4.43 (bs, 1H), 4.95 (m, 1H), 5.05 (m, 1H), 5.11 (d, 1H, J = 7.4 Hz), 5.97 (dd, 1H, J = 1.4, 5.9 Hz), 6.12 (dt, 1H, J = 2.8, 5.9 Hz), 6.29 (t, 1H, J = 6.9 Hz), 7.23 (d, 1H, J = 4.6 Hz), 7.65 (d, 1H, J = 0.9 Hz), 7.80 (d, 1H, J = 1.4), 8.08 (dd, 1H, J = 2.3, 7.3 Hz), 11.73 (bs, 1H). ^{13}C-NMR (DMSO-d_6): δ 67.6, 73.1, 78.6, 112.2, 112.5, 115.1, 132.9, 133.7, 136.2, 136.9, 140.1, 146.5, 156.5 HRMS calculated for $C_{13}H_{13}N_3O_3$ [M+H]$^+$ 260.1035, found 260.1033.

3.3. Anti-Trypanosome Activity

In vitro activity against *Trypanosoma brucei* was determined using the Alamar blue (resazurin) assay for cell viability exactly as described [52]. Briefly, serial dilutions of test compounds were made in 96-well plates by serial passage of 100 µL of test compound (usually at 2 mM) to 100 µL of HMI9 medium containing 10% fetal bovine serum (Invitrogen), using 2 rows, with the negative control values obtained from wells with 100 µL of medium without test compound. Serial dilutions with pentamidine isethionate (Sigma) were used as positive control. To each well, 100 µL of medium, containing 10^4 culture-adapted bloodstream *T. b. brucei* (strain Lister 427), was added and the plates were incubated at 37 °C for 48 h after which 20 µL 5 mM resazurin solution was added. Following a further incubation of 24 h at 37 °C, fluorescence was determined in a FLUOstar OPTIMA (BMG Labtech, Aylesbury, UK) fluorimeter with excitation and emission filters at 544 nm and 620 nm, respectively. EC$_{50}$ values (the effective concentration reducing specific fluorescence by 50%) were calculated by nonlinear regression using the Prism 5 software package (GraphPad, La Jolla, CA, USA).

3.4. Transport Assays

Transport assays with bloodstream forms of *T. b. brucei* were performed exactly as described previously [53,54]. Briefly, transport was initiated by the addition of 100 µL of *T. b. brucei* bloodstream forms (10^7 cells/mL in assay buffer [53]) to 100 µL of [2,8,5'-^{3}H]-adenosine (PerkinElmer, Waltham, MA, USA; 54.4 Ci/mmol) pre-mixed with up to 1 mM of test inhibitor in assay buffer. After exactly 10 s the mixture was centrifuged through an oil layer in a microfuge (13,000× *g*) and the microfuge tubes were flash-frozen in liquid nitrogen. Pellets were cut off and collected in scintillation tubes; after solubilisation in 2% SDS and addition of scintillation fluid, radioactivity was determined in a liquid scintillation counter. Inhibition data were fitted to a sigmoidal curve with variable slope (GraphPad Prism 5.0), allowing for the determination of EC$_{50}$ values, from which inhibition constants (K_i) were calculated using the Cheng-Prusoff equation, and Gibbs Free Energy using ΔG^0 = $-$RTln (K_i), as described [52].

Molecules **2014**, *19*, 21200–21214

4. Conclusions

The strategy of the work presented herein was to potentially synthesize new and more potent inhibitors of SAHases, thereby disrupting mRNA capping in protozoa as a strategy towards new antiparasitic therapeutics. To this end, characteristics of known SAHase inhibitors such as neplanocin and Aristeromycin were combined, and the nucleoside analogue was given enhanced flexibility using the "fleximer" approach, and added specificity by omitting the N3 equivalent nitrogen residue in the pyrimidine half of the fleximer base group. In addition, the 4'-CH$_2$OH moiety was omitted to reduce general cytotoxicity [10,14]. The data, however, show that the resulting 3-deazaneplanocin fleximers (1–3) displayed only moderate activity in a standardized anti-protozoal test, against *Trypanosoma brucei*, despite the possibility of this species being vulnerable to inhibition of SAHase [55].

We have previously shown that the trypanocidal action of nucleoside and nucleobase analogues is either enabled or limited by their rate of uptake by specific transport proteins [52–54,56,57], and therefore investigated the effect of the fleximer modification on nucleoside transport. We found that the introduction of this modification of the purine ring reduces affinity, and thus presumably translocation rates, for both of the transport systems expressed in bloodstream *T. brucei*, and conclude that the lack of suitable transporters for these molecules causes (or at least contributes to) the observed lack of trypanocidal potency. However, we have also shown that purine transporters in other protozoan parasites, e.g., *Toxoplasma gondii* [58], *Plasmodium falciparum* [59], *Leishmania donovani* [48], and *Trichomonas vaginalis* (Natto and De Koning, unpublished data) all have very different substrate-specificity characteristics. Further studies with additional parasites, and the optimization of the inhibitors for enhanced uptake by the parasites, are in progress.

Acknowledgments: This work was supported by the National Institutes of Health (NIH) (R01 CA97634 to KSR, NIH T32 GM066706 CBI Fellowship to SCZ) and Wellcome Trust (grant to HPdK). We are also grateful to Phil Mortimer (Johns Hopkins Mass Spectrometry Facility) for his invaluable assistance with the HRMS analysis. We also thank Therese Ku for her help with compound characterizations and editorial assistance.

Author Contributions: KLS was responsible for the oversight of the project and supervision of the students, SCZ and EO, while HdK was responsible for supervision of GUE and LW. KLS and SCZ designed the synthesis; SCZ and EO performed the synthesis and characterization of the compounds; GUE, LW and HdK were responsible for the parasite testing and adenosine uptake assays. All authors read and approved the final manuscript.

Conflicts of Interest: The authors have no conflict of interest.

References

1. Tseng, C.K.; Marquez, V.E.; Fuller, R.W.; Goldstein, B.M.; Haines, D.R.; McPherson, H.; Parsons, J.L.; Shannon, W.M.; Arnett, G.; Hollingshead, M.; *et al.* Synthesis of 3-deazaneplanocin A, a powerful inhibitor of *S*-adenosylhomocysteine hydrolase with potent and selective *in vitro* and *in vivo* antiviral activities. *J. Med. Chem.* **1989**, *32*, 1442–1446. [CrossRef]

2. Cantoni, G. The centrality of *S*-adenosylhomocysteinase in the regulation of the biological utilization of *S*-adenosylmethionine. In *Biological Methylation and Drug Design*; Borchardt, R.T., Creveling, C.R., Ueland, P.M., Eds.; Humana Press: Clifton, NJ, USA, 1986; pp. 227–238.

3. Cantoni, G.L.; Scarano, E. The formation of *S*-adenosylhomocysteine in enzymatic transmethylation reactions. *J. Am. Chem. Soc.* **1954**, *76*, 4744. [CrossRef]

4. Borchardt, R.T.; Keller, B.T.; Patel-Thombre, U.; Neplanocin, A. A potent inhibitor of *S*-adenosylhomocysteine hydrolase and of vaccinia virus multiplication in mouse L929 cells. *J. Biol. Chem.* **1984**, *259*, 4353–4358. [PubMed]

5. Bujnicki, J.M.; Prigge, S.T.; Caridha, D.; Chiang, P.K. Structure, evolution, and inhibitor interaction of *S*-adenosyl-L-homocysteine hydrolase from Plasmodium falciparum. *Proteins* **2003**, *52*, 624–632. [CrossRef] [PubMed]

6. Chiang, P.K. Biological effects of inhibitors of *S*-adenosylhomocysteine hydrolase. *Pharmacol. Ther.* **1998**, *77*, 115–134. [CrossRef] [PubMed]

7. De Clercq, E. John Montgomery's legacy: Carbocyclic adenosine analogues as SAH hydrolase inhibitors with broad-spectrum antiviral activity. *Nucleosides Nucleotides Nucleic Acids* **2005**, *24*, 1395–1415.

8. Marquez, V.E. Carbocyclic nucleosides. *Adv. Antivir. Drug Des.* **1996**, *2*, 89–146.
9. Borchardt, R.T.; Wu, Y.-S. S-Aristeromycinyl-L-homocysteine, a potent inhibitor of S-adenosylmethionine-dependent transmethylations. *J. Med. Chem.* **1976**, *19*, 197–198. [CrossRef] [PubMed]
10. Guranowski, A.; Montgomery, J.A.; Cantoni, G.L.; Chiang, P.K. Adenosine analogues as substrates and inhibitors of S-adenosylhomocysteine hydrolase. *Biochemistry* **1981**, *20*, 110–115. [CrossRef] [PubMed]
11. Van Brummelen, A.C.; Olszewski, K.L.; Wilinski, D.; Llinas, M.; Louw, A.I.; Birkholtz, L.M. Co-inhibition of Plasmodium falciparum S-adenosylmethionine decarboxylase/ornithine decarboxylase reveals perturbation-specific compensatory mechanisms by transcriptome, proteome, and metabolome analyses. *J. Biol. Chem.* **2009**, *284*, 4635–4646. [CrossRef] [PubMed]
12. Wolfe, M.S.; Borchardt, R.T. S-adenosyl-L-homocysteine hydrolase as a target for antiviral chemotherapy. *J. Med. Chem.* **1991**, *34*, 1521–1530. [CrossRef] [PubMed]
13. Hasobe, M.; Liang, H.; Ault-Riche, D.B.; Borcherding, D.R.; Wolfe, M.S.; Borchardt, R.T. (1'R,2'S,3'R)-9-(2',3'-Dihydroxycyclopentan-1'-yl)-adenine and -3-deaza-adenine: Analogs of aristeromycin which exhibit potent antiviral activity with reduced cytotoxicity. *Antivir. Chem. Chemother.* **1993**, *4*, 245–248.
14. Wolfe, M.S.; Lee, Y.; Bartlett, W.J.; Borcherding, D.R.; Borchardt, R.T. 4'-Modified analogs of aristeromycin and neplanocin A: Synthesis and inhibitory activity toward S-adenosyl-L-homocysteine hydrolase. *J. Med. Chem.* **1992**, *35*, 1782–1791. [CrossRef] [PubMed]
15. Richards, H.H.; Chiang, P.K.; Cantoni, G.L. Adenosylhomocysteine hydrolase. Crystallization of the purified enzyme and its properties. *J. Biol. Chem.* **1978**, *253*, 4476–4480. [PubMed]
16. Hasobe, M.; McKee, J.G.; Borcherding, D.R.; Borchardt, R.T. 9-(trans-2',trans-3'-Dihydroxycyclopent-4'-enyl)-adenine and -3-deazaadenine: Analogs of neplanocin A which retain potent antiviral activity but exhibit reduced cytotoxicity. *Antimicrob. Agents Chemother.* **1987**, *31*, 1849–1851. [CrossRef] [PubMed]
17. Ruan, J.P.; Shen, S.; Ullu, E.; Tschudi, C. Evidence for a capping enzyme with specificity for the trypanosome spliced leader RNA. *Mol. Biochem. Parasitol.* **2007**, *156*, 246–254. [CrossRef] [PubMed]
18. Mair, G.; Ullu, E.; Tschudi, C. Cotranscriptional Cap 4 Formation on the Trypanosoma brucei spliced leader RNA. *J. Biol. Chem.* **2000**, *275*, 28994–28999. [CrossRef] [PubMed]
19. Zamudio, J.R.; Mittra, B.; Campbell, D.A.; Sturm, N.R. Hypermethylated cap 4 maximizes Trypanosoma brucei translation. *Mol. Microbiol.* **2009**, *72*, 1100–1110. [CrossRef] [PubMed]
20. Seley, K.L.; Zhang, L.; Hagos, A. "Fleximers". Design and synthesis of two novel split nucleosides. *Org. Lett.* **2001**, *3*, 3209–3210. [CrossRef] [PubMed]
21. Seley, K.L.; Zhang, L.; Hagos, A.; Quirk, S. "Fleximers". Design and synthesis of a new class of novel shape-modified nucleosides. *J. Org. Chem.* **2002**, *67*, 3365–3373. [CrossRef] [PubMed]
22. Seley, K.L.; Quirk, S.; Salim, S.; Zhang, L.; Hagos, A. Unexpected inhibition of S-adenosyl-L-homocysteine hydrolase by a guanosine nucleoside. *Bioorg. Med. Chem. Lett.* **2003**, *13*, 1985–1988. [CrossRef] [PubMed]
23. Polak, M.; Seley, K.L.; Plavec, J. Conformational properties of shape modified nucleosides—Fleximers. *J. Am. Chem. Soc.* **2004**, *126*, 8159–8166. [CrossRef] [PubMed]
24. Seley, K.L.; Salim, S.; Zhang, L. "Molecular chameleons". Design and synthesis of C-4-substituted imidazole fleximers. *Org. Lett.* **2005**, *7*, 63–66. [CrossRef] [PubMed]
25. Seley, K.L.; Salim, S.; Zhang, L.; O'Daniel, P.I. "Molecular chameleons". Design and synthesis of a second series of flexible nucleosides. *J. Org. Chem.* **2005**, *70*, 1612–1619. [CrossRef] [PubMed]
26. Zimmermann, S.C.; Sadler, J.M.; Andrei, G.; Snoeck, R.; Balzarini, J.; Seley-Radtke, K.L. Carbocyclic 5'-nor "reverse" fleximers. Design, synthesis, and preliminary biological activity. *Med. Chem. Commun.* **2011**, *2*, 650–654. [CrossRef]
27. Wauchope, O.R.; Velasquez, M.; Seley-Radtke, K. Synthetic routes to a series of proximal and distal 2'-deoxy fleximers. *Synthesis (Stuttg.)* **2012**, *44*, 3496–3504. [CrossRef]
28. Zimmermann, S.C.; Sadler, J.M.; O'Daniel, P.I.; Kim, N.T.; Seley-Radtke, K.L. "Reverse" carbocyclic fleximers: Synthesis of a new class of adenosine deaminase inhibitors. *Nucleosides Nucleotides Nucleic Acids* **2013**, *32*, 137–154. [CrossRef] [PubMed]
29. Quirk, S.; Seley, K.L. Identification of catalytic amino acids in the human GTP fucose pyrophosphorylase active site. *Biochemistry* **2005**, *44*, 13172–13178. [CrossRef] [PubMed]
30. Quirk, S.; Seley, K.L. Substrate discrimination by the human GTP fucose pyrophosphorylase. *Biochemistry* **2005**, *44*, 10854–10863. [CrossRef] [PubMed]

31. Williamson, J.; Scott-Finnigan, T.J. Trypanocidal activity of antitumor antibiotics and other metabolic inhibitors. *Antimicrob. Agents Chemother.* **1978**, *13*, 735–744. [CrossRef] [PubMed]

32. Sufrin, J.R.; Rattendi, D.; Spiess, A.J.; Lane, S.; Marasco, C.J., Jr.; Bacchi, C.J. Antitrypanosomal activity of purine nucleosides can be enhanced by their conversion to *O*-acetylated derivatives. *Antimicrob. Agents Chemother.* **1996**, *40*, 2567–2572. [PubMed]

33. Cai, S.; Li, Q.S.; Borchardt, R.T.; Kuczera, K.; Schowen, R.L. The antiviral drug ribavirin is a selective inhibitor of *S*-adenosyl-L-homocysteine hydrolase from Trypanosoma cruzi. *Bioorg. Med. Chem.* **2007**, *15*, 7281–7287. [CrossRef] [PubMed]

34. Seley, K.L.; Schneller, S.W.; Rattendi, D.; Bacchi, C.J. (+)-7-deaza-5'-noraristeromycin as an anti-trypanosomal agent. *J. Med. Chem.* **1997**, *40*, 622–624. [CrossRef] [PubMed]

35. Yang, M.; Ye, W.; Schneller, S.W. Preparation of carbocyclic *S*-adenosylazamethionine accompanied by a practical synthesis of (−)-aristeromycin. *J. Org. Chem.* **2004**, *69*, 3993–3996. [CrossRef] [PubMed]

36. Luche, J.-L. Lanthanides in organic chemistry. 1. Selective 1,2 reductions of conjugated ketones. *J. Am. Chem. Soc.* **1978**, *100*, 2226–2227. [CrossRef]

37. Swamy, K.C.; Kumar, N.N.; Balaraman, E.; Kumar, K.V. Mitsunobu and related reactions: Advances and applications. *Chem. Rev.* **2009**, *109*, 2551–2651. [CrossRef] [PubMed]

38. Moussa, Z. The Hendrickson 'POP' reagent and analogues thereof: Synthesis, structure, and application in organic synthesis. *ARKIVOC* **2012**, 432–490.

39. Kim, J.-H.; Kim, H.O.; Lee, K.M.; Chun, M.W.; Moon, H.R.; Jeong, L.S. Asymmetric synthesis of homo-apioneplanocin A from D-ribose. *Tetrahedron* **2006**, *62*, 6339–6342. [CrossRef]

40. Hassan, J.; Sevignon, M.; Gozzi, C.; Schulz, E.; Lemaire, M. Aryl-aryl bond formation one century after the discovery of the Ullmann reaction. *Chem. Rev.* **2002**, *102*, 1359–1469. [CrossRef] [PubMed]

41. Vo, G.D.; Hartwig, J.F. Palladium-catalyzed coupling of ammonia with aryl chlorides, bromides, iodides, and sulfonates: A general method for the preparation of primary arylamines. *J. Am. Chem. Soc.* **2009**, *131*, 11049–11061. [CrossRef] [PubMed]

42. Huang, X.; Buchwald, S.L. New ammonia equivalents for the Pd-catalyzed amination of aryl halides. *Org. Lett.* **2001**, *3*, 3417–3419. [CrossRef] [PubMed]

43. Benkeser, R.A.; Buting, W.E. The preparation of aromatic amines with sodium amide in liquid ammonia. *J. Am. Chem. Soc.* **1952**, *74*, 3011–3014. [CrossRef]

44. Zhang, Y.; Tang, Q.; Luo, M. Reduction of hydrazines to amines with aqueous solution of titanium(iii) trichloride. *Org. Biomol. Chem.* **2011**, *9*, 4977–4982. [CrossRef] [PubMed]

45. Matovu, E.; Stewart, M.L.; Geiser, F.; Brun, R.; Maser, P.; Wallace, L.J.; Burchmore, R.J.; Enyaru, J.C.; Barrett, M.P.; Kaminsky, R.; *et al.* Mechanisms of arsenical and diamidine uptake and resistance in Trypanosoma brucei. *Eukaryot. Cell* **2003**, *2*, 1003–1008. [CrossRef] [PubMed]

46. Munday, J.C.; Eze, A.A.; Baker, N.; Glover, L.; Clucas, C.; Aguinaga Andres, D.; Natto, M.J.; Teka, I.A.; McDonald, J.; Lee, R.S.; *et al.* Trypanosoma brucei aquaglyceroporin 2 is a high-affinity transporter for pentamidine and melaminophenyl arsenic drugs and the main genetic determinant of resistance to these drugs. *J. Antimicrob. Chemother.* **2014**, *69*, 651–663. [CrossRef] [PubMed]

47. De Koning, H.P.; Jarvis, S.M. Adenosine transporters in bloodstream forms of Trypanosoma brucei brucei: Substrate recognition motifs and affinity for trypanocidal drugs. *Mol. Pharmacol.* **1999**, *56*, 1162–1170. [PubMed]

48. De Koning, H.P.; Bridges, D.J.; Burchmore, R.J.S. Purine and pyrimidine transport in pathogenic protozoa: From biology to therapy. *FEMS Microbiol. Rev.* **2005**, *29*, 987–1020. [CrossRef] [PubMed]

49. Al-Salabi, M.I.; Wallace, L.J.; Luscher, A.; Maser, P.; Candlish, D.; Rodenko, B.; Gould, M.K.; Jabeen, I.; Ajith, S.N.; de Koning, H.P. Molecular interactions underlying the unusually high adenosine affinity of a novel Trypanosoma brucei nucleoside transporter. *Mol. Pharmacol.* **2007**, *71*, 921–929. [CrossRef] [PubMed]

50. De Koning, H.P. Transporters in African trypanosomes: Role in drug action and resistance. *Int. J. Parasitol.* **2001**, *31*, 512–522. [CrossRef] [PubMed]

51. Munday, J.C.; Rojas Lopez, K.E.; Eze, A.A.; Delespaux, V.; van den Abbeele, J.; Rowan, T.; Barrett, M.P.; Morrison, L.J.; de Koning, H.P. Functional expression of TcoAT1 reveals it to be a P1-type nucleoside transporter with no capacity for diminazene uptake. *Int. J. Parasitol. Drugs Drug Resist.* **2013**, *3*, 69–76. [CrossRef] [PubMed]

52. Wallace, L.J.; Candlish, D.; de Koning, H.P. Different substrate recognition motifs of human and trypanosome nucleobase transporters. Selective uptake of purine antimetabolites. *J. Biol. Chem.* **2002**, *277*, 26149–26156. [CrossRef] [PubMed]

53. Natto, M.J.; Wallace, L.J.; Candlish, D.; Al-Salabi, M.I.; Coutts, S.E.; de Koning, H.P. Trypanosoma brucei: Expression of multiple purine transporters prevents the development of allopurinol resistance. *Exp. Parasitol.* **2005**, *109*, 80–86. [CrossRef] [PubMed]

54. Ali, J.A.; Creek, D.J.; Burgess, K.; Allison, H.C.; Field, M.C.; Maser, P.; de Koning, H.P. Pyrimidine salvage in Trypanosoma brucei bloodstream forms and the trypanocidal action of halogenated pyrimidines. *Mol. Pharmacol.* **2013**, *83*, 439–453. [CrossRef] [PubMed]

55. Parker, N.B.; Yang, X.; Hanke, J.; Mason, K.A.; Schowen, R.L.; Borchardt, R.T.; Yin, D.H. Trypanosoma cruzi: Molecular cloning and characterization of the S-adenosylhomocysteine hydrolase. *Exp. Parasitol.* **2003**, *105*, 149–158. [CrossRef] [PubMed]

56. Geiser, F.; Luscher, A.; de Koning, H.P.; Seebeck, T.; Maser, P. Molecular pharmacology of adenosine transport in Trypanosoma brucei: P1/P2 revisited. *Mol. Pharmacol.* **2005**, *68*, 589–595. [PubMed]

57. Vodnala, S.K.; Lundbäck, T.; Yeheskieli, E.; Sjöberg, B.; Gustavsson, A.-L.; Svensson, R.; Olivera, G.C.; Eze, A.A.; de Koning, H.P.; Hammarström, L.G.J.; et al. Structure-activity relationships of synthetic cordycepin analogues as experimental therapeutics for African Trypanosomiasis. *J. Med. Chem.* **2013**, *56*, 9861–9873. [CrossRef] [PubMed]

58. De Koning, H.P.; Al-Salabi, M.I.; Cohen, A.M.; Coombs, G.H.; Wastling, J.M. Identification and characterisation of high affinity nucleoside and nucleobase transporters in Toxoplasma gondii. *Int. J. Parasitol.* **2003**, *33*, 821–831. [CrossRef] [PubMed]

59. Quashie, N.B.; Dorin-Semblat, D.; Bray, P.G.; Biagini, G.A.; Doerig, C.; Ranford-Cartwright, L.C.; de Koning, H.P. A comprehensive model of purine uptake by the malaria parasite Plasmodium falciparum: Identification of four purine transport activities in intraerythrocytic parasites. *Biochem. J.* **2008**, *411*, 287–295. [CrossRef] [PubMed]

Sample Availability: *Sample Availability:* Samples of the compounds are not available from the authors at this time.

molecules

MDPI

Article

Development of New 1,3-Diazaphenoxazine Derivatives (ThioG-Grasp) to Covalently Capture 8-Thioguanosine

Yasufumi Fuchi, Hideto Obayashi and Shigeki Sasaki *

Graduate School of Pharmaceutical Sciences, Kyushu University, 3-1-1 Maidashi, Higashi-ku, Fukuoka 812-8582, Japan; fuchi@phar.kyushu-u.ac.jp (Y.F.); 2PS14001P@s.kyushu-u.ac.jp (H.O.)
* Correspondence: sasaki@phar.kyushu-u.ac.jp; Tel.: +81-92-642-6615

Academic Editor: Fumi Nagatsugi
Received: 9 December 2014; Accepted: 7 January 2015; Published: 9 January 2015

Abstract: The derivatives of 8-thioguanosine are thought to be included in the signal transduction system related to 8-nitroguanosine. In this study, we attempted to develop new 1,3-diazaphenoxazine (G-clamp) derivatives to covalently capture 8-thioguanosine (thioG-grasp). It was expected that the chlorine atom at the end of the linker would be displaced by the nucleophilic attack by the sulfur atom of 8-thioguanosine via multiple hydrogen-bonded complexes. The thioG-grasp derivative with a propyl linker reacted efficiently with 8-thioguanosine to form the corresponding adduct.

Keywords: oxidative damage; oxidized nucleoside; 8-oxoguanosine; 8-nitroguanosine; 8-thioguanosine

---

1. Introduction

Reactive oxygen species (ROS) and reactive nitrogen oxide species (RNOS) are the chemical sources of oxidative stress, and they oxidize nucleic acids to produce the 8-oxoguanosine and 8-nitroguanosine derivatives [1–3]. The oxidized nucleosides/nucleotides are highly mutagenic and are regarded as biomarkers of oxidative stress [4]. On the other hand, recent studies have revealed that nitrated-guanosine derivatives also play important roles as signal messengers; 8-nitroguanosine-3′,5′-cyclic monophosphate (8-nitro-cGMP) is generated from guanosine- 5′-triphosphate (GTP) in response to the production of peroxynitrite (ONOO⁻) and reacts with the sulfhydryl groups of proteins [5], H_2S/HS^- [6] or persulfide [7] to form 8-adduct-cGMP (S-guanylation) or 8-thio-cGMP. These metabolic cycles have led to the proposal of a new signaling pathway mediated by 8-nitro-cGMP. 8-Thio-cGMP may be returned to cGMP by ROS such as hydrogen peroxide in cells, but its biochemical role is not well understood. Thus, selective molecules that can form a covalently bonded complex with the 8-thioG derivative are required to further understand their biological functions. However, there is no specific recognition compound for 8-thioguanosine derivatives. In this study, we reported new 1,3-diazaphenoxazine nucleoside derivatives (thioG-grasp) that exhibit an efficient covalent capture of 8-thioguanosine via the formation of multiple hydrogen-bonded complexes.

We have focused on the development of recognition molecules for the 8-oxidized guanosine derivative based on the tricyclic cytosine analog "G-clamp" [8–12], and have reported the "8-oxoG-clamp" derivatives for the selective fluorescent detection of 8-oxo-dG [13–15]. Most recently, a new 1,3-diazaphenoxazine nucleoside derivative bearing a thiol arm, nitroG-grasp (1), has demonstrated the efficient capture of 8-nitroguanosine via multiple hydrogen-bonded complexes [16]. The displacement reactivity of nitroG-grasp (1) depends on the thiol pK_a and the length of the alkyl linker between the urea and the thiol group. For the new capture molecules for 8-thioguanosine, we based the design on a hydrogen bonded complex, such as that between the 8-nitroguanine portion and the 1,3-diazaphenoxazine portion connecting the urea-linker (Figure 1A). A nucleophilic attack of the 8-sulfur atom on the chloride leaving group was expected to form the corresponding covalent bond. The urea-type linker (X=NH) and the carbamate-type linker (X=O) were anticipated to form a suitable hydrogen bond with the thioenolate form or with the thioamide form, respectively (Figure 1B, nitroG-grasp, **2a–c**, **3**). In this study, we report the synthesis of 8-thioG-grasp derivatives and evaluated their reactivity with 8-thioguanosine.

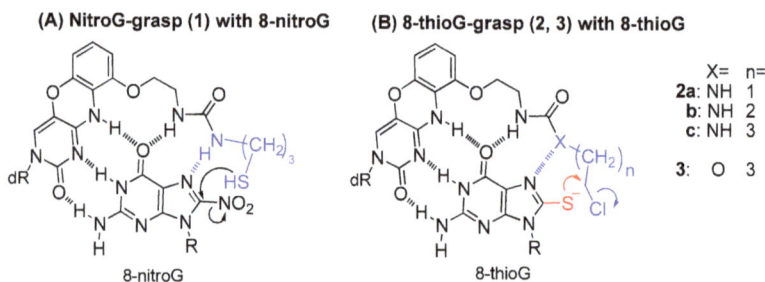

Figure 1. Molecular design of the selective capture molecules for 8-nitroG and 8-thioG. dR and R represent 3′,5′-diOTBS-2′-deoxyribosyl and triacetyl ribosyl groups, respectively.

2. Results and Discussion

2.1. Chemistry

In this study, 2′,3′,5′-tri-*O*-acetyl-8-thioguanosine (tri-Ac-8-thioG) was used as a substrate and was synthesized from 2′,3′,5′-tri-*O*-acetyl-8-bromoguanosine according to the literature [17]. The 8-thioG-grasp derivatives were synthesized through the reaction between the imidazole linker unit **5** and the amino group of the 3′,5′-*O*-diTBDMS-G-clamp unit **4** as a common intermediate (Scheme 1). The chloroalkylamine or chloropropanol were treated with carbonylimidazole in CH$_3$CN in the presence of triethylamine to form the corresponding imidazole intermediates **5**, which reacted with **4** to produce the desired 8-thioG-grasp derivatives. *N*-(3-Chloropropyl)-2-(pyrene-1-yl) acetamide **6** was synthesized from 1-pyrene-acetic acid as a control compound with no complexation site for 8-thioG.

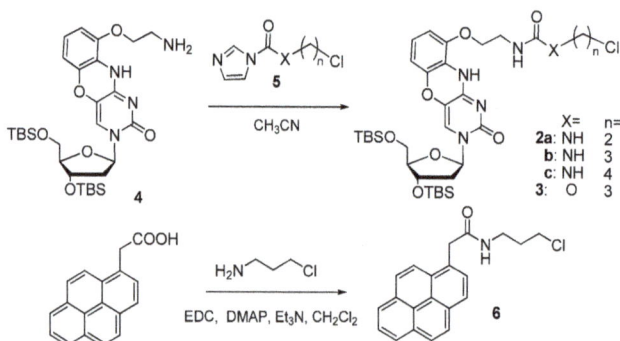

Scheme 1. The synthesis of the 8-thioG-grasp derivatives.

2.2. Reaction of 8-ThioG-Grasp with 8-ThioG

The reaction of the 8-thioG-grasp derivatives with triAc-8-thioG was performed at 50 °C in CH$_3$CN in the presence of Et$_3$N, and the reaction progress was monitored by HPLC. An equimolar mixture of 8-thioG-grasp (**2b**) and triAc-8-thioG formed an adduct in a nearly quantitative yield as a single product within 3 h (Figure 2).

This product was isolated, and its structure was confirmed as depicted in **7b** by ^{1}H-NMR, ESI-MS, and HMBC (see the Supporting Information). The HMBC spectrum indicated the distinct correlation between the C8 of the 8-thioguanine unit and the methylene protons next to the sulfur atom of **7b**.

2.3. Comparison of the Reactivity between 8-ThioG-Grasp and the Control Compound

The time courses of the reaction between the 8-thioG-grasp derivatives **2a–c**, **3** and triAc-8-thioG are compared in Figure 3. **2b** and **2c** exhibited efficient reactivity, and a relatively slow reaction was observed with **2a**. In the reaction of **2a**, the formation of the amino-oxazoline ring (**8**) as a byproduct decreased the adduct yield (Figure 3).

In contrast, the carbamate type **3** showed a significantly decreased efficiency compared with the compounds **2** with the urea-type linker. This is of great interest because 8-thioketo tautomer of 8-thioG is stable in neutral organic solvents [18] and forms more stable complexes with the carbamate type **9** than with the urea type **10** (K_s in CHCl$_3$, **9**: 7.6×10^6 M^{-1} *vs.* **10**: 1.7×10^6 M^{-1}) (Figure 4). Accordingly, it is reasonably explained that the 7N-H of 8-thioG is deprotonated by Et$_3$N to form the 8-thioenolate, thereby facilitating hydrogen-bonded complexes with the urea-type compounds (**2b** and **2c**) such as shown in Figure 1B to exhibit efficient reactivity. As the pK_a value of 7N-H of 8-thioguanosine is around 8.5 (see the Supporting Information), Et$_3$N is a suitable base for its deprotonation. No adduct was formed with N-(3-chloropropyl)-2-(pyrene-1-yl) acetamide (**6**), a control without a binding site, emphasizing the contribution of the hydrogen-bonded complexation of **2b** and **2c** for efficient reactivity. Among the urea-type 8-thioG-grasp derivatives, **2a** produced the adduct in a low yield. The HPLC monitoring showed that **2a** formed the byproduct **8** as a major peak, which structure was determined by H-NMR, ESI-MS and 2D-HMBC, as shown in Figure 3. The amino-oxazoline was formed by intramolecular displacement, which was faster than the intermolecular nucleophilic attack from 8-thioG. It should be emphasized that the displacement of the sulfur atom of 8-thioG in the complex with **2b** or **2c** is favorable compared with the intramolecular amino-oxazoline ring formation because this type of amino-oxazoline ring formation was not observed for **2b** or **2c**.

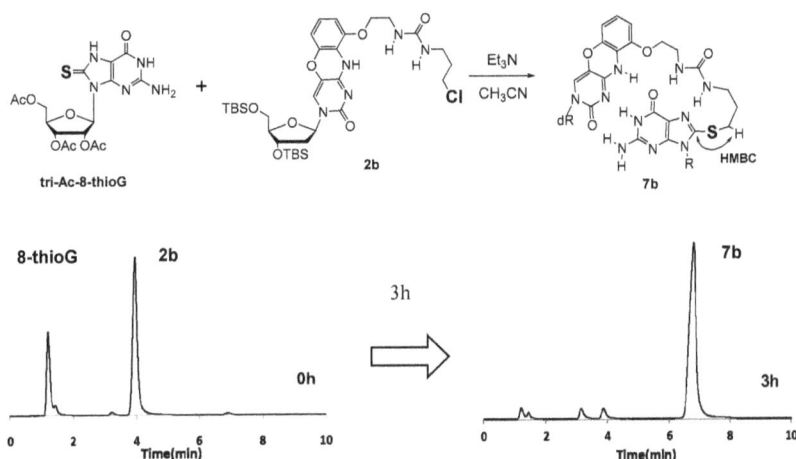

Figure 2. Reaction of 8-thioG-grasp with tri-Ac-8-thioG monitored by HPLC. dR and R represent 3′,5′-diOTBS-2′-deoxyribosyl and triacetyl ribosyl groups, respectively. The reaction was performed using 0.4 mM each of **2b** and 8-thioG in the presence of 20 mM Et$_3$N in CH$_3$CN at 50 °C. HPLC conditions: column: Xbridge C8 3.5 μm, 3.0 mm × 100 mm; solvents: (**A**) 0.1 M TEAA buffer at pH 7.0 and (**B**) CH$_3$CN, A/B = 20:80; flow rate: 0.5 mL/min; monitored by UV at 254 nm.

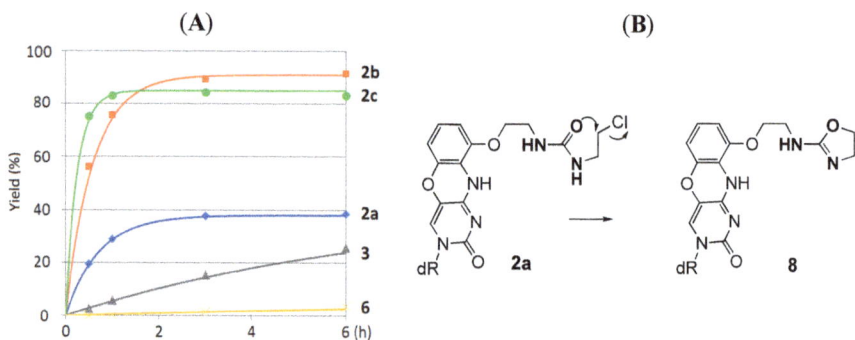

(A) **(B)**

Figure 3. (**A**) Comparison of the time courses of the reactions with 8-thioG-grasp (**2a–c** and **3**) and the control compound **6**. (**B**) Byproduct **8** formed via the intramolecular reaction of **2a**. Product yields were obtained at the indicated time points by HPLC analysis, as described in the footnote to Figure 2. dR represents the 3′,5′-diOTBS-2′-deoxyribosyl group.

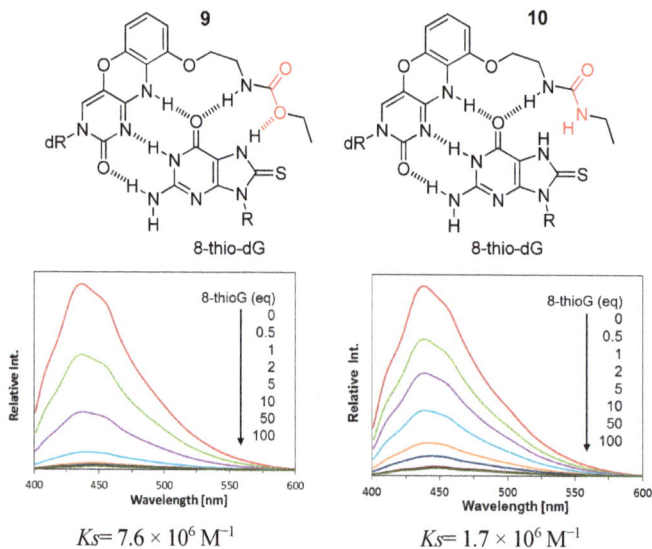

$Ks= 7.6 \times 10^6 \text{ M}^{-1}$ $Ks= 1.7 \times 10^6 \text{ M}^{-1}$

Figure 4. Proposed complexation of 8-thioG with the carbamate or the urea type. dR and R represent 3′, 5′-diOTBS-2′-deoxyribosyl and triacetyl ribosyl groups, respectively.

3. Experimental Section

3.1. General Information

The reagents and solvents were purchased from commercial suppliers and were used without purification. The ^{1}H- and ^{13}C- NMR spectra were recorded on a Bruker Avance III spectrometer. The 2D-NMR spectra were measured on a Varian Inova 500 instrument. The IR spectra were recorded on a Perkin Elmer Spectrum One FT-IR spectrometer. The ESI-HRMS spectra were measured on an Applied Biosystems Mariner Biospectrometry Workstation using neurotensin, angiotensin I, bradykinin and picolinic acid as the internal standards.

3.2. Chemistry

3.2.1. General Synthesis of 8-ThioG-Grasp Derivatives

1,1'-Carbonyldiimidazole (5 equiv.) was added to a solution of chloroalkylamine hydrochloride (5 equiv.) in anhydrous CH_3CN (0.05 M in G-clamp unit) under an argon atmosphere. The reaction mixture was stirred at room temperature for 30 min, followed by the addition of the G-clamp unit **4** (1 equiv.) and Et_3N (8 equiv.). After stirring overnight at room temperature, saturated aqueous $NaHCO_3$ solution was added to the reaction mixture, which was extracted with $CHCl_3$. The organic layer was washed with brine, dried over Na_2SO_4 and evaporated *in vacuo*. The resulting residue was purified by silica gel column chromatography (CH_2Cl_2/MeOH =100:0 to 95:5) to give **2a–c** and **3** as light yellow foams.

1-(2-((3-((2R,4S,5R)-4-((tert-Butyldimethylsilyl)oxy)-5-(((tert-butyldimethylsilyl)oxy)methyl)-tetrahydrofuran-2-yl)-2-oxo-2,10-dihydro-3H-benzo[b]pyrimido[4,5-e][1,4]oxazin-9-yl)oxy)ethyl)-3-(2-chloroethyl)urea. (C2-8-ThioG-grasp, **2a**): **2a** was synthesized using G-clamp unit **4** and 2-chloroethylamine in 48% as a light yellow foam. ^{1}H-NMR (400 MHz, CD_3OD) δ (ppm) 7.62 (1H, s), 6.82 (1H, t, *J* = 8.6 Hz), 6.60 (1H, d, *J* = 8.6 Hz), 6.35 (1H, d, *J* = 8.6 Hz), 6.17 (1H, t, *J* = 6.4 Hz), 4.46 (1H, dt, *J* = 6.4, 3.4 Hz), 4.05 (2H, t, *J* = 4.9 Hz), 3.96–3.93 (2H, m), 3.82 (1H, dd, *J* = 12.1, 3.4 Hz), 3.57 (2H, t, *J* = 6.0 Hz), 3.54 (2H, t, *J* = 5.2 Hz), 3.45 (2H, t, *J* = 6.4 Hz), 2.33 (1H, ddd, *J* = 13.4, 6.1, 4.6 Hz), 2.11 (1H, dt, *J* = 13.4, 6.4 Hz), 0.97 (9H, s), 0.91 (9H, s,), 0.18 (3H, s), 0.16 (3H, s), 0.11 (6H, s). ^{13}C-NMR (125 MHz, $CDCl_3$) δ (ppm) 158.53, 152.89, 146.88, 143.06, 127.52, 124.49, 122.97, 115.06, 108.21, 107.54, 87.75, 86.00, 70.84, 69.22, 62.32, 42.09, 41.89, 39.33, 29.28, 26.04, 25.71, 18.49, 17.94, −4.58, −4.90, −5.45, −5.52. IR (cm^{-1}): 2929, 1673, 1557, 1473. HR ESI-MS (*m/z*): Calcd. for $[C_{32}H_{53}ClN_5O_7Si_2]^+$: 710.3167 ([M+H]$^+$), found 710.3176.

1-(2-((3-((2R,4S,5R)-4-((tert-Butyldimethylsilyl)oxy)-5-(((tert-butyldimethylsilyl)oxy)methyl)-tetrahydrofuran-2-yl)-2-oxo-2,10-dihydro-3H-benzo[b]pyrimido[4,5-e][1,4]oxazin-9-yl)oxy)ethyl)-3-(3-chloropropyl)urea. (C3-8-ThioG-grasp, **2b**): **2b** was synthesized using G-clamp unit **4** and 3-chloropropylamine in 46% as a light yellow foam. ^{1}H-NMR (400 MHz, CD_3OD) δ (ppm) 7.62 (1H, s), 6.82 (1H, t, *J* = 8.2 Hz), 6.60 (1H, dd, *J* = 8.4, 1.2 Hz), 6.35 (1H, dd, *J* = 8.2, 1.2 Hz), 6.17 (1H, t, *J* = 6.4 Hz), 4.47 (1H, dt, *J* = 5.7, 3.7 Hz), 4.05 (2H, t, *J* = 5.2 Hz), 3.95 (1H, t, *J* = 3.1 Hz), 3.94 (1H, t, *J* = 3.1 Hz), 3.82 (1H, dd, *J* = 12.2, 3.4 Hz), 3.57 (2H, t, *J* = 6.7 Hz), 3.53 (2H, t, *J* = 5.2 Hz), 3.27 (2H, t, *J* = 6.7 Hz), 2.33 (1H, ddd, *J* = 13.2, 6.2, 4.0 Hz), 2.12 (1H, dt, *J* = 13.2, 6.4 Hz), 1.92 (2H, quin, *J* = 6.4 Hz), 0.97 (9H, s), 0.91 (9H, s,), 0.18 (3H, s), 0.16 (3H, s), 0.11 (6H, s). ^{13}C-NMR (125 MHz, $CDCl_3$) δ (ppm) 158.99, 152.94, 152.67, 146.79, 143.04, 127.61, 124.38, 122.87, 115.26, 108.27, 107.74, 87.71, 85.94, 70.82, 69.45, 62.31, 42.74, 41.88, 39.39, 37.45, 33.21, 26.04, 25.71, 18.48, 17.93, −4.58, −4.90, −5.45, −5.52. IR (cm^{-1}): 2931, 1670, 1558, 1499. HR ESI-MS (*m/z*): Calcd. for $[C_{33}H_{55}ClN_5O_7Si_2]^+$: 724.3323 ([M+H]$^+$), found 724.3340.

1-(2-((3-((2R,4S,5R)-4-((tert-butyldimethylsilyl)oxy)-5-(((tert-butyldimethylsilyl)oxy)methyl)tetrahydro-furan-2-yl)-2-oxo-2,10-dihydro-3H-benzo[b]pyrimido[4,5-e][1,4]oxazin-9-yl)oxy)ethyl)-3-(4-chloro-butyl)urea. (C4-8-ThioG-grasp, **2c**): **2c** was synthesized using G-clamp unit **4** and 4-chlorobutylamine [19] in 82% as a light yellow foam. ^{1}H-NMR (400 MHz, CD_3OD) δ (ppm) 7.62 (1H, s), 6.81 (1H, t, *J* = 8.6 Hz), 6.59 (1H, d, *J* = 8.6 Hz), 6.24 (1H, t, *J* = 8.2 Hz), 6.16 (1H, t, *J* = 6.1 Hz), 4.46 (1H, br), 4.03 (2H, t, *J* = 5.2 Hz), 3.94 (2H, dd, *J* = 11.6, 2.4 Hz), 3.82 (1H, d, *J* = 9.5 Hz), 3.61 (2H, t, *J* = 6.4 Hz) 3.53 (2H, t, *J* = 6.4 Hz), 3.12 (2H, t, *J* = 6.7 Hz), 2.35–2.29 (1H, m), 2.11 (1H, dt, *J* = 13.1, 6.4 Hz), 1.88–1.73 (4H, m), 0.97 (9H, s), 0.91 (9H, s,), 0.18 (3H, s), 0.16 (3H, s), 0.11 (6H, s). ^{13}C-NMR (125 MHz, CD_3OD) δ (ppm) 161.16, 156.45, 155.65, 148.10, 144.31, 129.56, 125.02, 123.52, 109.17, 108.64, 89.33, 87.51, 72.88, 70.17, 63.73, 45.49, 42.80, 40.43, 40.32, 31.10, 28.78, 26.60, 26.25, 19.37, 18.86, −4.45, −4.65, −5.25, −5.30. IR (cm^{-1}): 2953, 1670, 1558. HR ESI-MS (*m/z*): calcd. for $[C_{34}H_{57}ClN_5O_7Si_2]^+$, 738.3480 ([M+H]$^+$); found 738.3452.

(2-((3-((2R,4S,5R)-4-((tert-Butyldimethylsilyl)oxy)-5-(((tert-butyldimethylsilyl)oxy)methyl)tetrahydro-furan-2-yl)-2-oxo-2,10-dihydro-3H-benzo[b]pyrimido[4,5-e][1,4]oxazin-9-yl)oxy)ethyl)-3-chloropropyl carbamate. (C3 (O)-8-ThioG-grasp, **3**): **3** was synthesized using G-clamp unit **4** and 3-chloropropanol in 82% as a

light yellow foam. ^{1}H-NMR (400 MHz, CD$_3$OD)　(ppm) 7.64 (1H, s), 6.83 (1H, t, *J* = 8.2 Hz), 6.61 (1H, dd, *J* = 8.2, 0.9 Hz), 6.37 (1H, dd, *J* = 8.2, 0.9 Hz), 6.18 (1H, t, *J* = 6.1 Hz), 4.47 (1H, dt, *J* = 5.8, 4.0 Hz), 4.19 (2H, t, *J* = 6.1 Hz), 4.06 (1H, t, *J* = 5.2 Hz), 3.96 (1H, t, *J* = 2.75 Hz), 3.94 (1H, t, *J* = 3.1 Hz), 3.83 (1H, dd, *J* = 12.2, 3.4 Hz), 3.63 (2H, t, *J* = 6.7 Hz), 3.52 (2H, t, *J* = 5.2 Hz), 3.27 (2H, t, *J* = 6.7 Hz), 2.34 (1H, ddd, *J* = 13.4, 6.1, 4.3 Hz), 2.12 (1H, dt, *J* = 13.4, 6.1 Hz), 2.06 (2H, quin, *J* = 6.1 Hz), 0.97 (9H, s), 0.92 (9H, s,), 0.18 (3H, s), 0.17 (3H, s), 0.11 (6H, s). ^{13}C-NMR (125 MHz, CDCl$_3$) δ (ppm) 156.62, 152.69, 152.02, 146.63, 143.16, 127.26, 124.14, 122.85, 115.58, 108.41, 107.15, 87.63, 86.02, 70.71, 68.38, 62.26, 41.89, 41.38, 40.39, 32.05, 26.04, 25.72, 18.48, 17.94, −4.56, −4.90, −5.46, −5.52. IR (cm^{-1}): 2953, 1679, 1556, 1474. HR ESI-MS (*m/z*): calcd. for [C$_{33}$H$_{54}$ClN$_4$O$_8$Si$_2$]$^+$, 725.3163 ([M+H]$^+$); found 725.3142.

3.2.2. Synthesis of 8-ThioG Adduct **7b**

Et$_3$N (100 µL, 0.72 mmol) was added to a solution of **2b** (11 mg, 0.015 mmol) and tri-Ac-8-thioG (14 mg, 0.031 mmol) in anhydrous CH$_3$CN (1 mL) under an argon atmosphere. The reaction mixture was stirred at 50 °C for 12 h, and evaporated *in vacuo*. The resulting residue was purified by silica gel column chromatography (CHCl$_3$/MeOH =70:1) to give **5b** as a colorless oil (9 mg, 50%). ^{1}H-NMR (400 MHz, CD$_3$OD) δ (ppm) 7.51 (1H, s), 6.70 (1H, t, *J* = 8.2 Hz), 6.41 (1H, d, *J* = 8.6 Hz), 6.18 (1H, t, *J* = 4.9 Hz), 6.15 (1H, d, *J* = 6.1 Hz), 6.04–6.03 (1H, m), 5.84 (1H, d, *J* = 4.6 Hz), 5.77 (1H, t, *J* = 6.1 Hz), 4.51 (1H, dd, *J* = 7.6, 2.4 Hz), 4.41–4.31 (3H, m), 3.89–3.85 (4H, m), 3.76 (1H, d, *J* = 9.5 Hz), 3.60 (2H, br), 3.29–3.21 (4H, m), 2.29 (1H, ddd, *J* = 13.0, 6.5 Hz), 2.13–2.06 (1H, m), 1.96 (2H, br), 0.93 (9H, s), 0.91 (9H, s,), 0.14 (3H, s), 0.13 (3H, s), 0.09 (6H, s) ^{13}C-NMR (125 MHz, CD$_3$OD) δ (ppm) 172.40, 171.44, 171.20, 161.36, 160.17, 156.20, 156.01, 155.09, 155.04, 148.62, 147.36, 144.32, 129.27, 125.38, 123.73, 117.51, 115.86, 107.64, 88.03, 87.47, 81.00, 72.63, 71.79, 64.20, 42.70, 40.58, 39.76, 31.34, 29.52, 26.66, 26.34, 20.42, 19.37, 18.88, −4.29, −4.54, −5.14. IR (cm^{-1}): 2928, 1749, 1684. HR ESI-MS (*m/z*): calcd. for [C$_{49}$H$_{73}$N$_{10}$O$_{15}$SSi$_2$]$^+$, 1129.4511 ([M+H]$^+$); found 1129.4469.

3.2.3. N-(3-Chloropropyl)-2-(pyren-1-yl)acetamide (**6**)

3-Chloropropylamine hydrochloride (12 mg, 0.093 mmol), 1-(3-dimethylaminopropyl)-3-ethyl-carbodiimide hydrochloride (EDC, 17 mg, 0.089 mmol), 4,4-dimethylaminopyridine (DMAP, 1 mg 0.008 mmol) and Et$_3$N (50 µL, 0.359 mmol) were added to a solution of 1-pyreneacetic acid (20 mg, 0.077 mmol) in anhydrous CH$_2$Cl$_2$ (1 mL) under an argon atmosphere. The reaction mixture was stirred at room temperature for 7 h, quenched with aqueous saturated NH$_4$Cl solution, and extracted with AcOEt. The organic layer was washed with brine, dried over Na$_2$SO$_4$ and evaporated *in vacuo*. The resulting residue was purified by silica gel column chromatography (CH$_2$Cl$_2$) to give **6** as a pale yellow solid (13 mg, 50%). ^{1}H-NMR (400 MHz, CDCl$_3$) δ (ppm) 8.20 (2H, d, *J* = 7.6 Hz), 8.15 (3H, d, *J* = 7.9 Hz), 8.08 (1H, d, *J* = 9.2 Hz), 8.04 (1H, d, *J* = 9.2 Hz), 8.02 (1H, t, *J* = 7.9 Hz), 7.88 (1H, d, *J* = 7.9 Hz), 4.28 (2H, s), 3.31 (2H, t, *J* = 6.4 Hz), 3.25 (2H, q, *J* = 6.4 Hz), 1.78 (2H, quin, J = 6.4 Hz). ^{13}C-NMR (125 MHz, CDCl$_3$) δ (ppm) 171.23, 131.28, 131.16, 130.76, 129.53, 128.50, 128.16, 127.69, 127.27, 126.28, 125.60, 125.48, 125.20, 125.12, 124.62, 122.76, 42.22, 42.08, 37.28, 31.86. IR (cm^{-1}): 3293, 1646, 1544. HR ESI-MS (*m/z*): calcd. for [C$_{21}$H$_{19}$ClNO]+, 336.1150 ([M+H]$^+$); found 336.1161.

3.2.4. Determination of the Structure of the Byproduct **8**

To a solution of **2a** (70 mg, 0.099 mmol) in MeOH (4 mL) was added NaHCO$_3$ (90 mg, 1.07 mmol) under an argon atmosphere. The reaction mixture was refluxed for 25 h, filtered, and evaporated *in vacuo*. The resulting residue was purified by silica gel column chromatography (CHCl$_3$/Acetone =100:0 to 50:50) to give **8** as light yellow foams (26 mg, 39%). ^{1}H-NMR (400 MHz, CD$_3$OD) δ (ppm) 7.60 (1H, s), 6.81 (1H, t, *J* = 8.2 Hz), 6.58 (1H, d, *J* = 8.2 Hz), 6.35 (1H, d, *J* = 8.2 Hz), 6.17 (1H, t, *J* = 6.4 Hz), 4.47 (1H, dt, *J* = 6.1, 3.5 Hz), 4.35 (2H, t, *J* = 8.2 Hz), 4.07 (2H, *J* = 4.9 Hz), 3.96–3.93 (2H, m), 3.82 (1H, dd, *J* = 12.2, 3.1 Hz), 3.78 (2H, t, *J* = 8.2 Hz), 3.55 (2H, t, *J* = 5.2 Hz), 2.33 (1H, ddd, *J* = 13.3, 6.1, 4.1 Hz), 2.12 (1H, dt, *J* = 13.3, 6.3 Hz), 0.97 (9H, s), 0.91 (9H, s,), 0.18 (3H, s), 0.16 (3H, s), 0.11 (6H, s). ^{13}C-NMR (125 MHz, CD$_3$OD) δ (ppm) 164.20, 156.51, 155.66, 147.98, 144.31, 129.55, 124.95, 123.42, 117.14, 109.25,

108.51, 89.34, 87.51, 72.97, 69.64, 69.37, 63.77, 43.22, 42.77, 29.54, 26.60, 26.26, 19.36, 18.85, -4.46, -4.65, -5.26, -5.30. IR (cm^{-1}): 2931, 1669, 1557, 1497, 1473. HR ESI-MS (*m/z*): calcd. for $[C_{32}H_{52}N_5O_7Si_2]^+$, 674.3400 ([M+H]$^+$); found 674.3438.

3.2.5. General Procedure of Reaction Monitoring by HPLC

Reaction was initiated by the addition of Et$_3$N (20 mM) to a solution of 8-thioG-grasp derivative (0.4 mM) and triAc-8-thioG) (0.4 mM) in CH$_3$CN at 50 °C. The reaction progress was monitored by HPLC at 0.5, 1, 3 and 5 h. Product yield with time course of reaction were obtained from reverse-phase HPLC analysis. (Column: Xbridge C8 3.5 μm, 3.0 × 100 mm; Solvent: A: 0.1 M TEAA buffer at pH 7.0, B: CH$_3$CN, A/B= 20: 80; Flow rate: 0.5 mL/min; monitored by UV detector at 254 nm).

4. Conclusions

In this study, we designed new recognition molecules to covalently capture 8-thioguanosine based on the G-clamp skeleton by introducing chloroalkyl urea linker. 8-ThioG-grasp **2b** and **2c** with chloropropyl urea and chlorobutyl linker exhibited the most efficient reactivity for tri-Ac-8-thioG. It has been shown from the comparison with control compounds that the multiple hydrogen-bonded complexes contribute to the efficient reactivity. There is increasing interest in 8-thioguanosine derivatives for their biological roles in signal transduction pathways [7], the 8-thioG-grasp derivatives are expected to be a potential platform to develop specific molecules. For example, an 8-thioG-grasp derivative with a phosphate binding unit are expected to covalently trap phosphate derivatives of 8-thioguanosine and interfere their biological functions. Systematic studies are now ongoing in this line in our group, which will be reported in due course.

Supplementary Materials: Supplementary materials can be accessed at: http://www.mdpi.com/1420-3049/20/01/1078/s1.

Acknowledgments: We are grateful for support provided by a Grant-in-Aid for Challenging Exploratory Research (No. 26670004) from the Japan Society for the Promotion of Science (JSPS). S.S. and Y.F. also acknowledge the Astellas Foundation for Research on Metabolic Disorders and a Research Fellowship for Young Scientists from JSPS, respectively.

Author Contributions: Y. Fuchi and H. Obayashi performed chemistry experiments and analyzed the data. Y. Fuchi and S. Sasaki designed the capture molecules, and wrote the paper.

Conflicts of Interest: The authors declare no conflict of interest.

References

1. Halliwell, B. Oxidative stress and cancer: Have we moved forward? *Biochem. J.* **2007**, *401*, 1–11. [CrossRef] [PubMed]
2. Yermilov, V.; Rubio, J.; Ohshima, H. Formation of 8-nitroguanine in DNA treated with peroxynitrite in vitro and its rapid removal from DNA by depurination. *FEBS Lett.* **1995**, *376*, 207–210. [CrossRef] [PubMed]
3. Kasai, H. Analysis of a form of oxidative DNA damage, 8-hydroxy-2'-deoxyguanosine, as a marker of cellular oxidative stress during carcinogenesis. *Mutat. Res.* **1997**, *387*, 147–163. [CrossRef] [PubMed]
4. Maeda, H. The link between infection and cancer: Tumor vasculature, free radicals, and drug delivery to tumors via the EPR effect. *Cancer Sci.* **2013**, *104*, 779–789. [CrossRef] [PubMed]
5. Sawa, T.; Zaki, M.H.; Okamoto, T.; Akuta, T.; Tokutomi, Y.; Kim-Mitsuyama, S.; Ihara, H.; Kobayashi, A.; Yamamoto, M.; Fujii, S.; *et al.* Protein S-guanylation by the biological signal 8-nitroguanosine 3',5'-cyclic monophosphate. *Nat. Chem. Biol.* **2007**, *3*, 727–735. [CrossRef] [PubMed]
6. Nishida, M.; Sawa, T.; Kitajima, N.; Ono, K.; Inoue, H.; Ihara, H.; Motohashi, H.; Yamamoto, M.; Suematsu, M.; Kurose, H.; *et al.* Hydrogen sulfide anion regulates redox signaling via electrophile sulfhydration. *Nat. Chem. Biol.* **2012**, *8*, 714–724. [CrossRef] [PubMed]
7. Ida, T.; Sawa, T.; Ihara, H.; Tsuchiya, Y.; Watanabe, Y.; Kumagai, Y.; Suematsu, M.; Motohashi, H.; Fujii, S.; Matsunaga, T.; *et al.* Reactive cysteine persulfides and S-polythiolation regulate oxidative stress and redox signaling. *Proc. Natl. Acad. Sci. USA* **2014**, *111*, 7606–7611. [CrossRef] [PubMed]

8. Lin, K.; Matteucci, M.D. A Cytosine analogue capable of clamp-like binding to a guanine in helical nucleic acids. *J. Am. Chem. Soc.* **1998**, *120*, 8531–8532. [CrossRef]

9. Flanagan, W.M.; Wolf, J.J.; Olson, P.; Grant, D.; Lin, K.Y.; Wangner, R.W.; Matteucci, M.D. A cytosine analog that confers enhanced potency to antisense oligonucleotides. *Proc. Natl. Acad. Sci. USA* **1999**, *96*, 3513–3518. [CrossRef] [PubMed]

10. Maier, M.A.; Leeds, J.M.; Balow, G.; Springer, R.H.; Bharadwaj, R.; Manoharan, M. Nuclease resistance of oligonucleotides containing the tricyclic cytosine analogues phenoxazine and 9-(2-aminoethoxy)-phenoxazine ("G-clamp") and origins of their nuclease resistance properties. *Biochemistry* **2002**, *41*, 1323–1327. [CrossRef] [PubMed]

11. Wilds, C.J.; Maier, M.A.; Tereshko, V.; Manoharan, M.; Egli, M. Direct observation of a cytosine analogue that forms five hydrogen bonds to guanosine: guanidino G-clamp. *Angew. Chem. Int. Ed.* **2002**, *41*, 115–117. [CrossRef]

12. Holmes, S.C.; Arzumanov, A.A.; Gait, M.J. Steric inhibition of human immunodeficiency virus type-1 Tat-dependent trans-activation in vitro and in cells by oligonucleotides containing 2'-O-methyl G-clamp ribonucleoside analogues. *Nucleic Acids Res.* **2003**, *31*, 2759–2768. [CrossRef] [PubMed]

13. Nakagawa, O.; Ono, S.; Li, Z.; Tsujimoto, A.; Sasaki, S. Specific fluorescent probe for 8-Oxoguanosine. *Angew. Chem. Int. Ed. Engl.* **2007**, *119*, 4584–4587. [CrossRef]

14. Li, Z.; Nakagawa, O.; Koga, Y.; Taniguchi, Y.; Sasaki, S. Synthesis of new derivatives of 8-oxoG-clamp for better understanding the recognition mode and improvement of selective affinity. *Bioorg. Med. Chem.* **2010**, *18*, 3992–3998. [CrossRef] [PubMed]

15. Koga, Y.; Fuchi, Y.; Nakagawa, O.; Sasaki, S. Optimization of fluorescence property of the 8-oxodGclamp derivative for better selectivity for 8-oxo-2'-deoxyguanosine. *Tetrahedron* **2011**, *67*, 6746–6752. [CrossRef]

16. Fuchi, Y.; Sasaki, S. Efficient covalent capture of 8-nitroguanosine via a multiple hydrogen-bonded complex. *Org. Lett.* **2014**, *16*, 1760–1763. [CrossRef] [PubMed]

17. Holmes, E.; Robins, K. Purine nucleosides. VII. Direct bromination of adenosine, deoxyadenosine, guanosine, and related purine nucleosides. *J. Am. Chem. Soc.* **1964**, *86*, 1242–1245. [CrossRef]

18. Strassmaier, T.; Karpen, J.W. Novel N7- and N1-substituted cGMP derivatives are potent activators of cyclic nucleotide-gated channels. *J. Med. Chem.* **2007**, *50*, 4186–4194. [CrossRef] [PubMed]

19. Song, A.; Walker, S.G.; Parker, K.A.; Sampson, N.S. Antibacterial studies of cationic polymers with alternating, random, and uniform backbones. *ACS Chem. Biol.* **2011**, *6*, 590–599. [CrossRef] [PubMed]

Sample Availability: *Sample Availability:* Samples are available from authors.

molecules

MDPI

Article

Hybridisation Potential of 1',3'-Di-*O*-methylaltropyranoside Nucleic Acids

Akkaladevi Venkatesham †, Dhuldeo Kachare †, Guy Schepers, Jef Rozenski, Mathy Froeyen and Arthur Van Aerschot *

Medicinal Chemistry, Rega Institute for Medical Research, KU Leuven, Minderbroedersstraat 10,
Leuven BE-3000, Belgium; akkaladevi.venkatesham@rega.kuleuven.be (A.V.); ddkachare@gmail.com (D.K.);
Guy.Schepers@rega.kuleuven.be (G.S.); Jef.Rozenski@rega.kuleuven.be (J.R.);
Mathy.Froeyen@rega.kuleuven.be (M.F.);

* Correspondence: Arthur.Vanaerschot@rega.kuleuven.be; Tel.: +32-16-372-624
† These authors contributed equally to this work.

Academic Editor: Mahesh K. Lakshman
Received: 22 January 2015; Accepted: 24 February 2015; Published: 3 March 2015

Abstract: In further study of our series of six-membered ring-containing nucleic acids, different 1',3'-di-*O*-methyl altropyranoside nucleoside analogs (DMANA) were synthesized comprising all four base moieties, adenine, cytosine, uracil and guanine. Following assembly into oligonucleotides (ONs), their affinity for natural oligonucleotides was evaluated by thermal denaturation of the respective duplexes. Data were compared with results obtained previously for both anhydrohexitol (HNAs) and 3'-*O*-methylated altrohexitol modified ONs (MANAs). We hereby demonstrate that ONs modified with DMANA monomers, unlike some of our previously described analogues with constrained 6-membered hexitol rings, did not improve thermodynamic stability of dsRNA complexes, most probably in view of an energetic penalty when forced in the required 1C4 pairing conformation. Overall, a single incorporation was more or less tolerated or even positive for the adenine congener, but incorporation of a second modification afforded a slight destabilization (except for A), while a fully modified sequence displayed a thermal stability of −0.3 °C per modification. The selectivity of pairing remained very high, and the new modification upon incorporation into a DNA strand, strongly destabilized the corresponding DNA duplexes. Unfortunately, this new modification does not bring any advantage to be further evaluated for antisense or siRNA applications.

Keywords: modified oligonucleotides; hexitol nucleic acids; pairing behaviour; hybridisation; constrained oligonucleotides

1. Introduction

Gene silencing has become a standard technique for studying gene functions or in trying to obtain therapeutic effects and theoretically can be attained by interfering with transcription (via formation of triple stranded complexes [1] or translation processes. The latter can be obtained via steric blocking antisense oligonucleotides (ASOs) or via mRNA cleavage of double stranded complexes with RNAseH activating ASOs [2–4]. However, the vast majority of researchers nowadays have turned to the use of RNA interference (RNAi)-based strategies, which has recently become the technique of choice to silence gene expression in mammalian cell culture and is envisaged as a first choice for therapeutic treatment as well [5–7]. However, we need to point out that at the moment only one aptamer [8] and one antisense oligonucleotide have been effectively FDA approved for gene silencing [9], and also exon-skipping oligonucleotides are receiving considerable attention [10].

In cell culture in general unmodified siRNAs are highly efficient, however for *in vivo* application some chemical modifications are warranted to stabilise the siRNAs and to increase their selectivity and to promote delivery [11,12]. In the past, we and others have studied a wide variety of strategies for both ASO and siRNA modification as reviewed several times [13–15].

Our group has been very successful in increasing the affinity for RNA using modified building blocks based on 6-membered hexitol rings which resulted in the hexitol nucleic acids series [16–19]. However, the LNA monomers of the Wengel group [20] consistently showed the strongest affinity for RNA, and the series comprises many alternative structures [21]. Overall, both our hexitol nucleic acids and the LNA series of compounds take on a pre-organized conformation, fitting the A-form of dsRNA and rationalizing the strong hybridization characteristics noticed.

Herein, hexitol nucleic acids (HNA, Figure 1) are composed of 2,3-dideoxy-D-*arabino*-hexitol units with a nucleobase situated in the 2-(S)-position (in *cis* to the hydroxymethyl substituent as in natural nucleosides). Addition of a supplementary hydroxyl at the 3'-α-position resulted in D-altritol nucleic acid (ANA, Figure 1) analogs with increased affinity for RNA strands [18,22]. More recently, we finally reported on the 3'-O-methylated ANA congeners (MANA, Figure 1) resulting in a further increase of 0.5 °C/modification when evaluating melting temperatures (T_m) upon hybridisation to RNA [23]. Herein, both the heterocyclic base and the 3'-O-methyl moiety are located in an axial position with a $^4C_1$ conformation. However, assembly of the hexitol series of nucleosides is long-routed starting with synthesis of the 1,5-anhydrohexitol ring. In view of the positive results obtained for the MANA series of congeners and the abundance of cheap α-D-methylglucoside, we now planned to prepare and evaluate "bis-methylated altritol nucleosides", or more correctly di-O-methylated altropyranoside nucleic acids (DMANA, Figure 1).

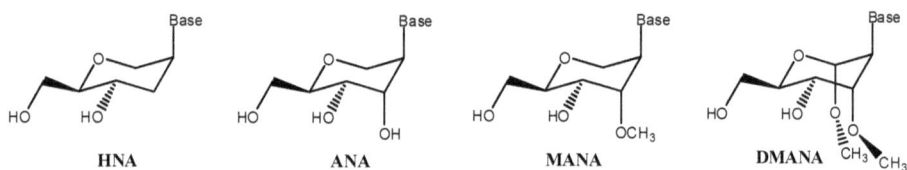

Figure 1. Structures of the different hexitol based nucleic acids (HNA, ANA and MANA) as discussed above and the newly envisaged structure DMANA, based on methyl-altropyranoside.

2. Results and Discussion

2.1. Chemical Synthesis of the New Building Blocks

The 1'-O-methylglycosidic protected analogues **4**, **6** and **9** (Scheme 1) were obtained starting from ubiquitous methyl glucopyranoside **1**, which in three steps was converted to **2** in 51.4% overall yield according to literature procedures [24]. Herein, regioselective epoxide ring opening of **2** with the sodium salts of uracil or adenine in DMF at 120–130 °C afforded the corresponding altrohexitol derivatives **3** and **7** in 66%–85% yield. Chemoselective methylation of **7** and **3** was accomplished using NaH, MeI in dry THF at low temperature for 1 h to afford the methylated nucleosides **4** and **8** in 65% and 75% yield, respectively. The selective O- *vs.* N-methylation mainly depends upon the dielectric constant of the solvent [23,25] and the stability of sodium-enolate chelation [26]. Hence, low dielectric constant and high chelation stabilizing capacity of THF afforded a higher O-selectivity compared to DMF. The one pot conversion of compound **3** to the triazolide derivative [27] using 1,2,4-triazole, POCl₃, and triethylamine, and subsequent treatment with aqueous ammonia/dioxane (1:1) at ambient temperature for 18 h yielded the cytidine derivative (**5**) with 58% (overall yield for 2 steps). Base protection of **5** and **8** using benzoylchloride in pyridine at rt for 3 h afforded **6** and **9** in 83 and 88% yield, respectively.

Due to the scalability, solubility and reproducibility of the reaction it proved advantageous to use the guanine derivative **10** for the epoxide ring opening reaction. The latter was obtained via Mitsunobu reaction of N^2-acetylguanine [28] and 2-(trimethylsilyl)ethanol in analogy with the previously described protocol for O^6-[2-(p-nitrophenyl)ethyl]guanine [29]. Selective epoxide ring opening was accomplished with the lithium salt of **10** (Scheme 2) in DMF at 130 °C and afforded 41% of **11** along with 26% of recovered **10**. Remarkably, the acetyl protection was lost in **11** upon the prolonged heating in DMF. Further chemoselective methylation using NaH (60%) and MeI in DMF and DCM at low temperature for 4 h gave **12** in 94% yield. Deprotection of **12** was done by 1 M TBAF in THF at rt for 2 h to yield 75% of **13**. The more base labile N^2-dimethylformamidine (dmf) [30–32] group was introduced using N,N-dimethylformamide diethylacetal in methanol under reflux for 12 h to afford **14** with 85% yield.

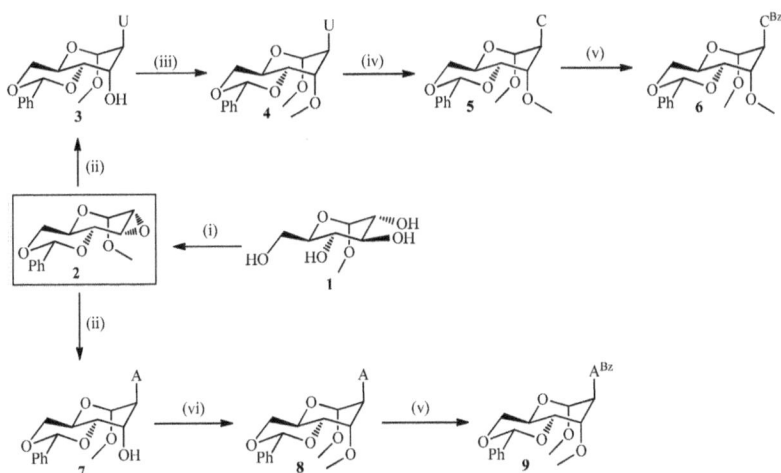

Scheme 1. Synthetic scheme of the protected DMANA congeners for uracil, cytosine and adenine. *Reagents and Conditions*: (i) (a) C_6H_5CHO, $ZnCl_2$, 72 h (66%); (b) 6 eq. $CH_3C_6H_4SO_2Cl$, pyridine, 60 °C, 72 h (78%); (c) 5.3 M in MeOH, CH_2Cl_2, rt, 12 h (99%); (ii) NaH (60%), DMF, 20 °C, 12 h (33.3% of **3** along with 40.1% recovery of **2** and 75% of **7**); (iii) NaH (60%), CH_3I, THF, 0 °C, 1 h (64.7%); (iv) (a) POCl$_3$, 1H-1,2,4-triazole, Et$_3$N, pyridine, rt, 2 h; (b) 1,4 dioxane, aq. NH$_3$, rt, 12 h (58.17% overall 2 steps); (v) Benzoyl chloride, pyridine, rt, 3 h (87.6% of **6** and 83% of **9**); (vi) NaH (60%), CH_3I, THF, −78 °C, 4.5 h to −30 °C, 1 h (75%).

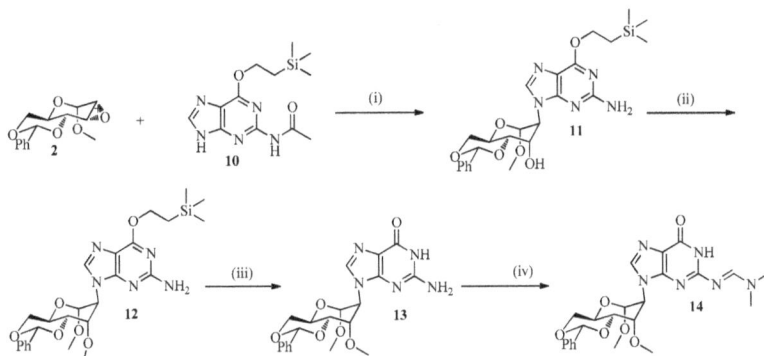

Scheme 2. Synthetic scheme for the protected guanine containing analog. *Reagents and Conditions*: (i) LiH, DMF, 130 °C, 18 h (42% of **11** along with 26% recovery of **10**); (ii) NaH (60%), CH_3I, DMF, DCM, −30 °C to −20 °C, 4 h (94%); (iii) TBAF, THF, rt, 2 h (75%); (iv) Me$_2$NCH(OEt)$_2$, MeOH, reflux for 12 h (85%).

Deprotection of the benzylidene protecting group under mild conditions using AcOH:H$_2$O (3:1) at 45 °C for 12 h afforded **15a–d** in 50%–94% yield (Scheme 3), which was followed by classical dimethoxytritylation and phosphitylation. Hereto, the nucleosides **15a–d** were selectively protected at the 6'-OH by reaction with DMTrCl in pyridine at rt for 3 h to furnish the corresponding protected derivatives **16a–d** in 74%–93% yield. Phosphitylation of **16a–d** at the 4'-OH with 2-cyanoethyl N,N-diisopropylchlorophosphoramidite in anhydrous CH$_2$Cl$_2$ at 0 °C for 1.5 h afforded the corresponding phosphoramidite building blocks **17a–d** in 65%–93% yield, to be used for oligomer assembly. Assembly of all oligonucleotides and purification was carried out as described before [33].

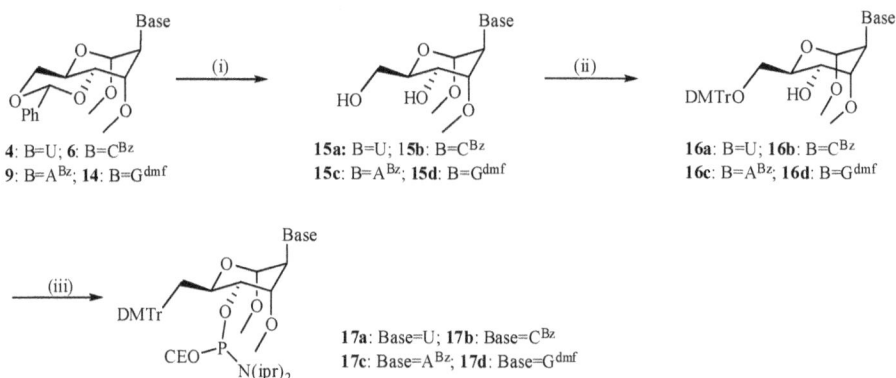

Base   (i)   Base   (ii)   Base

4: B=U; **6**: B=C^{Bz}
9: B=A^{Bz}; **14**: B=G^{dmf}

15a: B=U; **15b**: B=C^{Bz}
15c: B=A^{Bz}; **15d**: B=G^{dmf}

16a: B=U; **16b**: B=C^{Bz}
16c: B=A^{Bz}; **16d**: B=G^{dmf}

(iii)   Base

17a: Base=U; **17b**: Base=C^{Bz}
17c: Base=A^{Bz}; **17d**: Base=G^{dmf}

Scheme 3. Scheme for assembly of the different phosporamidites. *Reagents and Conditions*: (i) AcOH: H$_2$O (3:1), 45 °C, 12 h (50% to 94%); (ii) DMTrCl, pyridine, rt, 3 h (74% to 93%); (iii) (iPr)$_2$N(OCE)PCl, DIPEA, DCM, 0 °C, 1.5 h (65% to 93%).

2.2. Oligonucleotide Affinity Measurements

First, following incorporation of a single DMANA modification into an dsRNA nonamer sequence [5'-GCGU-X*-UGCG/5'-CGCAYACGC], the respective affinities for complementary RNA were studied in a 0.1 M NaCl buffer and were compared with the melting temperatures (T$_m$) of previously studied six-membered ring structures substituting for the ribose ring (Table 1). Within the context of all four natural bases, the HNA (anhydrohexitol) and ANA (altritol) substitutions proved advantageous and considerably stabilized the RNA helix. Where ANA modifications were either slightly less stabilizing (for pyrimidines) or more stabilizing (as with purines) *versus* HNA modifications, methylation of the 3'-hydroxyl moiety of ANA congeners further improved the affinity for RNA systematically with approximately 0.5 °C. However, converting the hexitol into a methyl hexopyranoside via attachment of a second "methoxy substituent" at the 1'-position as in our DMANA constructs, wiped out the advantage which was gained before in using constrained hexitol moieties. The obtained affinities of these DMANA containing constructs more or less matched those of the corresponding fully complementary RNA sequences. Incorporation of a single DMANA building block slightly destabilized the RNA duplex for substitution of a pyrimidine, but on contrast slightly stabilized the duplex in case of guanine and to a larger extent upon substitution of adenine within this sequence context (Table 1 and Figure 2, base matches).

As on average the DMNA pairing affinity to RNA was not advantageous nor really detrimental, we further studied their mismatch behavior within the same RNA nonamer constructs (5'-GCGU-X*-UGCG/5'-CGCAYACGC; Figure 2) in comparison to different constructs. The graphical chart displays the destabilization in relation to the respective matched sequence as obtained for DMANA building blocks in comparison to discrimination properties for HNA and MANA ([23] and within dsRNA duplexes. Analogous discrimination properties were noted, and especially for dmanaC very selective pairing to guanosine was obtained with −22 °C to −24 °C of destabilization for the different mismatches. However, no overall advantage can be seen in terms of either pairing selectivity or universal base pairing capabilities. Hence, normal WC pairing can be assumed.

Table 1. T_m values of complementary RNA duplexes [5'-GCGU-X*-UGCG/5'-CGCAYACGC].

X*	Structure	T_m (°C)	X*	Structure	T_m (°C)
	RNA	50.4 ± 0.0		RNA	60.4 ± 0.0
	HNA	53.4 ± 0.1		HNA	62.4 ± 0.1
U*	ANA	53.0 ± 0.2	G*	ANA	62.9 ± 0.1
	MANA	53.8 ± 0.2		MANA	63.4 ± 0.1
	DMANA	50.6 ± 0.2		DMANA	61.6 ± 0.1
	RNA	60.8 ± 0.1		RNA	52.5 ± 0.1
	HNA	62.0 ± 0.0		HNA	55.0 ± 0.2
C*	ANA	60.9 ± 0.1	A*	ANA	56.5 ± 0.1
	MANA	61.4 ± 0.0		MANA	57.0 ± 0.1
	DMANA	60.1 ± 0.2		DMANA	55.5 ± 0.1

Conditions: as determined in 100 mM NaCl buffer containing 20 mM KH_2PO_4 and 0.1 mM EDTA, pH 7.5, with a duplex concentration of 4 μM. Annotations U*, C*, A* and G* denote either a RNA, HNA, ANA, MANA or DMANA monomer respectively, *versus* the natural complementary base Y.

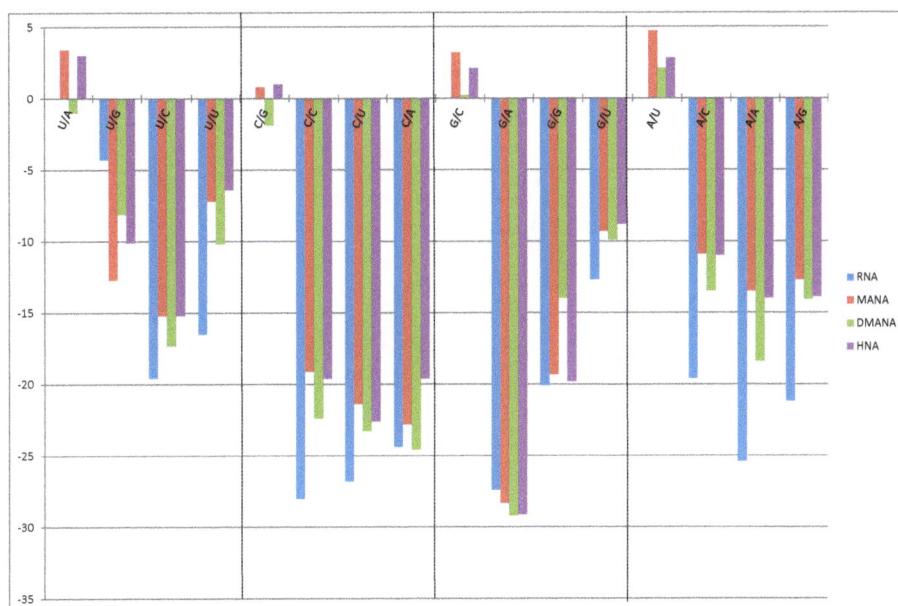

Figure 2. Base pairing selectivity for different 6-membered ring analogues. Conditions: graphical overview of the (de)stabilization of the matched and mismatched nonamer sequences following a single incorporation of a sugar modified nucleoside (MANA in red, DMANA in green and HNA in purple) wherein the selectivity of pairing is shown by destabilization of the respective mismatches. The pairing selectivity for RNA is shown in blue.

This picture was confirmed with a second substitution within different nonamer dsRNA sequences (Table 2), with the largest destabilization noted for dmanaC building blocks (−1.8 °C/modification) while still an increase in stability of 1.1 °C/modification was seen with incorporation of 2 dmanaA blocks. The stabilizing effects of the HNA and 3'-methylated ANA constructs (MANA) are included for comparative reasons. We therefore decided to prepare a fully modified DMANA octamer and hybridized it to the complementary RNA sequence (Table 2, bottom). Where the natural RNA duplex displayed a Tm of 40.6 °C, the DMANA construct still paired albeit with slightly lower affinity (−0.3 °C/modification). The different hexitol constructs on the other hand strongly increased the affinity for the RNA complement as documented before [23,34]. It hence can be concluded that the DMANA analogues are different from the previous hexitol series of compounds with a constrained 6-membered ring conformation fit for pairing to RNA.

Table 2. Thermal stability for RNA duplexes containing a double modification.

Sequences	X*	Tm (°C)	ΔTm/Modification (°C)
	RNA	55.6 ± 0.4	Reference
5'-GCU*GUGU*CG-3'	MANA	62.6 ± 0.2	3.5
	DMANA	54.5 ± 0.1	−0.5
	HNA	60.2 ± 0.1	2.3
	RNA	57.1 ± 0.1	Reference
5'-GCC*AUAC*CG-3'	MANA	59.3 ± 0.1	1.1
	DMANA	53.4 ± 0.1	−1.8
	HNA	58.2 ± 0.1	0.6
	RNA	51.3 ± 0.2	Reference
5'-GCG*UUUG*CG-3'	MANA	54.3 ± 0.1	1.5
	DMANA	51.1 ± 0.1	−0.1
	HNA	53.0 ± 0.2	0.8
	RNA	57.1 ± 0.1	Reference
5'-GCA*CUCA*CG-3'	MANA	63.5 ± 0.1	3.2
	DMANA	59.3 ± 0.1	1.1
	HNA	62.1 ± 0.1	2.5
	RNA	40.6 ± 0.1	Reference
	MANA	61.1 ± 0.3	2.6
5'-G*C*G*U*A*G*C*G*-3'	DMANA	38.1 ± 0.2	−0.3
	ANA	59.6 [34]	2.4
	HNA	52.0 [34]	1.4

Conditions: T_m as determined in 100 mM NaCl buffer containing 20 mM KH_2PO_4 and 0.1 mM EDTA, pH 7.5, with a duplex concentration of 4 μM. U*, C*, A* and G* denote either a RNA, MANA, DMANA, ANA or a HNA monomer, respectively. At the bottom the results are given for the fully modified strands paired to the complementary RNA strand.

Finally, incorporation of a single DMANA building block into dsDNA 13-mer sequences [5'-CACCGX*TGCTACC-3'/3'-GTGGCYACGATGG-5'] was evaluated at 0.1 M salt concentration (Table 3) for both match and mismatch sequences. However, a single DMANA incorporation already afforded respectively 9 °C (for dmanaU and dmanaC), 5 °C (for dmanaA) or 6 °C (for dmanaG) of destabilization for the different matched pairs within this sequence context. A slightly better result could be expected if we could compare the dmanaT construct instead of dmanaU in view of the stabilizing effect of a 5-methyl substituent on the base, but this still would have resulted in a destabilization of 7 to 8 °C. The selectivity of pairing is adequate but gives a mixed picture, with selectivity being dependent probably on the base and the sequence context. In view of the fairly strong destabilization *versus* DNA sequences it is clear however that these DMANA blocks do not have a DNA like conformation.

Table 3. Hybridization studies following incorporation into a DNA strand.

Y	A		T		G		C	
X* (X)	T_m	ΔT_m	T_m	ΔT_m	T_m	ΔT_m	T_m	ΔT_m
U*	47.7	-	40.9	−6.8	41.7	−6.0	38.3	−9.4
T	57.1	-	46.7	−10.4	50.3	−6.8	44.2	−12.9
A*	41.9	−10.8	52.7	-	45.3	−7.4	41.4	−11.3
A	46.6	−10.7	57.3	-	52.9	−4.4	44.4	−12.9
C*	41.8	−10.3	40.8	−11.3	52.1	-	38.5	−13.6
C	44.4	−16.5	45.6	−15.3	60.9	-	40.6	−20.3
G*	42.9	−11.0	48.9	−5.0	46.0	−7.9	53.9	-
G	51.6	−8.4	51.0	−9.0	53.5	−6.5	60.0	-

Conditions: T_m values are provided for a single incorporation of a DMANA (X*) modification into a mixed DNA sequence [5'-CACCGX*TGCTACC-3'/3'-GTGGCYACGATGG-5'] for the match and the different mismatch sequences (Y) at 0.1 M salt and 4 M of duplex concentration.

2.3. Discussion

The MedChem group of the Rega Institute has already been elaborating for many years on nucleoside analogues with a 6-membered ring system substituting for ribose for various applications. Especially the analogues with a 1,5-anhydrohexitol ring having the base at the C2' in syn orientation with the remaining hydroxymethyl substituent (HNA, ANA and MANA) turned out to be well pre-organized for pairing with RNA, and are thus strongly stabilizing for RNA duplexes. Several biological studies have been undertaken with both HNA and ANA [22,35,36] and interesting results were obtained more recently regarding their use as xeno-nucleic acids [37,38]. Highest affinities so far however were obtained with MANA building blocks [23]. We therefore started to study the influence of DMANA analogues carrying an additional methoxy substituent on the 6-membered ring system. However, as shown by the various T_m studies, incorporation of the new modification does not further increase the affinity for RNA, but overall rather tends to slightly destabilize dsRNA complexes, while strongly destabilizing DNA upon incorporation. Therefore the modification still resembles more closely RNA monomers with their 3'-endo conformation as in RNA duplexes.

These findings are further corroborated by inserting *in silico* a modified building block into a dsRNA. As can be seen in Figure 3, no steric hindrance occurs when substituting a DMANA residue having two OMe groups at the sugar 1' and 3' positions for a uridine into an RNA duplex for which the model of Mooers was used [39]. The HNA sugar conformation upon pairing with RNA is $^1C_4$ having an axial base orientation. Likewise, for the DMANA structure incorporated into the dsRNA, both methoxy substituents likewise are oriented axially with no apparent steric clashes. However, it is well known that apart from forces like the anomeric effect, substituents in 6-membered rings prefer the equatorial orientation to avoid 1,3-diaxial interactions. Energy calculations for the monomers using Amber force field [40] indeed show a slight preference for having the base and both methoxy groups in an all-equatorial configuration (with 4'-OH and 6'-CH$_2$OH in axial orientation), opposite to what is found for HNA and ANA building blocks with an axial oriented heterocyclic base. Increasing the number of OMe substituents therefore may destabilize the $^1C_4$ chair conformation giving preference to $^4C_1$, which is less compatible with the RNA duplex. Hence, the energy penalty to preserve the DMANA monomer in a $^1C_4$ conformation to allow for pairing within a dsRNA strand, might upset the entropic gain as expected of a pre-organized monomer.

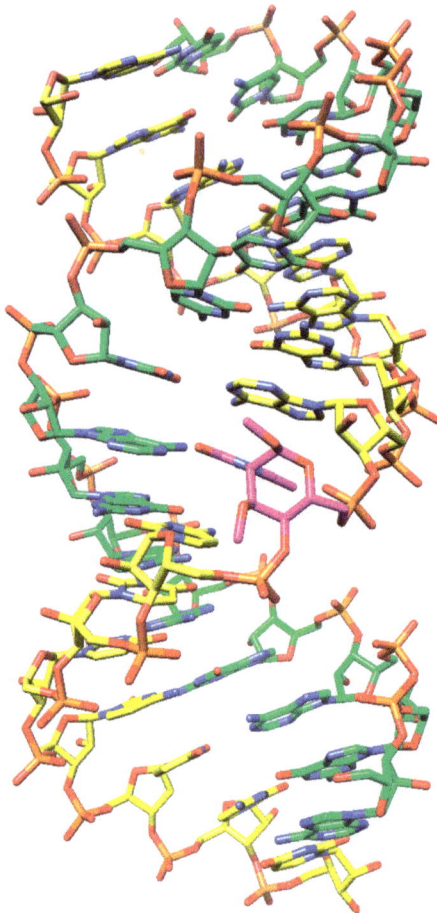

Figure 3. RNA duplex following insertion of a DMANA modification (DMANA modification in purple; picture generated using Chimera (UCSF Chimera—a visualization system for exploratory research and analysis [41]).

Table 4 indeed shows the largest energy difference for both chair conformations for DMANA constructs. This energetic penalty for a forced change in conformation could be the basis of the reduced fitness of DMANA analogues for pairing with RNA.

Table 4. Energy calculations for monomers with different scaffold.

Base and OMe Orientation	Axial	Equatorial	Base and OMe Orientation	Axial	Equatorial
HNA	−205.31	−205.79	MANA	−195.22	−200.35
ANA	−200.58	−198.19	DMANA	−187.62	−194.82
Chair	$^1C_4$	$^4C_1$	Chair	$^1C_4$	$^4C_1$

Conditions: Final energies in kcal/mol are shown following minimization using Amber force field. Nucleotides all have a thymine base, uncharged residues (phosphate groups protonated), with parametrization via antechamber, Gaff force field, 5000 steps of energy minimization, born solvation energy model.

3. Experimental Section

3.1. General

All chemicals including methylglucopyranoside were provided by Sigma-Aldrich (Diegem, Belgium) or Acros Organics (Geel, Belgium) and were of the highest quality. ^{1}H and ^{13}C-NMR spectra were determined with a 300, 500 and 600 MHz Varian Gemini apparatus (currently Agilent Technologies, Santa Clara, CA, USA) with tetramethylsilane as internal standard for the ^{1}H NMR spectra (s = singlet, d = doublet, dd = double doublet, t = triplet, br. s = broad signal, m = multiplet) and the solvent signal; CD$_3$OD-d_4 (δ = 48.9 ppm), DMSO-d_6 (δ = 39.6 ppm) or CDCl$_3$ (δ = 76.9 ppm) for the ^{13}C-NMR spectra. Exact mass measurements were performed with a quadrupole/orthogonal acceleration time-of-flight tandem mass spectrometer (qTOF2, Micromass, Manchester, UK) fitted with a standard electrospray ionization (ESI) interface. All solvents were carefully dried or bought as such.

3.2. 1',5';2',3'-Dianhydro-4',6'-O-benzylidene-1'-O-methyl-D-allopyranoside (2)

To a solution of the 4',6'-O-benzylidene-2',3'-ditosyl intermediate (10.0 g, 17 mmol) in CH$_2$Cl$_2$ (80 mL) was added NaOMe (5.3 M in MeOH, 12.8 mL, 67.8 mmol). After stirring for overnight at room temperature, the reaction mixture was concentrated and the residue was dissolved in CH$_2$Cl$_2$ (150 mL), and washed twice with brine (50 mL). The aqueous layer was again extracted with CH$_2$Cl$_2$ (2 × 80 mL). Combined organic layers were dried over anhydrous Na$_2$SO$_4$, filtered and evaporated to get pure epoxide **2** (4.45 g, 99%). ^{1}H-NMR (300 MHz, CDCl$_3$): δ 7.60–7.33 (m, 5H, Ar-H), 5.60 (d, J = 6.1 Hz, 1H, Ph-C_H_), 4.92 (dd, J = 6.1, 2.7 Hz, 1H, 1'-H), 4.32–4.22 (m, 1H, 6'-He), 4.18–4.05 (m, 1H, 5-H), 3.98 (dd, J = 9.0, 6.2 Hz, 1H, 4-H), 3.71 (m, 1H, 6-Ha), 3.58–3.45 (m, 5H, 2'-H, O-Me). ^{13}C-NMR (75 MHz, CDCl$_3$): δ 137.11 (Ar-C$_i$); 129.19; 128.28 (Ar-C$_{p+o}$); 126.27 (Ar-C$_m$); 102.71 (Ph-C); 95.27 (C-1'); 77.83 (C-4'); 68.86 (C-6'); 59.99 (C-5'); 55.82 (OMe); 53.07 (C-3'); 50.66 (C-2'); HRMS calcd. for C$_{14}$H$_{16}$O$_5$Na$^+$ [M+Na]$^+$ 287.0890, found 287.0891.

3.3. 4',6'-O-Benzylidene-1'-O-methyl-2'-deoxy-2'-(uracil-l-yl)-D-altropyranoside (3)

To a solution of uracil (4.62 g, 38.6 mmol) in dry DMF (30 mL) was added NaH (60% dispersion in oil, 1.29 g, 41.2 mmol). The reaction mixture was heated at 120 °C under an argon atmosphere for 1 h and to this reaction mixture epoxide **2** (3.4 g, 12.9 mmol) in dry DMF (20 mL) was added and stirring was continued for overnight at same temperature. The reaction mixture was then cooled and evaporated to dryness. The residue was dissolved in ethyl acetate (150 mL) and the organic layer was washed with a saturated aqueous NaHCO$_3$ solution (2 × 50 mL). The aqueous layer was again extracted with EtOAc (3 × 50 mL). The combined organic layers were washed with brine (2 × 50 mL), dried over Na$_2$SO$_4$, and concentrated under *vacuo* and purification by silica gel column chromatography (elution with 2% MeOH in DCM) afforded **3** (1.61 g, 33%) as a white foam while recovering a large part of the starting epoxide **2** (1.94 g, 40% yield). ^{1}H-NMR (500 MHz, CDCl$_3$): δ 7.75 (d, J = 8.1 Hz, 1H, 6-H), 7.49–7.41 (m, 2H, Ar-H), 7.39–7.30 (m, 3H, Ar-H), 5.77 (d, J = 8.1 Hz, 1H, 5-H), 5.62 (s, 1H, Ph-C_H_), 4.83 (s, 1H, 1'-H), 4.81 (s, 1H, 2'-H), 4.48 – 4.38 (m, 2H, 5'-H, 6'-He), 4.16 (brs, 1H, 3'-H), 3.81 (t, J = 9.9 Hz, 1H, 6'-Ha), 3.69 (dd, J = 9.9, 2.4 Hz, 1H, 4'-H), 3.41–3.51 (m, 4H, OMe, -OH). ^{13}C-NMR (125 MHz, CDCl$_3$): δ 163.08 (C-4); 150.60 (C-2); 141.14 (C-6); 136.81 (Ar-C$_i$); 129.26 (Ar-C$_p$); 128.28 (Ar-C$_m$); 126.18 (Ar-C$_o$); 102.91 (C-5); 102.27 (Ph-C); 98.99 (C-1'); 75.61 (C-3'); 68.94 (C-4'); 67.02 (C-6'); 58.19 (C-5'); 57.91 (OMe); 55.97(C-2'). HRMS calcd. for C$_{18}$H$_{20}$N$_2$O$_7$Na$^+$ [M+Na]$^+$ 399.1163, found 399.1156.

3.4. 4',6'-O-Benzylidene-1',3'-di-O-methyl-2'-deoxy-2'-(uracil-l-yl)-D-altropyranoside (4)

To a solution of **3** (1.51 g, 4.0 mmol) in dry THF (15 mL) was added NaH (60% dispersion in oil, 404 mg, 12.04 mmol) at 0 °C and the reaction mixture was stirred at 0 °C for 1h under argon atmosphere. Methyl iodide (0.37 mL, 6.02 mmol) in dry THF (1 mL) was added and stirring was continued for another 1 h at the same temperature. The reaction mixture was quenched with 5 mL

Molecules **2015**, *20*, 4020–4041

MeOH. The solution was concentrated and dissolved in ethyl acetate (150 mL) and washed with saturated aqueous NaHCO₃ (150 mL). The aqueous layer was again extracted with ethyl acetate (3 × 50 mL). The combined the organic layers were dried over Na_2SO_4, filtered, concentrated under *vacuo*, and purification by silica gel column chromatography (elution with 2% MeOH in DCM) afforded the dimethylated nucleoside **4** (1.01 g, 65%). ^{1}H-NMR (500 MHz, CDCl₃): δ 9.41 (s, 1H, N³-H), 7.80 (d, *J* = 8.2 Hz, 1H, 6-H), 7.47–7.35 (m, 5H, Ar-H), 5.80 (dd, *J* = 8.1, 1.7 Hz, 1H, 5-H), 5.55 (s, 1H, Ph-C̲H̲), 4.90 (d, *J* = 1.7 Hz, 1H, 1'-H), 4.80 (s, 1H, 2'-H), 4.48–4.36 (m, 2H, 6'-He, 3'-H), 3.77 (t, *J* = 10.3 Hz, 1H, 6'-Ha),3.72–3.67 (m, 2H, 4'-H, 5'-H), 3.62 (s, 3H, OMe), 3.46 (s, 3H, OMe). ^{13}C-NMR (125 MHz, CDCl₃): δ 162.98 (C-4); 150.39 (C-2); 141.18 (C-6); 137.06 (Ar-C$_i$); 129.20 (Ar-C$_p$); 128.28 (Ar-C$_m$); 126.21 (Ar-C$_o$); 102.86 (C-5); 102.48 (Ph-C̲); 98.84 (C-1'); 76.06 (C-3'); 75.95 (C-4'); 69.11 (C-5'); 59.59 (C-6'); 58.56 (OMe); 55.94 (OMe); 55.81(C-2'). HRMS calcd. for $C_{19}H_{23}N_2O_7$ [M+H]⁺ 391.1500, found 391.1494.

3.5. 4',6'-O-Benzylidene-1',3'-di-O-methyl-2-deoxy-2'-(cytosin-1-yl)-D-altropyranoside (5)

A solution of triazole (1.05 g, 15.2 mmol) and phosphorus oxychloride (0.3 mL, 3.18 mmol) was prepared in pyridine (8 mL) at 0 °C. Triethylamine (2.03 mL, 14.5 mmol) was added dropwise at 0 °C and the solution was stirred 30 min. The uracil derivative **4** (0.650 g, 1.66 mmol) dissolved in dry pyridine (8 mL) was added at 0 °C and the solution was stirred for 2 h at room temperature and concentrated and co-evaporated with toluene (2 × 20 mL). The crude product was dissolved with DCM (150 mL) and washed twice with brine (60 mL). The aqueous layer was extracted with DCM (30 mL). Combined organic layers were dried over anhydrous Na_2SO_4, filtered and evaporated. The residue was dissolved in 1,4-dioxane (30 mL), cooled to 0 °C and aqueous ammonia 25% (13 mL) were added. The solution was left overnight at RT. The solution was evaporated and co-evaporated with toluene (3 × 20 mL). The residue was purified by silica gel column chromatography (elution with 10% MeOH/dichloromethane) and afforded the dimethylated cytidine analog **5** (377 mg, 58%) as a white foam. ^{1}H-NMR (500 MHz, CD₃OD): δ 7.91 (d, *J* = 7.5 Hz, 1H, 6-H), 7.50–7.42 (m, 2H, Ar-H), 7.38–7.26 (m, 3H, Ar-H), 5.95 (d, *J* = 7.5 Hz, 1H, 5-H), 5.63 (s, 1H, Ph-C̲H̲), 4.93 (s, 1H, 2'-H), 4.91 (d, *J* = 2.1 Hz, 1H, 1'-H), 4.34–4.30 (m, 1H, 6'-He), 4.28 (dd, *J* = 9.9, 5.3 Hz, 1H, 5'-H), 3.84 (dd, *J* = 11.0, 2.3 Hz, 1H, 6'-Ha), 3.81 (dd, *J* = 6.6, 2.3 Hz, 1H, 4'-H), 3.67 (t, *J* = 2.3 Hz, 1H, 3'-H), 3.57 (s, 3H, OMe), 3.42 (s, 3H, OMe). ^{13}C-NMR (125 MHz, CD₃OD): δ 167.49 (C-4); 158.15 (C-2); 144.10 (C-6), 139.09 (Ar-C$_i$), 129.97 (Ar-C$_p$), 129.07 (Ar-C$_m$), 127.44 (Ar-C$_o$), 103.41 (Ph-C̲), 100.28 (C-5), 96.46 (C-1'), 77.40 (C-3'), 76.86 (C-4'), 70.05 (C-6'), 59.93 (C-5'), 59.13 (OMe), 57.45 (OMe), 55.84 (C-2'). HRMS calcd. for $C_{19}H_{24}N_3O_6$⁺ [M+H]⁺ 390.1654, found 390.1653.

3.6. 4',6'-O-Benzylidene-1',3'-O-methyl-2'-deoxy-2'-(N⁶-benzoylcytosin-1-yl)-D-altropyranoside (6)

The analog **5** (0.35 g, 0.9 mmol) was co-evaporated with dry pyridine (6 mL), dissolved in dry pyridine (4 mL), and cooled at 0 °C. Benzoyl chloride (0.314 mL, 2.7 mmol) was added and the solution was allowed to come to RT. The solution was stirred for 3 h at RT. The mixture was cooled to 0 °C and water (0.25 mL) was added. Then, aqueous ammonia 25% (2 mL) was added and the solution was stirred for 30 min at RT. The volatiles were removed under reduced pressure and co-evaporated, each time with toluene (3 × 5 mL). The residue was adsorbed on silica by co-evaporation from DCM and purified by silica column chromatography (elution with 2% MeOH/DCM) affording the protected nucleoside **6** (370 mg, 83% yield) as a white foam. ^{1}H-NMR (300 MHz, CDCl₃): δ 8.24 (d, *J* = 7.5 Hz, 1H, 6-H), 7.94 (d, *J* = 7.3 Hz, 2H, Bz-H), 7.70–7.33 (m, 9H, 5-H, Ar-H, Bz-H), 5.56 (s, 1H, Ph-C̲H̲), 5.12 (d, *J* = 1.6 Hz, 1H, 1'-H), 4.90 (s, 1H, 2'-H), 4.55–4.37 (m, 2H, 5'-H, 6'-He), 3.86–3.71 (m, 2H, 6'-Ha, 3'-H), 3.74 (dd, *J* = 9.5, 3.1 Hz, 1H, 4'-H), 3.68 (s, 3H, OMe), 3.49 (s, 3H, OMe). ^{13}C-NMR (75 MHz, CDCl₃): δ 167.76 (PhCONH); 162.14 (C-4); 137.11 (C-2); 133.34 (C-6); 129.18, 129.08, 128.28, 127.60, 126.23 (Ar-C, Bz-C); 102.48 (Ph-C̲); 99.00 (C-5); 97.14 (C-1'); 78.35 (C-3'), 75.77 (C-4'); 75.33 (C-5'); 69.20 (C-6'); 59.51 (OMe); 58.72 (OMe); 56.03 (C2'). HRMS calcd. for $C_{26}H_{28}N_3O_7$⁺ [M+H]⁺ 494.1922, found 494.1919.

3.7. 4',6'-O-Benzylidene-1'-O-methyl-2'-deoxy-2'-(adenin-9-yl)-D-altropyranoside (7)

To a solution of adenine (3.22 g, 23.85 mmol) in dry DMF (40 mL) was added NaH (60% dispersion in oil, 890 mg, 22.26 mmol). The reaction mixture was heated at 90 °C under argon atmosphere for 1 h, after which the epoxide **2** (2.1 g, 7.95 mmol) in dry DMF (15 mL) was added and stirring was continued overnight at 120 °C. The reaction mixture was quenched with MeOH (15mL) and following evaporation the residue was partitioned between EtOAc and saturated aqueous NaHCO₃ solution. The organic layer was separated and the aqueous layer was extracted with EtOAc (3 × 50 mL). The combined the organic layers were dried over Na₂SO₄, filtered, concentrated *in vacuo*, and purification by normal silica gel column chromatography (elution with 3% MeOH in DCM) afforded compound **7** (2.37 g, 75%). ¹H-NMR (300 MHz, CDCl₃): δ 8.30 (s, 1H, 8-H), 8.18 (s, 1H, 2-H), 7.50–7.29 (m, 5H, Ar-H), 6.00 (s, 2H, -NH₂), 5.55 (s, 1H, Ph-C$\underline{H}$), 5.15 (s, 1H, 1'-H), 5.07 (s, 1H, 2'-H), 4.64– 4.52 (m, 1H, 5'-H), 4.46 (dd, *J* = 9.9, 4.7 Hz, 1H, 6'-He), 4.33 (brs, 1H, 3'-H), 3.86 (t, *J* = 9.9 Hz, 1H, 6'-Ha), 3.73 (d, *J* = 8.3 Hz, 1H, 4'-H), 3.54 (s, 3H, OMe). ¹³C-NMR (75 MHz, CDCl₃): δ 155.60 (C-6); 153.44 (C-2); 149.77 (C-4); 138.45 (C-8); 136.86 (C-5); 129.28, 128.26, 126.15, 118.89 (Ar-C); 102.35 (Ph-$\underline{C}$); 99.66 (C-1'); 75.87 (C-5'); 69.14 (C-3'); 67.03 (C-4'); 58.52 (C-6'); 57.18 (OMe); 56.13 (C-2'). HRMS calcd. for C₁₉H₂₂N₅O₅⁺ [M+H]⁺ 400.1615, found 400.1613.

3.8. 4',6'-O-Benzylidene-1',3'-di-O-methyl-2'-deoxy-2'-(adenin-9-yl)-D-altropyranoside (8)

The adenine analog **7** (2.36 g, 5.9 mmol) was co-evaporated with dry DMF (40 mL).The foam was dissolved in dry DMF (70 mL), cooled to −78 °C and NaH (330 mg, 8.28 mmol) was added and the mixture was stirred at −78 °C for 30 min under argon. Subsequently, methyl iodide (0.590 mL, 9.46 mmol) was dissolved in dry DCM (20 mL) and the solution was added drop wise over 30 min to the reaction mixture at the same temperature. Stirring was continued for another 4.5 h at −78 °C and at −30 °C for an additional 1 h before quenching of the reaction with MeOH (10 mL). The solution was warmed to RT and concentrated to dryness under vacuum. The residue was dissolved in ethyl acetate (150 mL) and washed with aqueous saturated NaHCO₃ (130 mL). The organic layer was then washed with brine (130 mL), and the combined aqueous layers were again extracted with ethyl acetate (2 × 150 mL). The combined organic layers were dried over Na₂SO₄, filtered, and concentrated. Purification by silica column chromatography (0%–3% MeOH/DCM) afforded the methylated nucleoside **8** (1.84 g, 75.4%). ¹H-NMR (500 MHz, CDCl₃): δ 8.39 (s, 1H, 8-H), 8.20 (s, 1H, 2-H), 7.47–7.40 (m, 2H, Ar-H), 7.37–7.31 (m, 3H, Ar-H), 5.92 (s, 2H, -NH₂), 5.49 (s, 1H, Ph-C$\underline{H}$), 5.15 (d, *J* = 2.5 Hz, 1H, 1'-H), 5.07 (s, 1H, 2'-H), 4.54 (td, *J* = 10.1, 5.3 Hz, 1H, 5'-H), 4.44 (dd, *J* = 10.5, 5.3 Hz, 1H, 6-He), 3.88 (t, *J* = 2.6 Hz, 1H, 3'-H), 3.83 (t, *J* = 10.5 Hz, 1H, 6'-Ha), 3.79 (dd, *J* = 9.8, 2.5 Hz, 1H, 4'-H), 3.71 (s, 3H, OMe), 3.52 (s, 3H, OMe). ¹³C-NMR (125 MHz, CDCl₃): δ 155.57 (C-6); 153.54 (C-2); 149.99 (C-4); 138.56 (C-8); 137.11, 129.15, 128.26, 126.16 (Ar-C); 119.10 (C-5); 102.48 (Ph-$\underline{C}$); 99.47 (C-1'); 76.25 (C-5'); 76.19 (C-3'); 69.26 (C-4'); 60.06 (C-6'); 58.81 (OMe); 56.06 (OMe); 55.05 (C-2'). HRMS calcd. for C₂₀H₂₄N₅O₅⁺ [M+H]⁺ 414.1772, found 414.1767.

3.9. 4',6'-O-Benzylidene-1',3'-di-O-methyl-2'-deoxy-2'-(N⁶-benzoyladenin-9-yl)-D-altropyranoside (9)

The obtained analog **8** (1.59 g, 3.85 mmol) was co-evaporated with dry pyridine (2 mL), dissolved in dry pyridine (16 mL), and cooled at 0 °C. Benzoyl chloride (1.34 mL, 11.54 mmol) was added and the solution was allowed to come to RT. The solution was stirred 3 h at RT. The mixture was cooled to 0 °C and water (4 mL) was added. Then, aqueous ammonia 25% (8 mL) was added and the solution was stirred 30 min at RT. The volatiles were removed under reduced pressure and co-evaporated, each time with toluene (3 × 25 mL). The residue was adsorbed on silica by co-evaporation and purified by silica column chromatography (elution with 2% MeOH/DCM) affording the protected nucleoside **9** (1.70 g, 83% yield) as a white foam. ¹H-NMR (500 MHz, CDCl₃): δ 9.43 (s, 1H, NH), 8.78 (s, 1H, 8-H), 8.43 (s, 1H, 2-H), 8.05 (d, *J* = 7.5 Hz, 2H, H-Bz), 7.66–7.29 (m, 8H, Bz-H, Ar-H), 5.49 (s, 1H, Ph-C$\underline{H}$), 5.22 (d, *J* = 2.1 Hz, 1H, 1'-H), 5.09 (s, 1H, 2'-H), 4.55 (td, *J* = 10.0, 5.3 Hz, 1H, 5'-H), 4.44 (dd, *J* = 10.4,

5.2 Hz, 1H, 6'-He), 3.88 (brs, 1H, 3'-H), 3.85–3.76 (m, 2H, ,6'-Ha, 4'-H), 3.71 (s, 3H, OMe), 3.52 (s, 3H, OMe). ^{13}C-NMR (125 MHz, CDCl$_3$): δ 164.86 (C=O); 152.77 (C-6); 152.06 (C-2); 149.78 (C-4); 141.06 (C-8); 137.01, 133.35, 132.75, 129.09, 128.72, 128.19, 128.00, 126.11 (Ar-C, Bz-C); 122.75 (C-5); 102.43 (Ph-C); 99.19 (C-1'); 76.11 (C-5'); 76.08 (C-3'); 69.16 (C-4'); 59.98 (C-6'); 58.83 (OMe); 56.01 (OMe); 55.13 (C-2'). HRMS calcd. for C$_{27}$H$_{28}$N$_5$O$_6$$^+$ [M+H]$^+$ 518.2034, found 518.2031.

3.10. 4',6'-O-Benzylidene-1'-O-methyl-2'-deoxy-2'-(2-amino-6-(2-trimethylsilyl-ethoxy)purin-9-yl)-D-altropyranoside (11)

A mixture of 4.5 g (15.4 mmol) of **10** and 0.266 g (33.5 mmol) of lithium hydride in 60 mL of dry DMF was stirred under nitrogen at 90 °C for 2 h. After addition of 2.6 g (9.8 mmol) of the epoxide **2** dissolved in 20 mL of dry DMF, stirring was continued for 18 h at 130 °C. The reaction mixture was cooled and evaporated to dryness. The residue was dissolved in ethyl acetate (100 mL) and washed with brine (100 mL). A small part of the base precipitated during extraction and was filtered off and kept for recycling. The aqueous layer was again extracted with EtOAc (3 × 50 mL). The combined organic layers were dried over Na$_2$SO$_4$, filtered, concentrated under *vacuo*, and purification by silica gel column chromatography (elution with 3% MeOH in DCM) afforded **11** (2.11 g, 42%) while recovering a large point of the starting **10** (1.5 g, 26%). ^{1}H-NMR (300 MHz, CDCl$_3$): δ 8.03 (s, 1H, 8-H), 7.39 (m, 5H, Ar-H), 5.55 (s, 1H, Ph-CH), 5.27 (d, *J* = 1.8 Hz, 1H, 1'-H), 5.04 (s, 1H, 2'-H), 4.83–4.60 (m, 3H, -NH$_2$, 5'-H), 4.59–4.38 (m, 3H, 6'-He, 6'-Ha, 3'-H), 4.36–4.29 (m, 1H, 4'-H), 3.89–3.73 (m, 2H, OCH$_2$-), 3.55 (s, 3H, OMe), 1.21–1.10 (m, 9H, -SiC$_3$H$_9$). ^{13}C-NMR (75 MHz, CDCl$_3$): δ 161.74 (C-6); 159.43 (C-2); 152.79 (C-4); 137.10 (C-8); 136.53 (Ar-C$_i$); 129.53 (Ar-C$_p$); 128.34 (Ar-C$_m$); 126.46 (Ar-C$_o$); 115.06 (C-5); 102.58 (Ph-C); 100.05 (C-1'); 76.14 (C-4'); 69.25 (C-6'); 66.10 (C-5'); 65.09 (C-3'); 58.63 (OCH$_2$-); 57.19 (OMe); 56.40 (C-2'); 17.49 (CH$_2$Si); −1.46 (-SiC$_3$H$_9$).

3.11. 4',6'-O-Benzylidene-1',3'-di-O-methyl-2'-deoxy-2'-(2-amino-6-(2-trimethylsilylethoxy)-purin-9-yl)-D-altropyranoside (12)

The obtained foam **11** (2.11 g, 4.1 mmol) was dried overnight under vacuum. The foam was dissolved in dry DMF (42 mL), cooled to −30 °C and NaH (1.3 g, 32 mmol) was added and the mixture was stirred at −30 °C for 1 h under argon. Subsequently, methyl iodide (1 mL, 16 mmol) was dissolved in dry DCM (10.5 mL) and the solution was added drop wise over 30 min to the reaction mixture at −30 °C which was stirred further at −30 °C for 1 h. The reaction was finally running 3 h more at −30 °C to −20 °C before quenching with 10 mL of MeOH for 20 min. The solution was concentrated and dissolved in DCM (50 mL) and washed with brine (50 mL). The aqueous layer was again extracted with DCM (3 × 50 mL). The combined organic layers were dried over Na$_2$SO$_4$, filtered and concentrated under *vacuo*, and purification by silica gel column chromatography (elution with 2% MeOH in DCM) afforded methylated nucleoside **12** (2.03 g, 94%) as a white foam. ^{1}H-NMR (500 MHz, CDCl$_3$): δ 7.98 (s, 1H, 8-H), 7.49–7.40 (m, 1H, 2H, Ar-H), 7.38–7.30 (m, 3H, Ar-H), 5.47 (s, 1H, Ph-CH), 5.01 (s, 1H, 1'-H), 4.98–4.92 (m, 1H, 5'-H), 4.66–4.55 (m, 1H, 3'-H), 4.53–4.46 (m, 1H, 4'-H), 4.41 (dd, *J* = 10.4, 5.3 Hz, 1H, 6'-Ha), 1.26 (t, *J* = 12.1 Hz, 1H). ^{13}C-NMR (126 MHz, CDCl$_3$): δ 161.59 (C-6); 159.61 (C-2); 153.62 (C-4); 137.21 (C-8); 136.93 (Ar-C$_i$); 129.07 (Ar-C$_p$); 128.21 (Ar-C$_m$); 126.16 (Ar-C$_o$); 115.15 (C-5); 102.42 (Ph-C); 99.45 (C-1'); 76.26 (C-4'); 76.06 (C-6'); 69.26 (C-5'); 65.04 (C-3'); 59.98 (OCH$_2$-); 58.73 (OMe); 55.97 (OMe); 54.82 (C-2'); 17.54 (CH$_2$Si); −1.41 (3C, Si(CH$_3$)$_3$). HRMS calcd. For C$_{25}$H$_{36}$N$_5$O$_6$Si$^+$ [M+H]$^+$ 530.2429, found 530.2430.

3.12. 4',6'-O-Benzylidene-1',3'-di-O-methyl-2'-deoxy-2'-(guanin-9-yl)-D-altropyranoside (13)

A 1 M solution of tetrabutylammonium fluoride in dry THF (15.10 mL) was added to the guanine derivative **12** (2.0 g , 3.77 mmol) and the mixture was stirred at room temperature under N$_2$ for 2 h after which water (15 mL) was added. The pH was adjusted to 5 with acetic acid and the mixture was evaporated. The residue was purified by silica gel column chromatography (elution with 10% MeOH in DCM) affording **13** (1.25 g, 75%). ^{1}H-NMR (500 MHz, CDCl$_3$): δ 11.99 (s, 1H, NH), 7.86 (s, 1H, 8-H),

7.42 (m, 2H, Ar-H), 7.31 (m, 3H, Ar-H), 6.73 (s, H, 2H, 2-NH_2), 5.49 (s, 1H, Ph-C$\underline{H}$), 5.07 (s, 1H, 1'-H), 4.81 (s, 1H, 2'-H), 4.50–4.42 (m, 1H, 5'-H), 4.37 (dd, J = 10.1, 5.0 Hz, 1H, 6'-He), 3.78 (s, 3H, 3'-H, 6-Ha, 4'-H), 3.58 (s, 3H, OMe), 3.53 (s, 3H, OMe). ^{13}C-NMR (125 MHz, CDCl$_3$): δ 159.03 (C-6); 154.05 (C-2); 151.58 (C-4); 137.20 (Ar-C$_i$); 135.42 (C-8); 129.08, 128.22, 126.19 (Ar-C$_{m+o+p}$); 116.24 (C-5); 102.40 (Ph-C$\underline{\ }$); 99.23 (C-1'); 76.11 (C-4'); 76.04 (C-6'); 69.19 (C-5'); 59.76 (C-4'); 58.68 (OMe); 55.98 (OMe); 54.79 (C-2'). HRMS calcd. For C$_{20}$H$_{24}$N$_5$O$_6$$^+$ [M+H]$^+$ 430.1721, found 430.1711.

3.13. 4',6'-O-Benzylidene-1',3'-di-O-methyl-2'-deoxy-2'-(N^2-(dimethylamino)methylene-guanin-9-yl))-D-altropyranoside (14)

An amount of **13** (0.66 g, 1.54 mmol) was co evaporated three times with pyridine, dissolved in 30 mL of dry MeOH (20 mL) and *N,N*-dimethylformamide diethyl acetal (0.824 mL, 6.16 mmol) was added. The mixture was stirred at reflux for 2 h under argon, after which it was evaporated and co-evaporated with toluene (3 × 30 mL). The residue was purified by silica gel column chromatography (2%–5% MeOH/dichloromethane) affording the analog **14** (0.634 g, 85%). ^{1}H-NMR (500 MHz, CDCl$_3$): δ 9.78 (s, 1H, N^1-H), 8.65 (s, 1H, N=C$\underline{H}$-N), 7.97 (s, 1H, 8-H), 7.45 (m, 2H, Ar-H), 7.40–7.30 (m, 3H, Ar-H), 5.51 (s, 1H, Ph-C$\underline{H}$), 5.02 (s, 1H, 2'-H), 4.99 (d, J = 2.2 Hz, 1H, 1'-H), 4.50 (td, J = 10.4, 5.3 Hz, 1H, 5'-H), 4.41 (dd, J = 10.4, 5.3 Hz, 1H, 6'-He), 3.90–3.80 (m, 3H, 3'-H, 6-Ha, 4'-H), 3.68 (s, 3H, OMe), 3.50 (s, 3H, OMe), 3.18 (s, 3H, NMe), 3.15 (s, 3H, NMe). ^{13}C-NMR (125 MHz, CDCl$_3$): δ 158.12 (N=C$\underline{H}$-N); 157.97 (C-6); 157.04 (C-2); 150.18 (C-4); 137.20 (C-8); 136.08, 129.05, 128.19, 126.14 (Ar-C); 119.85 (C-4); 102.39 (Ph-C$\underline{H}$); 99.31 (C-1'); 76.47 (C-4'); 76.12 (C-6'); 69.17 (C-3'); 60.14 (C-5'); 58.67 (OMe); 55.89 (OMe); 55.13 (C-2'); 41.44 (NMe); 35.23 (NMe). HRMS calcd. for C$_{23}$H$_{29}$N$_6$O$_6$$^+$ [M+H]$^+$ 485.2143, found 485.2145.

3.14. 1',3'-Di-O-methyl-2'-deoxy-2'-(uracil-1-yl)-D-altropyranoside (15a)

Compound **4** (0.9 g, 2.31 mmol) was dissolved in 60 mL of AcOH-H$_2$O (3:1) at rt. The reaction mixture was slowly heated at 45 °C and reaction progress was monitored using TLC. After 12 h, the mixture was concentrated and co evaporated with toluene (30 mL). The crude residue was purification by flash silica gel column chromatography (elution with 5% MeOH in DCM) afforded compound **15a** (0.66 g, 94%). ^{1}H-NMR (600 MHz, DMSO-d_6): δ 11.15 (s, 1H, NH), 7.71 (s, 1H, 6-H), 5.56 (d, J = 5.5 Hz, 1H, 5-H), 4.93 (d, J = 3.9 Hz, 1H, 1'-H), 4.88 (t, J = 5.0 Hz, 1H, 2'-H), 4.84–4.60 (m, 1H, 3'-H), 4.03 (dd, J = 9.8, 5.5 Hz, 1H, 5'-H), 3.77 (q, J = 5.5 Hz, 1H, 4'-H), 3.65–3.53 (m, 2H, 6'-H), 3.26, (s, 3H, OMe), 3.22 (s, 3H, OMe). ^{13}C-NMR (150 MHz, DMSO-d_6): δ 170.83 (C-4); 163.61 (C-2); 151.66 (C-6); 101.62 (C-5); 99.07 (C-1'); 77.52 (C-5'); 76.08 (C-3); 63.67 (C-4'); 61.12 (C-6'); 60.23 (OMe); 56.65 (OMe); 55.39 (C-2'). HRMS calcd. For C$_{12}$H$_{19}$N$_2$O$_7$$^+$ [M+H]$^+$ 325.1006, found 325.1005.

3.15. 1',3'-Di-O-methyl-2'-deoxy-2'-(N^6-benzoylcytosin-1-yl)-D-altropyranoside (15b)

Compound **6** (0.36 g, 0.73 mmol) was dissolved in 24.1 mL of AcOH:H$_2$O (3:1) at rt. The reaction mixture was slowly heated at 45 °C and further treated as per the synthesis of **15a** affording the nucleoside analog **15b** (0.28 g, 50%). ^{1}H-NMR (300 MHz, CD$_3$OD): 8.23–7.94 (m, 3H, 6-H, Bz-H), 7.68–7.51 (m, 4H, 5-H, Bz-H), 5.26 (brs, 1H, 1'-H), 4.87–4.81 (m, 1H, 2'-H), 4.23–4.09 (m, 1H, 5'-H), 4.03 (q, 1H, J = 5.5 Hz, 4'-H), 3.80–3.66 (m, 2H, 3'-H, 6'-He), 3.30 (s, 6H, 2OMe), 3.26–3.20 (m, 1H, 6'-Ha). ^{13}C-NMR (150 MHz, CD$_3$OD): 169.07 (Ph$\underline{C}$ONH), 164.69 (C-4), 158.61 (C-2), 134.65 (C-6), 134.13, 130.47, 129.84, 129.59, 129.19 (Bz-C), 100.35 (C-5), 98.59 (C-1'), 64.94 (C-3'), 64.74 (C-5'), 63.46 (C-4'), 62.47 (C-6'), 57.75 (OMe), 56.05 (OMe), 49.85 (C-2'). HRMS calcd. for C$_{19}$H$_{24}$N$_3$O$_7$$^+$ [M+H]$^+$ 406.1609, found 406.1608.

3.16. 1',3'-Di-O-methyl-2'-deoxy-2'-(N^6-benzoyladenin-9-yl)-D-altropyranoside (15c)

Following the procedure for **15a**, compound **9** (1.65 g, 3.19 mmol) was treated 82.85 mL of AcOH:H$_2$O (3:1) at rt. The reaction mixture was slowly heated at 45 °C and reaction progress was monitored using TLC. After 12 h, the mixture was concentrated and co evaporated with toluene, and after purification afforded (0.78 g, 57%) of **15c**.

^{1}H-NMR (300 MHz, CD$_3$OD): δ 8.76 (s, 1H, 8-H), 8.52 (s, 1H, 2-H), 8.13 (d, *J* = 7.4 Hz, 2H, Bz-H), 7.77–7.53 (m, 3H, Bz-H), 5.39 (d, *J* = 5.5 Hz, 1H, 1'-H), 4.88–4.82 (m, 1H, 2'-H), 4.31 (dd, *J* = 8.8, 3.7 Hz, 1H, 5'-H), 4.26–4.21 (m, 1H, 3'-H), 4.13 (q, *J* = 5.5 Hz, 1H, 4'-H), 4.00–3.9 (m, 2H, 6'-H), 3.42 (s, 3H, OMe), 3.36 (s, 3H, OMe). ^{13}C-NMR (75 MHz, DMSO-d_6): δ 165.20 (C=O); 152.25 (C-6); 150.77 (C-2); 149.68 (C-4); 143.85 (C-8); 132.92 (C-5); 131.89, 127.95, 127.91, 125.07 (Bz-C); 97.91 (C-1'); 77.10 (C-5'); 75.44 (C-3'); 62.37 (C-4'); 59.78 (C-6'); 55.75 (OMe); 55.53 (OMe); 54.53 (C-2'). HRMS calcd. for C$_{20}$H$_{24}$N$_5$O$_6$$^+$ [M+H]$^+$ 430.1721, found 430.1716.

3.17. 1',3'-Di-O-methyl-2'-deoxy-2'-(N^2-(dimethylamino)methylene-guanin-9-yl))-D-altro-pyranoside (**15d**)

Compound **14** (0.58 g, 1.2 mmol) was dissolved in 30 mL of AcOH:H$_2$O (3:1) at rt. The reaction mixture was slowly heated at 45 °C and reaction progress was monitored using TLC. After 12 h, the mixture was concentrated and co-evaporated with toluene. The crude residue was purified by flash silica gel column chromatography (elution with 10% MeOH in DCM) afforded compound **15d** (0.4 g, 81%). ^{1}H-NMR (500 MHz, CD$_3$OD): δ 8.65 (s, 1H, N=CH-N), 7.85 (s, 1H, 8-H), 5.21 (d, *J* = 5.3 Hz, 1H, 2'-H), 4.64–4.57 (m, 1H, 1'-H), 4.10 (m, 2H, 4'-H, 3'-H), 4.00 (dd, *J* = 8.8, 4.1 Hz, 1H, 5'-H), 3.84 (d, *J* = 4.6 Hz, 2H, 6'-H), 3.33 (s, 3H, OMe), 3.30 (s, 3H, OMe), 3.16 (s, 3H, OMe), 3.07 (s, 3H, OMe). ^{13}C-NMR (125 MHz, CD$_3$OD): δ 160.31 (N=C-N); 159.83 (C-6); 158.79 (C-2); 152.31 (C-4); 140.37 (C-8); 120.48 (C-5); 100.37 (C-1'); 78.41 (C-5'); 76.47 (C-3'); 65.24 (C-4'); 62.66 (C-6'); 57.94 (OMe); 57.74 (OMe); 56.17 (C-2'); 41.53 and 35.34 (NMe$_2$). HRMS calcd. for C$_{16}$H$_{25}$N$_6$O$_6$$^+$ [M+H]$^+$ 397.1830, found 397.1827.

3.18. 6'-O-Dimethoxytrityl-1',3'-di-O-methyl-2'-deoxy-2'-(uracil-1-yl)-D-altropyranoside (**16a**)

To a solution of the nucleoside **15a** (0.4 g, 1.32 mmol) in anhydrous pyridine (10 mL) and under argon atmosphere, dimethoxytrityl chloride (0.49 g, 1.45 mmol) was added under stirring on an ice bath. After stirring at room temperature for 3 h, 5% aqueous NaHCO$_3$ solution (1 mL) was added, the reaction solvent was evaporated, diluted with CH$_2$Cl$_2$ (50 mL) and washed with 5% aqueous NaHCO$_3$ (2 × 30 mL). The aqueous layers were back extracted once with 30 mL CH$_2$Cl$_2$. The combined organic layer was dried (Na$_2$SO$_4$), evaporated under reduced pressure and the crude residue was purified by flash chromatography (CH$_2$Cl$_2$/MeOH 96:4) affording the corresponding dimethoxytritylated nucleoside **16a** (0.745 g, 93% yield) as a white foam. ^{1}H-NMR (125 MHz, CDCl$_3$): δ 9.31 (brs, 1H, NH), 7.48–7.43 (m, 2H, 6-H, Ar-H), 7.37–7.33 (m, 4H, Ar-H), 7.31–7.26 (m, 2H, Ar-H), 7.24–7.19 (m, 1H, Ar-H), 6.85–6.82 (m, 4H, Ar-H), 5.69 (dd, *J* = 8.0, 1.6 Hz, 1H, 5-H), 5.30 (brs, 1H, 1'-H), 4.94 (brs, 1H, 2'-H), 4.08–3.95 (m, 2H, 5'-H, 3'-H), 3.79 (s, 3H, OMe), 3.78–3.77 (m, 4H, OMe, 4'-H), 3.50–3.42 (brs, s, 6H, 2OMe), 3.42–3.40 (m, 2H, 6'-H). ^{13}C-NMR (125 MHz, CDCl$_3$): δ 163.28 (C-4); 158.50 (C-2); 150.68 (C-6); 144.85, 135.84, 130.09, 130.06, 128.10, 127.80, 126.81, 113.11 (Ar-C); 102.35 (C-5); 98.36 (C-1'); 86.20 (C^{Tr}-O); 77.14 (C-5'); 76.87 (C-3'); 64.05 (C-4'); 62.88 (C-6'); 57.93 (OMe); 55.69 (OMe); 55.19 (2OMe, C-2'). HRMS calcd. for C$_{33}$H$_{36}$N$_2$O$_9$Na$^+$ [M+Na]$^+$ 627.23186, found 627.2311.

3.19. 6'-O-Dimethoxytrityl-1',3'-di-O-methyl-2'-deoxy-2'-(N^6-benzoylcytosin-1-yl)-D-altropyranoside (**16b**)

Compound **16b** (0.346 g, 87% yield) was synthesized from compound **15b** (0.228 g, 0.56 mmol) using dimethoxytrityl chloride (0.09 g, 1.29 mmol) in anhydrous pyridine (10 mL) according to the procedure used for the synthesis of compound **16a**. ^{1}H-NMR (500 MHz, CDCl$_3$) : δ 8.75 (brs, 1H, NH), 7.90 (d, *J* = 7.3 Hz, 2H, 6-H, Ar-H), 7.64–7.45 (m, 6H, Ar-H), 7.40–7.19 (m, 8H, 5-H, Ar-H), 6.86 (d, *J* = 8.8 Hz, 4H, Ar-H), 5.29 (s, 1H, 1'-H), 5.03 (brs, 1H, 2'-H), 4.11–3.95 (m, 2H, 5'-H, 3'-H), 3.83–3.77 (m, 7H, 2OMe, 4'-H), 3.60–3.37 (m, 8H, 6'-H, 2OMe), 2.50 (s, 1H, 4-OH). ^{13}C-NMR (126 MHz, CDCl$_3$): δ 166.13 (C=O), 162.13 (C-4), 158.47 (C-4), 144.71 (C-6), 136.02, 133.17, 130.03, 129.01, 128.20, 127.79, 127.50, 126.80, 113.10 (Ar-C), 98.06 (C-5), 96.85 (C-1'), 86.11 (C^{Tr}-O), 77.13 (C-5'), 76.88 (C-3'), 65.62 (C-4'), 62.73 (C-6'), 58.04 (OMe), 55.72 (OMe), 55.18 (C-2, 2OMe). HRMS calcd. For C$_{40}$H$_{42}$N$_3$O$_9$$^+$ [M+H]$^+$ 708.29208, found 708.2909.

*3.20. 6'-O-Dimethoxytrityl-1',3'-di-O-methyl-2'-deoxy-2'-(N⁶-benzoyladenin-9-yl)-D-altropyranoside (**16c**)*

Compound **16c** (0.77 g, 91% yield) was synthesized from compound **15c**(0.5 g, 1.16 mmol) using dimethoxytrityl chloride (0.44 g, 1.27 mmol) in anhydrous pyridine (10 mL) according to the procedure used for the synthesis of compound **16a**. ^{1}H-NMR (500 MHz, CDCl$_3$): δ 9.15 (brs, 1H, 2-H), 8.12 (s, 1H, 8-H),8.03 (d, J = 8.0 Hz, 2H, Ar-H), 7.60 (t, J = 7.4 Hz, 1H, Ar-H), 7.52–7.49 (m, 3H, Ar-H), 7.40 (d, J = 8.8 Hz, 4H, Ar-H), 7.34–7.21 (m, 5H, Ar-H), 6.85 (d, J = 8.8 Hz, 3H, Ar-H), 5.29 (t, J = 1.1 Hz, 1H, 1'-H), 5.20 (d, J = 4.8 Hz, 1H, 2'-H), 4.79–4.68 (m, 1H, 6'-He), 4.27 (q, J = 5.5 Hz, 1H, 5'-H), 4.18 (dd, J = 7.9, 3.8 Hz, 1H, 4'-H), 4.07 (t, J = 4.7 Hz, 1H, 3'-H), 3.79 (s, 6H, OMe), 3.53 – 3.47 (m, 2H, 6'-Ha, OH), 3.41 (s, 3H, OMe), 3.29 (s, 3H, OMe). ^{13}C-NMR (125 MHz, CDCl$_3$): δ 164.63 (C-6); 158.52 (Ar-C); 152.46 (C-2); 151.78 (C-4); 149.56 (C-8); 144.66, 143.58, 135.84, 133.67, 132.75, 130.05, 128.84, 128.17, 127.85, 126.84, 123.11, 113.16 (Ar-C); 98.31 (C-1'); 86.33 (C^{Tr}-O); 76.26 (C-5'); 72.89 (C-3'); 64.68 (C-4'); 62.90 (C-6'); 57.89 (OMe); 56.87 (OMe); 55.94 (OMe); 55.19 (OMe); 53.38 (C-2'). HRMS calcd. for C$_{41}$H$_{42}$N$_5$O$_8$$^+$ [M+H]$^+$ 732.30332, found 732.3030.

*3.21. 6'-O-Dimethoxytrityl-1',3'-di-O-methyl-2'-deoxy-2'-(N²-(dimethylamino)methylene-guanin-9-yl))-D-altropyranoside (**16d**)*

Compound **16d** (0.51 g, 74% yield) was synthesized from compound **15d** (0.39 g, 0.974 mmol) using dimethoxytrityl chloride (0.362 g, 1.07 mmol) in anhydrous pyridine (10 mL) according to the procedure used for the synthesis of compound **16a**. ^{1}H-NMR (500 MHz, CDCl$_3$): δ 9.20 (s, 1H, NH), 8.61 (d, J = 4.2 Hz, 1H, N=CH-N), 8.52 (s, 1H, 8-H), 7.79 (s, 1H, Ar-H), 7.68 (tt, J = 7.7, 1.8 Hz, 1H, Ar-H), 7.52–7.48 (m, 2H, Ar-H), 7.41–7.36 (m, 4H, Ar-H), 7.33–7.20 (m, 5H, Ar-H), 6.87–6.82 (m, 4H, Ar-H), 5.29 (s, 1H, 1'-H), 5.09 (d, J = 3.1 Hz, 1H, 2'-H), 4.79 (dd, J = 6.4, 3.1 Hz, 1H, 3'-H), 4.17–4.11 (m, 1H, 5'-H), 4.01 (brs, 1H, -OH), 3.88 (dd, J = 6.4, 4.0 Hz, 1H, 4'-H), 3.52 (dd, J = 10.2, 2.9 Hz, 1H, 6-He), 3.46 (s, 3H, OMe), 3.44 (s, 3H, OMe), 3.35 (dd, J = 10.2, 5.5 Hz, 1H, 6'-Ha), 3.05 (s, 3H, OMe), 2.93 (s, 3H, OMe). ^{13}C-NMR (125 MHz, CDCl$_3$): δ 158.50 (N=CH-N); 157.92 (MMTr); 157.78 (C-6); 156.55(C-2); 150.39(C-4); 149.81 (C-8); 144.84 (Ar-C), 137.33, 135.98, 135.87, 130.06, 130.02, 128.10, 127.85, 126.81, 123.70, 120.37, 113.13 (Ar-C); 99.83 (C-1'); 86.17 (C^{Tr}-O); 78.04 (C-5'); 71.31 (C-3'); 64.79 (C-4'); 63.87 (C-6'); 58.37 (OMe); 55.65 (OMe); 54.56 (C-2'); 41.12 and 35.12 (-NMe$_2$). HRMS calcd. for C$_{37}$H$_{43}$N$_6$O$_8$$^+$ [M+H]$^+$ 699.3142, found 699.3121.

3.22. General Procedure for Nucleoside Phosphitylation

To a solution of the dimethoxytritylated nucleoside **16a** (0.73 g, 1.2 mmol) in anhydrous CH$_2$Cl$_2$ (6 mL) at 0 °C and under argon atmosphere, freshly dried diisopropylethylamine (0.063 mL, 3.6 mmol) and 2-cyanoethyl-N,N-diisopropylchlorophosphoramidite (0.040 mL, 1.8 mmol) were added. The reaction mixture was stirred at 0 °C for 90 min after which completeness of the reaction was indicated by TLC. Saturated NaHCO$_3$ solution (2 mL) was added, the solution was stirred for another10 min and partitioned between CH$_2$Cl$_2$ (50 mL) and aqueous NaHCO$_3$ (30 mL). The organic layer was washed with brine (3 × 30 mL) and the aqueous phases were back extracted with CH$_2$Cl$_2$ (30 mL). After solvent evaporation, the resulting oil was purified by flash chromatography (hexane/acetone/TEA = 62/36/2). The yellow solid was then dissolved in CH$_2$Cl$_2$ (2 mL) and precipitated twice in cold hexane (160 mL, −30 °C) to afford the desired corresponding phosphoramidite nucleoside **17a** (0.908 g, 93% yield) as a white powder. The obtained product was dried under vacuum and stored overnight under argon at −20 °C. ^{31}P-NMR (CDCl$_3$): δ = 150.84. HRMS calcd. for C$_{42}$H$_{54}$N$_4$O$_{10}$P$_1$$^+$ [M+H]$^+$ 805.35773, found 805.3557; ^{31}P-NMR (CDCl$_3$): δ = 150.84.

Compound **17b** (0.34 g, 79% yield) was synthesized from compound **16b** (0.34 g, 0.47 mmol), dry diisopropylethylamine (0.025 mL, 1.42 mmol), 2-cyanoethyl-N,N-diisopropylchloro-phosphoramidite (0.016 mL, 0.71 mmol) and anhydrous CH$_2$Cl$_2$ (10 mL) according to procedure used for the synthesis of compound **17a**. ^{31}P-NMR (CDCl$_3$): δ = 150.96. HRMS calcd. for C$_{49}$H$_{59}$N$_5$O$_{10}$P$_1$ [M+H]$^+$ 908.39993, found 908.3981.

Compound **17c** (0.83 g, 85% yield) was synthesized from compound **16c** (0.76 g, 1.03 mmol), dry diisopropylethylamine (0.054 mL, 3.09 mmol), 2-cyanoethyl-*N,N*-diisopropylchloro-phosphoramidite (0.034 mL, 1.54 mmol) and anhydrous CH_2Cl_2 (10 mL) according to procedure used for the synthesis of compound **17a**. ^{31}P-NMR (CDCl$_3$): δ = 150.287 and 151.231. HRMS calcd. for $C_{50}H_{59}N_7O_9P_1^+$ [M+H]$^+$ 932.41116, found 932.4103.

Compound **17d** (0.42 g, 65% yield) was synthesized from compound **16d** (0.5 g, 0.71 mmol), dry diisopropylethylamine (0.037 mL, 2.13 mmol), 2-cyanoethyl-*N,N*-diisopropylchloro-phosphoramidite (0.025 mL, 1.15 mmol) and anhydrous CH_2Cl_2 (10 mL) according to procedure used for the synthesis of compound **17a**. ^{31}P-NMR (CDCl$_3$): δ = 150.065 and 151.206. HRMS calcd. for $C_{46}H_{60}N_8O_9P_1^+$ [M+H]$^+$ 899.42206, found 899.4240.

4. Conclusions

A new nucleoside analogue scaffold for incorporation into oligonucleotides was developed and all four monomers with the natural heterocyclic bases have been prepared and evaluated on their hybridization potential with natural DNA and RNA. While it was anticipated that the constraint imposed by the 6-membered ring structure could afford the entropic advantage as seen with HNA and ANA constructs, an entropic penalty to preserve the $^1C_4$ conformation required for pairing with RNA annulated the affinity gain which one could expect from a pre-organized structure.

Acknowledgments: This work was supported by a FWO (Flemish Scientific Research) grant G.0784.11 and KU Leuven financial support (GOA/10/13). Mass spectrometry was made possible by the support of the Hercules Foundation of the Flemish Government (grant 20100225-7). We are indebted to Chantal Biernaux for final typesetting.

Author Contributions: A.V. and D.K. synthesized and analyzed the different monomers; G.S. assembled oligonucleotides and carried out Tm analysis; J.R. carried out all mass spectrometric analysis; M.F. performed the modeling experiments; A.V.A. conceived and supervised research and wrote the paper.

Conflicts of Interest: The authors declare no conflict of interest.

References

1. Helene, C. The anti-gene strategy: Control of gene expression by triplex-forming-oligonucleotides. *Anti-Cancer Drug Des.* **1991**, *6*, 569–584.
2. Wagner, R.W. Gene inhibition using antisense oligodeoxynucleotides. *Nature* **1994**, *372*, 333–335. [CrossRef] [PubMed]
3. Opalinska, J.B.; Gewirtz, A.M. Nucleic-acid therapeutics: Basic principles and recent applications. *Nat. Rev. Drug. Discov.* **2002**, *1*, 503–514. [CrossRef] [PubMed]
4. Castanotto, D.; Stein, C.A. Antisense oligonucleotides in cancer. *Curr. Opin. Oncol.* **2014**, *26*, 584–589. [CrossRef] [PubMed]
5. Dean, N.M.; Bennett, C.F. Antisense oligonucleotide-based therapeutics for cancer. *Oncogene* **2003**, *22*, 9087–9096. [CrossRef] [PubMed]
6. Yang, W.Q.; Zhang, Y. RNAi-mediated gene silencing in cancer therapy. *Expert Opin. Biol. Ther.* **2012**, *12*, 1495–1504. [CrossRef] [PubMed]
7. Gao, S.; Yang, C.; Jiang, S.; Xu, X.-N.; Lu, X.; He, Y.-W.; Cheung, A.; Wang, H. Applications of RNA interference high-throughput screening technology in cancer biology and virology. *Protein Cell* **2014**, *5*, 805–815. [CrossRef] [PubMed]
8. Sun, H.; Zhu, X.; Lu, P.Y.; Rosato, R.R.; Tan, W.; Zu, Y. Oligonucleotide aptamers: New tools for targeted cancer therapy. *Mol. Ther. Nucleic Acids* **2014**, *3*, e182. [CrossRef] [PubMed]
9. Jiang, K. Biotech Comes to Its "Antisenses" after Hard-Won Drug Approval. (Archived by WebCite®at http://www.webcitation.org/6WNb4M1rP). Available online: http://blogs.nature.com/spoonful/2013/02/biotech-comes-to-its-antisenses-after-hard-won-drug-approval.html (accessed on 6 January 2015).
10. Lu, Q.L.; Cirak, S.; Partridge, T. What Can We Learn From Clinical Trials of Exon Skipping for DMD? *Mol. Ther. Nucl. Acids* **2014**, *3*, e152. [CrossRef]

11. Soutschek, J.; Akinc, A.; Bramlage, B.; Charisse, K.; Constien, R.; Donoghue, M.; Elbashir, S.; Geick, A.; Hadwiger, P.; Harborth, J.; *et al.* Therapeutic silencing of an endogenous gene by systemic administration of modified siRNAs. *Nature* **2004**, *432*, 173–178. [CrossRef] [PubMed]

12. De Fougerolles, A.; Vornlocher, H.P.; Maraganore, J.; Lieberman, J. Interfering with disease: A progress report on siRNA-based therapeutics. *Nat. Rev. Drug. Discov.* **2007**, *6*, 443–453.

13. Herdewijn, P. Heterocyclic modifications of oligonucleotides and antisense technology. *Antisense Nucl. Acid. Drug Dev.* **2000**, *10*, 297–310. [CrossRef]

14. Herdewijn, P. Conformationally restricted carbohydrate-modified nucleic acids and antisense technology. *Biochim. Biophys. Acta* **1999**, *1489*, 167–179. [CrossRef] [PubMed]

15. Manoharan, M. RNA interference and chemically modified small interfering RNAs. *Curr. Opin. Chem. Biol.* **2004**, *8*, 570–579. [CrossRef] [PubMed]

16. Van Aerschot, A.; Verheggen, I.; Hendrix, C.; Herdewijn, P. 1,5-Anhydrohexitol nucleic acids, a new promising antisense construct. *Angew. Chem. Int. Ed. Engl.* **1995**, *34*, 1338–1339. [CrossRef]

17. Hendrix, C.; Rosemeyer, H.; Verheggen, I.; Seela, F.; van Aerschot, A.; Herdewijn, P. 1',5'-Anhydrohexitol oligonucleotides: Synthesis, base pairing and recognition by regular oligodeoxyribonucleotides and oligoribonucleotides. *Chem. Eur. J.* **1997**, *3*, 110–120. [CrossRef]

18. Allart, B.; Khan, K.; Rosemeyer, H.; Schepers, G.; Hendrix, C.; Rothenbacher, K.; Seela, F.; Van Aerschot, A.; Herdewijn, P. D-Altritol nucleic acids (ANA): Hybridisation properties, stability, and initial structural analysis. *Chem. Eur. J.* **1999**, *5*, 2424–2431. [CrossRef]

19. Wang, J.; Verbeure, B.; Luyten, I.; Lescrinier, E.; Froeyen, M.; Hendrix, C.; Rosemeyer, H.; Seela, F.; van Aerschot, A.; Herdewijn, P. Cyclohexene nucleic acids (CeNA): Serum stable oligonucleotides that activate RNase H and increase duplex stability with complementary RNA. *J. Am. Chem. Soc.* **2000**, *122*, 8595–8602. [CrossRef]

20. Singh, S.K.; Nielsen, P.; Koshkin, A.A.; Wengel, J. LNA (locked nucleic acids): Synthesis and high-affinity nucleic acid recognition. *Chem. Commun.* **1998**, 455–456. [CrossRef]

21. Veedu, R.N.; Wengel, J. Locked nucleic acids: Promising nucleic acid analogs for therapeutic applications. *Chem. Biodivers.* **2010**, *7*, 536–542. [CrossRef] [PubMed]

22. Fisher, M.; Abramov, M.; van Aerschot, A.; Xu, D.; Juliano, R.L.; Herdewijn, P. Inhibition of MDR1 expression with altritol-modified siRNAs. *Nucl. Acids Res.* **2007**, *35*, 1064–1074. [CrossRef] [PubMed]

23. Chatelain, G.; Schepers, G.; Rozenski, J.; van Aerschot, A. Hybridization potential of oligonucleotides comprising 3'-O-methylated altritol nucleosides. *Mol. Divers.* **2012**, *16*, 825–837. [CrossRef] [PubMed]

24. Richtmyer, N.K.; Hudson, C.S. Crystalline α-methyl-D-altroside and some new derivatives of D-altrose1. *J. Am. Chem. Soc.* **1941**, *63*, 1727–1731. [CrossRef]

25. Teste, K.; Colombeau, L.; Hadj-Bouazza, A.; Lucas, R.; Zerrouki, R.; Krausz, P.; Champavier, Y. Solvent-controlled regioselective protection of 5'-O-protected thymidine. *Carbohydr. Res.* **2008**, *343*, 1490–1495. [CrossRef] [PubMed]

26. Lucas, R.; Teste, K.; Zerrouki, R.; Champavier, Y.; Guilloton, M. Chelation-controlled regioselective alkylation of pyrimidine 2'-deoxynucleosides. *Carbohydr. Res.* **2010**, *345*, 199–207. [CrossRef] [PubMed]

27. Divakar, K.J.; Reese, C.B. 4-(1,2,4-Triazol-1-yl) and 4-(3-nitro-1,2,4-triazol-1-yl)-1-(beta-D-2,3,5-tri-O-acetylarabinofuranosyl)pyrimidin-2(1*H*)-ones—Valuable intermediates in the synthesis of derivatives of 1-(beta-D-arabinofuranosyl)cytosine (Ara-C). *J. Chem. Soc. Perk. Trans. 1* **1982**, 1171–1176. [CrossRef]

28. Pulido, D.; Sánchez, A.; Robles, J.; Pedroso, E.; Grandas, A. Guanine-containing DNA minor-groove binders. *Eur. J. Org. Chem.* **2009**, *2009*, 1398–1406. [CrossRef]

29. Jenny, T.F.; Schneider, K.C.; Benner, S.A. N-2-Isobutyryl-O-6-[2-(*para*-nitrophenyl)ethyl]guanine—A new building block for the efficient synthesis of carbocyclic guanosine analogs. *Nucleos. Nucleot.* **1992**, *11*, 1257–1261. [CrossRef]

30. Mcbride, L.J.; Kierzek, R.; Beaucage, S.L.; Caruthers, M.H. Amidine protecting groups for oligonucleotide synthesis 16. *J. Am. Chem. Soc.* **1986**, *108*, 2040–2048. [CrossRef]

31. Vu, H.; Mccollum, C.; Jacobson, K.; Theisen, P.; Vinayak, R.; Spiess, E.; Andrus, A. Fast Oligonucleotide deprotection phosphoramidite chemistry for DNA-synthesis. *Tetrahedron Lett.* **1990**, *31*, 7269–7272. [CrossRef]

32. Theisen, P.; McCollum, C.; Andrus, A. N-6-Dialkylformamidine-2'-deoxyadenosine phosphoramidites in oligodeoxynucleotide synthesis—Rapid deprotection of oligodeoxynucleotides. *Nucleos. Nucleot.* **1993**, *12*, 1033–1046. [CrossRef]

33. D'Alonzo, D.; van Aerschot, A.; Guaragna, A.; Palumbo, G.; Schepers, G.; Capone, S.; Rozenski, J.; Herdewijn, P. Synthesis and base pairing properties of 1′,5′-anhydro-L-hexitol nucleic acids (L-HNA). *Chem. Eur. J.* **2009**, *15*, 10121–10131. [CrossRef] [PubMed]

34. Van Aerschot, A.; Meldgaard, M.; Schepers, G.; Volders, F.; Rozenski, J.; Busson, R.; Herdewijn, P. Improved hybridisation potential of oligonucleotides comprising O-methylated anhydrohexitol nucleoside congeners. *Nucl. Acids Res.* **2001**, *29*, 4187–4194. [CrossRef] [PubMed]

35. Vandermeeren, M.; Preveral, S.; Janssens, S.; Geysen, J.; Saison-Behmoaras, E.; Van Aerschot, A.; Herdewijn, P. Biological activity of hexitol nucleic acids targeted at Ha-ras and intracellular adhesion molecule-1 mRNA. *Biochem. Pharmacol.* **2000**, *59*, 655–663. [CrossRef] [PubMed]

36. Fisher, M.; Abramov, M.; Van Aerschot, A.; Rozenski, J.; Dixit, V.; Juliano, R.L.; Herdewijn, P. Biological effects of hexitol and altritol-modified siRNAs targeting B-Raf. *Eur. J. Pharmacol.* **2009**, *606*, 38–44. [CrossRef] [PubMed]

37. Pinheiro, V.B.; Taylor, A.I.; Cozens, C.; Abramov, M.; Renders, M.; Zhang, S.; Chaput, J.C.; Wengel, J.; Peak-Chew, S.Y.; McLaughlin, S.H.; *et al.* Synthetic genetic polymers capable of heredity and evolution. *Science* **2012**, *336*, 341–344. [CrossRef] [PubMed]

38. Taylor, A.I.; Pinheiro, V.B.; Smola, M.J.; Morgunov, A.S.; Peak-Chew, S.; Cozens, C.; Weeks, K.M.; Herdewijn, P.; Holliger, P. Catalysts from synthetic genetic polymers. *Nature* **2015**, *518*, 427–430. [CrossRef] [PubMed]

39. Mooers, B.H.; Singh, A. The crystal structure of an oligo(U):pre-mRNA duplex from a trypanosome RNA editing substrate. *RNA* **2011**, *17*, 1870–1883. [CrossRef] [PubMed]

40. Salomon-Ferrer, R.; Case, D.A.; Walker, R.C. An overview of the Amber biomolecular simulation package. *Wires Comput. Mol. Sci.* **2013**, *3*, 198–210. [CrossRef]

41. Pettersen, E.F.; Goddard, T.D.; Huang, C.C.; Couch, G.S.; Greenblatt, D.M.; Meng, E.C.; Ferrin, T.E. UCSF Chimera—A visualization system for exploratory research and analysis. *J. Comput. Chem.* **2004**, *25*, 1605–1612. [CrossRef] [PubMed]

Sample Availability: *Sample Availability*: Only small amounts of the different amidites are still available from the authors.

MDPI

Article

Tethering in RNA: An RNA-Binding Fragment Discovery Tool

Kiet Tran [1], Michelle R. Arkin [2] and Peter A. Beal [1,*]

[1] Department of Chemistry, University of California, One Shields Ave, Davis, CA 95616, USA; ktutran@ucdavis.edu

[2] Small Molecule Discovery Center, Department of Pharmaceutical Chemistry, University of California, San Francisco, CA 94158, USA; Michelle.Arkin@ucsf.edu

* Correspondence: pabeal@ucdavis.edu; Tel.: +1-530-752-4132

Academic Editors: Mahesh K. Lakshman and Fumi Nagatsugi
Received: 5 December 2014; Accepted: 17 February 2015; Published: 4 March 2015

Abstract: Tethering has been extensively used to study small molecule interactions with proteins through reversible disulfide bond forming reactions to cysteine residues. We describe the adaptation of Tethering to the study of small molecule binding to RNA using a thiol-containing adenosine analog (A_{SH}). Among 30 disulfide-containing small molecules screened for efficient Tethering to A_{SH}-bearing RNAs derived from pre-miR21, a benzotriazole-containing compound showed prominent adduct formation and selectivity for one of the RNAs tested. The results of this screen demonstrate the viability of using thiol-modified nucleic acids to discover molecules with binding affinity and specificity for the purpose of therapeutic compound lead discovery.

Keywords: Tethering; miRNA; nucleoside analog

---

1. Introduction

The structural complexity and functional diversity of cellular RNAs make them attractive targets for high affinity small molecule ligands [1,2]. However, a significant challenge in the development of RNA-binding molecules is the difficulty in optimizing lead compounds from initial screening hits. This is due, at least in part, to difficulties in characterizing structures of RNA targets bound to weakly interacting lead compounds [3,4]. Here we describe the application of the Tethering screening method to the discovery of RNA-interacting fragments. In the Tethering method, lead molecules are covalently bound to the target such that future structural studies are not complicated by weak fragment interactions or multiple binding modes. Tethering has been successfully used in the generation of potent enzyme and protein-protein interaction inhibitors, but has yet to be applied to RNA [5]. For protein targets, Tethering involves the introduction of individual cysteine mutations surrounding the site of interest followed by screening against libraries of disulfide-containing fragments [6]. The disulfide exchange reactions allow identification of ligands since the reversible protein-fragment interactions increase the local concentration of the small molecules at the targeted site. Fragments having affinity for the protein, apart from the disulfide bond, are efficiently conjugated at equilibrium. Hits are then detected by mass spectrometry of the mixture. Fragment-target covalent conjugates can then be structurally characterized and guide the improvement of affinity and conversion to noncovalent inhibitors by subsequent modification [7]. To adapt Tethering to the discovery of RNA-binding ligands, one must introduce a thiol near the site of interest in the RNA. This can be achieved using thiol-bearing nucleoside analogs. In this initial report on Tethering for RNA, we endeavored to discover fragments that bind to a biosynthetic precursor to a biologically important miRNA, miR21.

MicroRNAs (miRNAs) are endogenously encoded ~23 nt RNA molecules that control gene expression at the post-transcriptional level [8]. Hundreds of human miRNAs are known with each miRNA regulating multiple targets. Furthermore, altered miRNA expression correlates with human disease states, including cancer [9]. Indeed, overexpression of certain miRNAs is known to cause malignant phenotypes and these miRNAs are targets for the development of cancer therapeutics. Among these is miR-21, whose expression is dramatically upregulated in many solid tumors [10]. MiRNAs are biosynthesized from hairpin precursors via the action of two duplex-specific endonucleases, Drosha and Dicer [11]. Given this biosynthetic pathway, small molecules that bind to functionally significant sites on miRNA precursors should block their conversion to the mature miRNA. Several recent reports highlight this attractive approach that will undoubtedly benefit from the discovery of new RNA-targeting compounds that have the requisite properties to be useful starting points for drug discovery efforts [12–15]. Here we use Tethering to identify molecular fragments with affinity for pre-miR-21 near the Dicer cleavage sites. We envision such fragments providing starting structures for the development of inhibitors of the biosynthesis of miR-21.

2. Results and Discussion

2.1. Synthesis of N^6-(2-(triphenylmethylthio)ethyl)adenosine Phosphoramidite and Reactivity of Modified RNAs

To introduce a thiol into RNA screening targets, we synthesized a derivative of adenosine (A_{SH}) that bears an ethane thiol substituent on N6 (Figure 1). This modification does not interfere with Watson-Crick hydrogen bonding and directs the reactive thiol into the major groove of an RNA duplex. Verdine and colleagues described the synthesis of RNA bearing this modification using a convertible nucleoside approach [16]. We chose to synthesize the S-trityl protected phosphoramidite, which had not been previously reported (Scheme 1). While many methods could be used to introduce a protected thiol moiety into a ribonucleoside [17,18], our previous work showed that a trityl-protected thiol group in RNA could be easily deprotected with $AgNO_{3(aq)}$ [19].

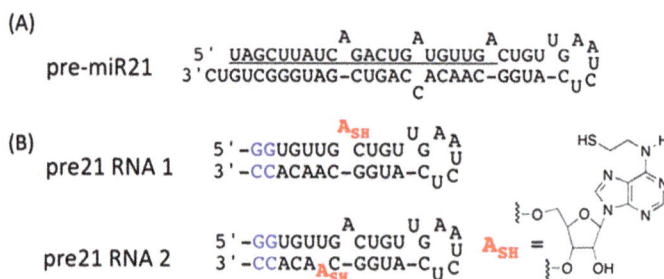

Figure 1. Pre-miR21 sequence-based RNAs bearing the thiol tether. (**A**) Sequence and secondary structure for pre-miR21. Underlined sequence is mature miR21; (**B**) Corresponding truncated pre21 RNA1 and pre21 RNA2 31-nucleotide oligomers (left) containing the adenosine modification A_{SH} (right). Two non-native G-C base pairs (blue) have been added to the duplex end for stability.

First, the tri-O-acetyl protected 6-chloropurine nucleoside was allowed to react with S-trityl protected 2-aminoethanethiol in an S_NAr reaction to give the N^6-alkylated adenosine analog. Removal of the acyl protecting groups with methanolic ammonia yielded free nucleoside **1**. Protection of the 5' and 2' hydroxyls of the ribose moiety was accomplished with DMTr-Cl and TBDMS-Cl to yield compounds **2** and **3**, respectively. Finally, the 3'-OH was allowed to react with β-cyanoethyl-(N,N-diisopropyl)chlorophosphite to give the corresponding phosphoramidite **4** which was then used to incorporate the nucleoside analog via solid phase synthesis into two different RNAs mimicking the hairpin loop of pre-miR21 (pre21 RNA 1 and pre21 RNA 2, Figure 1). These two RNAs differ in the positioning of the reactive adenosine analog near the Dicer cleavage sites on opposite sides of the duplex. The modified RNAs still containing S-trityl protecting groups were purified via denaturing polyacrylamide gel electrophoresis.

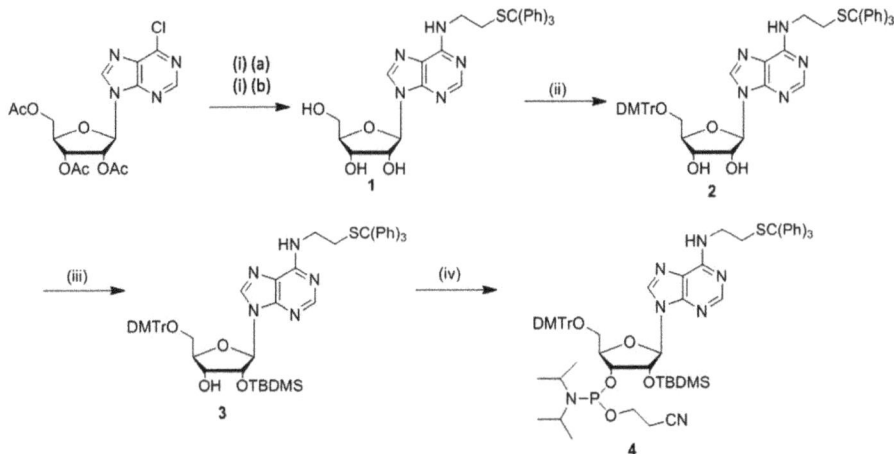

Scheme 1. Synthesis of N^6-(2-(triphenylmethylthio)ethyl)-adenosine phosphoramidite. *Reagents and conditions*: (i) (a) S-trityl-2-aminoethanethiol, DMF (b) NH$_3$(sat) in MeOH, 72%; (ii) DMTr-Cl, DMAP, pyridine, 88%; (iii) TBDMS-Cl, N(Et)$_3$, DMAP, THF, 43%; (iv) 2-cyanoethyl-(*N*,*N*-diisopropylamino) chlorophosphite, DIPEA, CH$_2$Cl$_2$, 82%.

Removal of the trityl protecting group on the sulfur was necessary before subjecting the RNA to Tethering. This was accomplished using aqueous silver nitrate giving the RNA free thiol. Deprotected pre21 RNA 2 (free thiol) and cystamine were then used to determine the appropriate conditions needed for the disulfide screen (Figure 2). Detection of reaction products was performed with ESI-MS and the relative abundances of species in a sample were determined from the areas of the peaks. The result clearly showed that the thiol-containing oligonucleotide can form a disulfide linkage with cystamine under the experimental conditions in Figure 2A, to give a product peak observed at 10,030 amu in the mass spectrum shown. The percent abundance of the cysteamine adduct was 33% in the presence of 0.1 mM β-mercaptoethanol (BME). The concentration of BME to be used in the screen should be high enough such that the majority of the small molecules would not conjugate efficiently with the RNA. Another experiment was performed using pre21 RNA 2 to determine the concentration of BME necessary to completely inhibit cystamine conjugation. The results in Figure 2B showed that no adduct could be detected at a concentration of 1.0 mM BME. Therefore, the screen was performed at a concentration of 0.5 mM BME.

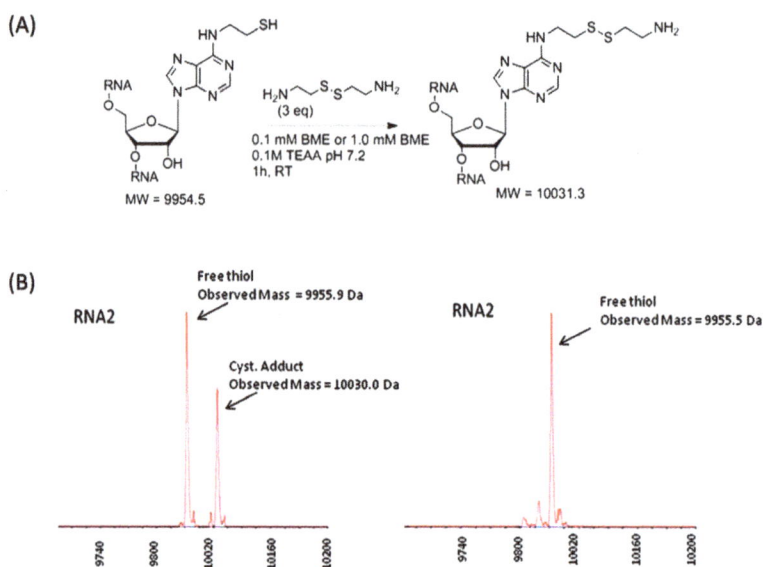

Figure 2. Testing the reactivity of a thiol-modified RNA. (**A**) Reaction of pre21 RNA 2 free-thiol with cystamine; (**B**) ESI-MS spectra of the disulfide conjugation reaction in 0.1 mM BME (left) and 1.0 mM BME (right).

2.2. RNA Tethering Screen with pre21 RNA 1 and pre21 RNA 2

A small library of thirty disulfide-containing small molecules, all containing an *N,N*-dimethylethylamino (*N,N*-DMA) moiety for solubility purposes, was obtained from the Small Molecule Discovery Center, UCSF for the Tethering screen (Figure 3) [20]. Each compound at 0.12 mM concentration was allowed to react with 20 μM of thiol-bearing RNA in 0.1 M TEAA buffer (pH 7.2) for 1 h at room temperature. The resulting sixty reaction mixtures were then analyzed by ESI-MS and percent relative abundance of the target adduct was used to assess small molecule binding efficiency. The results of the screen are shown in Figure 4, Supplementary Table S1 and Supplementary Table S2. In addition, the ESI-MS data for an example screening reaction is shown in Figure 5 with compound A05. The majority of the compounds in the library showed minimal to no target adduct formation (Figure 4). This was expected considering the concentration of reducing agent used. However, compounds A04 and A05 gave >30% target adduct formation for pre21 RNA 1 under the reducing conditions of the Tethering screen suggesting a binding interaction between the RNAs and these compounds. For compound A04, efficient conjugation was not specific to pre21 RNA 1 as pre21 RNA 2 also gave a high conjugate yield. This is perhaps not surprising considering the structure of A04 (Figure 3). 2-Phenylquinolines are known RNA ligands that bind by an intercalation mechanism [21]. Such a binding mode is possible at multiple sites on these RNAs consistent with efficient reaction independent of Tethering site. While this result is not desirable for the development of selective ligands, it does provide confirmation of the validity of this screening method as a means of discovering RNA-interacting ligands. In contrast to A04, compound A05 gave a high conjugate yield (31%) with pre21 RNA 1 and a lower yield (13%) with pre21 RNA 2 suggesting a selective binding site near the bulged adenosine for the benzotriazole appendage of A05 (Figures 4 and 5). Furthermore, when the reaction with A05 was carried out with an unrelated thiol-modified 16 bp duplex, a low conjugate yield was observed (<5%) while compound A04 produced conjugate with this RNA in 38% yield (Supplementary Information). Interestingly, the benzotriazole structure of A05 can be considered a base analog with the potential to stack and hydrogen bond like natural nucleobases.

Indeed, tetrabromobenzotriazole is a known inhibitor of nucleotide-dependent enzymes where the inhibitor interacts with nucleotide-binding sites [22]. A recent NMR study of the structure of this part of the pre-miR21 RNA indicates the bulged A is extrahelical and partially associated with a groove of the adjacent duplex [23]. It is possible that the benzotriazole of A05 contacts the RNA groove at this location. However, at this point, we cannot rule out the possibility that our covalent modification strategy altered the RNA structure, placing the bulged A in a different position. Nevertheless, if the binding site for the tethered benzotriazole is present in the native RNA, it can provide a useful starting point for the development of selective pre-miR21 ligands. Additional structural studies will be necessary to fully define this binding site.

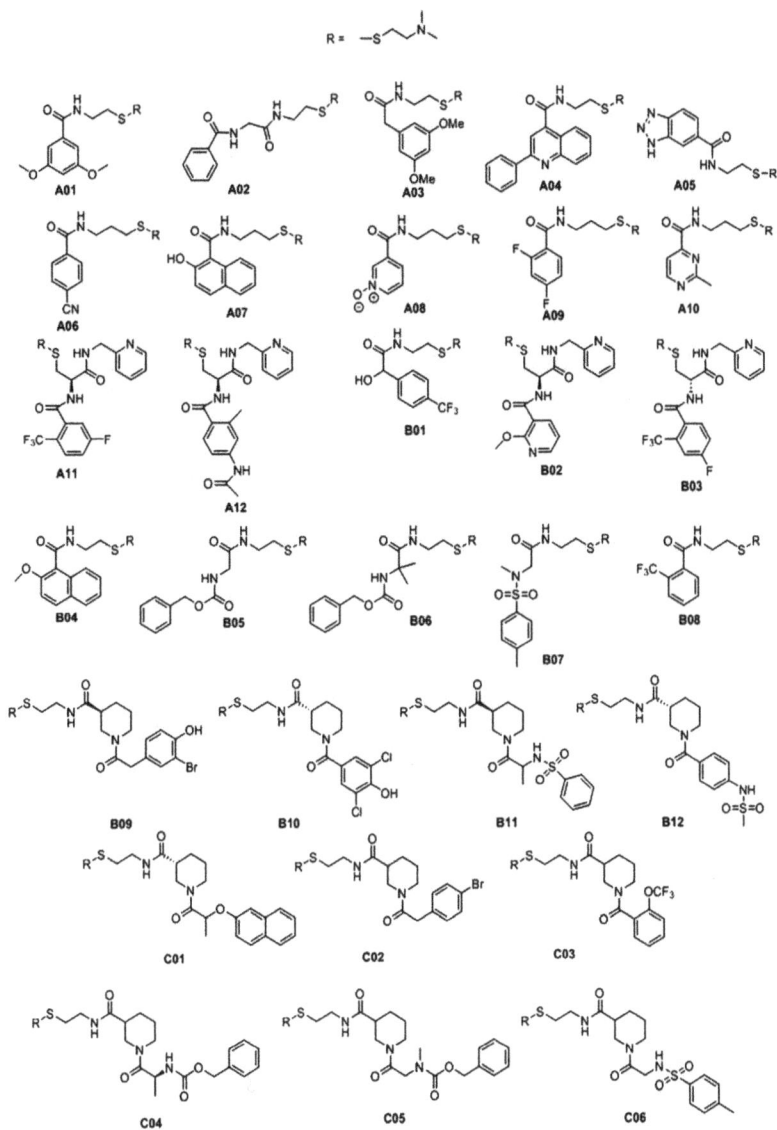

Figure 3. Structures of the thirty disulfide-containing small molecules used in RNA Tethering screen.

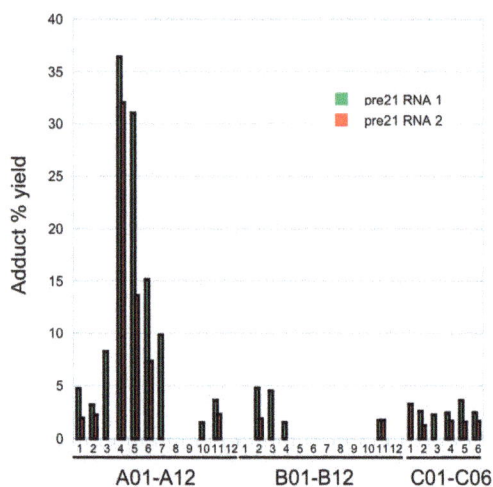

Figure 4. Results of an RNA Tethering screen with disulfide containing small molecules and pre21 RNA 1 and pre21 RNA 2. Plotted is the % yield of the target adduct *vs.* compound used (see Figure 3 for structures of disulfides and Experimental Section for reaction conditions.) No bar indicates no product detected in that reaction by ESI-MS.

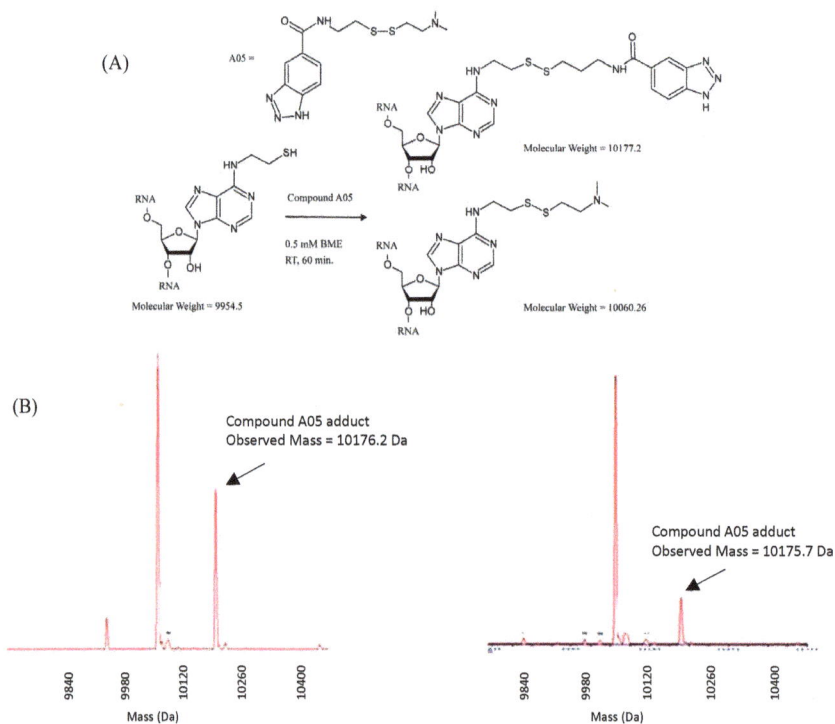

Figure 5. Selectivity was observed during the screen. (**A**) RNA Tethering reaction of pre21 RNA 1 and pre21 RNA 2 with compound A05; (**B**) Corresponding ESI-MS spectra for these reactions (pre21 RNA 1 (left) and pre21 RNA 2 (right)).

3. Experimental Section

3.1. General Procedures

All reagents and solvents were purchased from Sigma/Aldrich or Fischer Scientific and were used without additional purification, unless noted otherwise. Reactions using dry solvents were performed under an atmosphere of argon, in oven-dried glassware. Solvents were dried in a solvent purification system that passes solvent through two columns of dry neutral alumina. Column chromatography was conducted using silica gel (Sorbent Technologies, 60–200 mesh) and analytical TLC was performed on glass plates coated with 0.25 mm silica gel using UV for visualization. ^{1}H and ^{13}C magnetic resonance spectra were recorded with Varian VNMRS 600, Varian Mercury 300 spectrometers and referenced to the residual solvent peak. ^{31}P magnetic resonance spectra was recorded with a Varian Mercury 300 spectrometer. The abbreviations s, t, m, brs, dd, d stand for singlet, triplet, multiplet, broad singlet, doublet of doublets and doublet. High-resolution ESI mass spectra were obtained at the University of California, Davis Mass spectrometry facility, on a Thermo Electron LTQ-Orbitrap Hybrid Mass Spectrometer.

N^6-[2-(triphenylmethylthio)ethyl]-9-(β-D-ribofuranosyl)adenine (**1**):

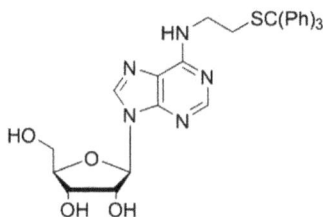

2',3',5'-Tri-O-acetyl-9-(β-D-ribofuranosyl)-6-chloropurine (100 mg, 0.218 mmol) was dissolved in 1.0 mL anhydrous DMF and treated with S-trityl-protected 2-aminoethanethiol [24] (161 mg, 2.3 equiv) in 1.0 mL DMF. After stirring for 18 h, NH$_3$/MeOH (5 mL, saturated solution) was added, and stirring continued for a further 15 h. The crude was then diluted with 10 mL CH$_2$Cl$_2$, extracted twice with saturated NaCl, and dried over Na$_2$SO$_4$. The crude mixture was then concentrated under reduced pressure, absorbed onto silica gel, and purified by column chromatography (0% → 6% MeOH in CH$_2$Cl$_2$), yielding a pale yellow solid (85 mg, 72%). ^{1}H-NMR (300 MHz, CD$_3$OD) δ 8.25 (s, 1H), 8.13 (s, 1H), 7.40-7.07 (m, 15H), 5.96 (d, J = 6.4 Hz, 1H), 5.49 (s, 1H), 4.76 (t, J = 6.0 Hz, 1H), 4.33 (d, J = 2.6 Hz, 1H), 4.18 (s, 1H), 3.90 (d, J = 12.5 Hz, 1H), 3.75 (d, J = 12.7 Hz, 1H), 3.49 (s, 1H), 2.54 (t, J = 6.5 Hz, 2H), 2.39 (dd, J = 14.5, 5.6 Hz, 4H). ^{13}C NMR (151 MHz, CDCl$_3$) δ 154.53, 152.15, 144.54, 139.92, 146.79, 129.49, 127.86, 127.20, 126.68, 91.10, 87.65, 73.63, 73.64, 72.59, 67.92, 39.22, 31.67, 29.64, 25.54. ESIHRMS: calcd for C$_{31}$H$_{31}$N$_5$O$_4$S (M+H)$^+$: 570.2169, obsd 570.2168.

N6-[2-(triphenylmethylthio)ethyl]-9-[5-O-(4,4′-dimethoxytrityl)-β-D-ribofuranosyl]adenine (**2**):

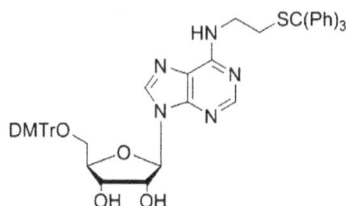

DMTrCl (229 mg, 1.1 eq) and DMAP (37 mg, 0.5 eq) were added to a solution of **1** (350 mg, 0.61 mmol) in 5 mL anhydrous pyridine. The reaction mixture was stirred at RT for 16 h.

Co-evaporation with 20 mL tolune and 20 mL CH_3CN was performed to remove pyridine and toluene, respectively. Afterwards, the mixture was diluted with CH_2Cl_2 (15 mL) and washed with saturated aqueous $NaHCO_3$ (2 × 10 mL). The organic layer was dried (Na_2SO_4), concentrated under reduced pressure and purified by column chromatography (1% Et_3N in CH_2Cl_2). Excess Et_3N in column fractions were removed via azeotrope formation with CH_3CN, yielding a white solid (467 mg, 88%). ^{1}H-NMR (300 MHz, CD_2Cl_2) δ 8.22 (s, 1H), 8.03 (s, 1H), 7.45–7.19 (m, 24H), 6.79 (m, 4H), 6.26 (d, J = 5.4 Hz, 1H), 5.99 (d, J = 5.2 Hz, 1H), 4.73 (t, J = 4.8 Hz, 1H), 4.40 (d, J = 4.5 Hz 1H), 3.75 (s, 6H), 3.47–3.29 (m, 4H), 2.62–2.51 (m, 2H), 2.31 (t, J = 6.4 Hz, 2H). 13C-NMR (75 MHz, CD_2Cl_2) δ 158.57, 152.29, 144,72, 144.59, 138.04, 135.54, 135.38, 129.91, 129.87, 129.49, 129.47, 127,90, 127.83, 127.75, 126.74, 126.66, 90.84, 86.32, 86.11, 76.02, 72.61, 66.76, 63.55, 55.14, 38.06, 31.85, 29.67, 22.92. ESIHRMS: calcd for $C_{52}H_{49}N_5O_6S$ $(M+H)^+$: 872.3476, obsd: 872.3473.

N^6-[2-(triphenylmethylthio)ethyl]-9-[2-O-(tert-butyldimethylsilyl)-5-O-(4,4'-dimethoxytrityl)-β-
D-ribofuranosyl]adenine (**3**):

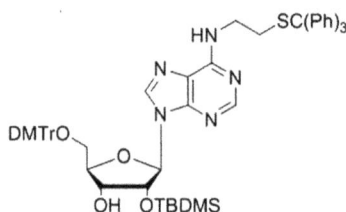

In a flame-dried 25 mL round-bottom flask, TBDMSCl (93 mg, 1.2 eq), freshly distilled triethylamine (145 uL, 2.0 eq) and DMAP (32 mg, 0.5 eq) were added to a solution 5'-O-DMTr protected derivative **2** (450 mg, 0.516 mmol, 1 eq) in 1 mL anhydrous THF. The reaction mixture was stirred at RT for 24 h. Upon completion of the reaction, the mixture was diluted with EtOAc (10 mL), filtered out white precipitate, and washed with 5% aqueous $NaHCO_3$ (2 × 10 mL). The organic layer was dried (Na_2SO_4) and concentrated under reduced pressure. The crude reaction mixture was purified by column chromatography (15% → 30% $(CH_3)_2CO$ in CH_2Cl_2) to afford a white foam (152 mg, 43%). ^{1}H-NMR (300 MHz, CDCl$_3$) δ 8.30 (s, 1H), 8.06 (s, 1H), 7.59–7.32 (m, 24H), 6.94 (d, J = 8.8 Hz, 4H), 6.09 (d, J = 5.0 Hz, 1H), 5.97 (s, 1H), 5.10 (s, 1H), 4.42 (d, J = 4.2 Hz, 1H), 4.32 (d, J = 4.2 Hz, 1H), 3.89 (s, 6H), 3.54 (m, 4H), 2.78 (d, J = 4.6 Hz, 1H), 2.68 (t, J = 6.6 Hz, 2H), 2.13 (s, 1H), 0.97 (s, 9H), 0.12 (s, 3H), 0.00 (s, 3H). 13C-NMR (75 MHz, CD_2Cl_2) δ 158.59, 144.80, 144.72, 138.73, 135.65, 135.62, 130.01, 130.00, 129.47, 128.01, 127.81, 127.77, 126.75, 126.61, 113.02, 88.37, 86.40, 83.81, 75.26, 71.35, 63.40, 55.12, 29.62, 25.29, 17.73. ESIHRMS: calcd for $C_{58}H_{63}N_5O_6SSi$ $(M + H)^+$: 986.4341, obsd: 986.4370.

N^6-[2-(triphenylmethylthio)ethyl]-9-[2-O-(tert-butyldimethylsilyl)-5-O-(4,4'-dimethoxytrityl)-β-
D-ribofuranosyl]adenine 3'-(2-Cyanoethyl)-N,N-diisopropylphosphoramidite (**4**):

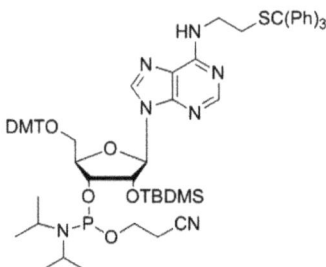

Dry diisopropylethylamine (70 uL, 6 eq) and 2-cyanoethyl-(N,N-diisopropylamino)chlorophosphite (29 uL, 2 eq) were added to a solution of **3** (74 mg, 0.075 mmol, 1 eq) in 1.0 mL dry DCM.

The reaction mixture was stirred for 1 h. Upon completion of the reaction, the mixture was diluted with EtOAc (5 mL) and washed with 5% (w/v) aqueous $NaHCO_3$ (2 × 5 mL). The organic layer was dried (Na_2SO_4) and concentrated under reduced pressure. Purification by column chromatography (30%–40% EtOAc in Hexane) yielded a white foam (62 mg, 82%). ^{31}P-NMR (121 MHz, CD_2Cl_2) δ 151.68, 150.24. ESIHRMS: calcd for $C_{67}H_{80}N_7O_7PSSi$ $(M+H)^+$: 1186.5420, obsd: 1186.5474.

3.2. RNA Synthesis, Purification and S-Tr Deprotection

RNA oligonucleotides were synthesized on an ABI 394 synthesizer (DNA/Peptide Core Facility, University of Utah) at 200 nmol and 1 umol scale using 5'-DMTr, 2'-OTBDMS protected β-cyanoethyl phosphoramidites. Conditions for synthesis, cleavage and standard deprotection are identical to those previously described [25]. Crude RNA synthesis products were purified by urea-polyacrylamide gel electrophoresis (19%), and desalted with Sep-Pak cartridges, as previously described. The S-Trityl RNA was stored as a lyophilized pellet at −70 °C. Identification and purity was determined by ESI-MS.

Prior to S-trityl deprotection [19], the RNA was allowed to fold into its native secondary structure by heating at 95 °C for 30 min. then slow cooling in the heat block until the sample has reached room temperature. To remove the S-Trityl protecting group, purified RNA (1.2 nmol) was dissolved in aqueous TEAA buffer (triethylammonium acetate, 15 uL, 0.1 M, pH 6.5) and treated with $AgNO_3$ (2 μL, 1 M in H_2O). After 30 min of gentle agitation by vortex at room temperature, DTT was added (2 μL, 1 M H_2O), and the reaction was allowed to proceed for a further 10 min. The mixture was diluted with aqueous 0.1 M TEAA (pH 6.5) and centrifuged (16,000× g, 4 °C, 5 min) to separate the precipitate. The supernatant was collected. The pellet was washed 3 times with 50 uL aqueous 0.1 M TEAA, to minimize the loss of the RNA. Each wash was followed by vigorous vortex agitation. The combined supernatants (~170 μL) were loaded in to a 3000 MWCO centrifugal concentrator (Microcon-3, Millipore) and centrifuged (11,000× g, 4 °C, approx. 30 min). The concentrated crude sample was then washed once with 150 μL of 0.1 M TEAA and centrifuged (11,000× g, 4 °C, approx. 30 min). A final wash was performed with 100 μL of nuclease-free water and centrifuged (11,000× g, 4 °C, approx. 30 min). After the buffer exchange into water, the deprotected RNA was quantified by the absorbance of the solution at 260 nm. Yields ranged from 340 pmol (28%) to 940 pmol (76%). ESI-MS confirmed quantitative conversion to the free thiol-RNA, and the purity of the product. The deprotected RNA was stored in aliquots of water at −20 °C if not used immediately for experimentation.

STr-protected pre21 RNA 1: 5'-GGU GUU GCX UGU UGA AUC UCA UGG CAA CAC C-3'

X = A_{SH}

Calcd mass (monoisotopic): 10,195.76

Obsd mass (ESI-MS): 10,193.56

Free thiol pre21 RNA 1: 5'- GGU GUU GCX UGU UGA AUC UCA UGG CAA CAC C -3'

X = A_{SH}

Calc mass (monoisotopic): 9954.5

Obsd mass (ESI-MS): 9955.5

STr-protected pre21 RNA 2: 5'-GGU GUU GCA UGU UGA AUC UCA UGG CXA CAC C-3'

X = A_{SH}

Calcd mass (monoisotopic): 10,195.8

Obsd mass (ESI-MS): 10,194.3

Free thiol pre21 RNA 2: 5'- GGU GUU GCX UGU UGA AUC UCA UGG CXA CAC C-3'

X = A_{SH}

Calc mass (monoisotopic): 9954.5

Obsd mass (ESI-MS): 9955.1

3.3. Mass Spectrometry Analysis of RNAs

Mass spectra were obtained on a Thermo Electron LTQ-Orbitrap Hybrid Mass Spectrometer. Free-thiol RNAs were desalted and purified prior to conjugate formation. Conjugate reaction samples were analyzed crude. The samples were run with a mobile phase consisting of 50% MeOH/H$_2$O w/0.2% triethylamine. Mass spectra were recorded in negative ionization mode. A 9 kDa DNA standard was used to tune the instrument before each run. Spectra were deconvoluted using ProMass and MassLynx software.

3.4. Conjugation Reaction Procedure

Free-thiol RNA (~200 pmol) was dissolved in 8 µL of 0.1 M TEAA (triethylammonium acetate, pH 7.2). Freshly prepared β-mercaptoethanol (1 µL, 0.5 M) was then added to the solution and pipet-mixed. The disulfide compound (1 µL, 1.25 mM in H$_2$O) was added to the reaction mixture immediately. The reaction was allowed to proceed for 1 h at room temperature under gentle agitation. Samples were then analyzed with ESI-MS immediately or stored at −20 °C.

3.5. Calculation of Reaction Product Abundances

Abundances were calculated using the area of the peaks in the deconvoluted spectrum of each sample reaction. All peaks were accounted for in each spectrum. The noteworthy peaks in each spectrum are as follows: (1) free thiol; (2) N,N-dimethylethylamino adduct (N,N-DMA); (3) product. Given their structural similarities, these species were assumed to have equivalent ionization efficiencies. However, due to possible variations in ionization behavior, quantification of conjugates in this manner may not always reflect exact yields. At an RNA concentration of 20 µM, no RNA dimers were observed in the spectra.

% abundance = (area of species peak/sum of the areas of all peaks in spectrum) × 100

4. Conclusions

We have demonstrated that oligonucleotides bearing the A$_{SH}$ nucleoside analog can effectively conjugate with disulfide-containing small molecules under varying reducing conditions. Among the library of 30 disulfide-containing compounds screened, a small molecule bearing a benzotriazole moiety displayed not only prominent yield of the adduct, but also selectivity for pre21 RNA 1 where the A$_{SH}$ analog replaces a bulged adenosine near the Dicer cleavage site. The result suggests a mode of binding to the RNA outside of the disulfide linkage and may provide a starting point for the development of pre-miRNA21 ligands.

Supplementary Materials: Supplementary materials can be accessed at: http://www.mdpi.com/1420-3049/20/03/4148/s1.

Acknowledgments: This work was supported by the University of California Cancer Research Coordinating Committee and a grant from the National Institutes of Health to PAB (R01GM061115). We also thank William Jewell, Core Mass Spectrometry Facility, UC Davis for technical assistance and Adam Renslo, Small Molecule Discovery Center, Department of Pharmaceutical Chemistry, UCSF for providing screening compounds.

Author Contributions: Peter A. Beal designed the research, analyzed the data, and wrote the paper; Kiet Tran performed the research, analyzed the data, and wrote the paper. Michelle Arkin assisted in the design of the research and editing of the paper. All authors read and approved the final transcript.

Conflicts of Interest: The authors declare no conflict of interest.

References

1. Gallego, J.; Varani, G. Targeting RNA with Small-Molecule Drugs: Therapeutic Promise and Chemical Challenges. *Acc. Chem. Res.* **2001**, *34*, 836–843. [CrossRef] [PubMed]

2. Guan, L.; Disney, M.D. Recent Advances in Developing Small Molecules Targeting RNA. *ACS Chem. Biol.* **2012**, *7*, 73–86. [CrossRef] [PubMed]
3. Griffey, R.H.; Sannes-Lowery, K.A.; Drades, J.J.; Venkatraman, M.; Swayze, E.E.; Hofstadler, S.A. Characterization of Low-Affinity Complexes between RNA and Small Molecule Using Electrospray Ionization Mass Spectrometry. *J. Am. Chem. Soc.* **2000**, *122*, 9933–9938. [CrossRef]
4. Jimenez-Moreno, E.; Gomez-Pinto, I.; Corzana, F.; Santan, A.G.; Revuelta, J.; Bastida, A.; Jimenez-Barbero, J.; Gonzalez, C.; Asensio, J.L. Chemical Interrogation of Drug/RNA Complexes: From Chemical Reactivity to Drug Design. *Angew. Chem. Int. Ed.* **2013**, *52*, 3148–3151. [CrossRef]
5. Hardy, J.A.; Lam, J.; Nguyen, J.; O'Brien, T.; Wells, J.A. Discovery of an allosteric site in the caspases. *Proc. Natl. Acad. Sci. USA* **2004**, *101*, 12461–12466. [CrossRef] [PubMed]
6. Yang, W.; Fucini, R.V.; Fahr, B.T.; Randal, M.; Lind, K.E.; Lam, M.B.; Lu, W.; Lu, Y.; Cary, D.R.; Romanowski, M.J.; *et al.* Fragment-based Discovery of Nonpeptidic BACE-1 Inhibitors Using Tethering. *Biochemistry* **2009**, *48*, 4488–4496. [CrossRef] [PubMed]
7. Erlanson, D.A.; Braisted, A.C.; Raphael, M.; Stroud, R.M.; Gordon, E.M.; Wells, J.A. Site-directed ligand discovery. *Proc. Natl. Acad. Sci. USA* **2000**, *97*, 9367–9372. [CrossRef] [PubMed]
8. Bartel, D.P. MicroRNAs: Target Recognition and Regulatory Functions. *Cell* **2009**, *136*, 215–233. [CrossRef] [PubMed]
9. Volinia, S.; Calin, G.; Liu, C.-G.; Ambs, S.; Cimmino, A.; Petrocca, F.; Visone, R.; Iorio, M.; Ferracin, M.; Prueit, R.L. A microRNA expression signature of human solid tumors defines cancer gene targets. *Proc. Natl. Acad. Sci. USA* **2006**, *103*, 2257–2261. [CrossRef] [PubMed]
10. Medina, P.P.; Nolde, M.; Slack, F.J. OncomiR addiction in an *in vivo* model of microRNA-21-induced pre-B-cell lymphoma. *Nature* **2010**, *467*, 86–91. [CrossRef] [PubMed]
11. Bartel, D.P. MicroRNAs: Genomics, Biogenesis, Mechanism, and Function. *Cell* **2004**, *116*, 281–297. [CrossRef] [PubMed]
12. Chirayil, S.; Chirayil, R.; Luebke, K.J. Discovering ligands for a microRNA precursor with peptoid microarrays. *Nucl. Acids Res.* **2009**, *37*, 5486–5497. [CrossRef] [PubMed]
13. Velagapudi, S.P.; Gallo, S.M.; Disney, M. Sequence-based Design of Bioactive Small Molecules that Target Precursor MicroRNAs. *Nat. Chem. Biol.* **2014**, *10*, 291–299. [CrossRef] [PubMed]
14. Bose, D.; Gopal, J.; Suryawanshi, H.; Agarwala, P.; Pore, S.K.; Banerjee, R.; Maiti, S. The Tuberculosis Drug Streptomycin as a Potential Cancer Therapeutic: Inhibition of miR-21 Function by Directly Targetting Its Precursor. *Angew. Chem. Int. Ed.* **2012**, *51*, 1019–1023. [CrossRef]
15. Maiti, M.; Nauwelaerts, K.; Herdewijn, P. Pre-mircoRNA binding aminoglycosides and antitumor drugs as inhibitors of Dicer catalyzed microRNA processing. *Bioorg. Med. Chem. Lett.* **2012**, *22*, 1709–1711. [CrossRef] [PubMed]
16. Allerson, C.; Swaine, L.; Verdine, G.L. A Chemical Method for Site-Specific Modification of RNA: The Convertible Nucleoside Approach. *J. Am. Chem. Soc.* **1997**, *119*, 7423–7433. [CrossRef]
17. Porcher, S.; Myeyyappan, M.; Pitsch, S. Spontaneous Aminoacylation of a RNA Sequence Containing a 3'-Terminal 2'-Thioadenosine. *Helvetica Chimica Acta* **2005**, *88*, 2897–2910. [CrossRef]
18. Geiermann, A.; Polacek, N.; Micura, R. Native Chemical Ligation of Hydrolysis-Resistant 3'-Peptidyl-RNA Mimics. *J. Am. Chem. Soc.* **2011**, *133*, 19068–19071. [CrossRef] [PubMed]
19. Peacock, H.; Bachu, R.; Beal, P.A. Covalent stabilization of a small molecule-RNA complex. *Bioorg. Med. Chem. Lett.* **2011**, *21*, 5002–5005. [CrossRef] [PubMed]
20. Burlingame, M.A.; Tom, C.T.; Renslo, A.R. Simple One-Pot Synthesis of Disulfide Fragments for Use in Disulfide-Exchange Screening. *ACS Comb. Sci.* **2011**, *13*, 205–208. [CrossRef] [PubMed]
21. Zhao, M.; Janda, L.; Nguyen, J.; Strekowski, L.; Wilson, W.D. The Interaction of Substituted 2-phenylquinoline Intercalators with Poly(A) • poly(U): Classical and Threading Intercalation Modes of RNA. *Biopolymers* **2004**, *34*, 61–73. [CrossRef]
22. Battistutta, R.; de Moliner, E.; Sarno, S.; Zanotti, G.; Pinna, L.A. Structural features underlying selective inhibition of protein kinase CK2 by ATP site-directed tetrabromo-2-benzotriazole. *Protein Sci.* **2001**, *10*, 2200–2206. [CrossRef] [PubMed]
23. Chirayil, S.; Wu, Q.; Amezcua, C.; Leubke, K.J. NMR Characterization of an Oligonucleotide Model of the MiR-21 Pre-Element. *PLoS One* **2014**, *9*, e108231. [CrossRef] [PubMed]

24. Li, M.; Yamato, K.; Ferguson, J.S.; Gong, B. Sequence-Specific Associaion in Aqueous Media by Integrating Hydrogen Bonding and Dynamic Covalent Interactions. *J. Am. Chem. Soc.* **2006**, *128*, 12628–12629. [CrossRef] [PubMed]
25. Hamm, M.L.; Piccirilli, J.A. Incorporation of 2'-deoxy-2'-mercaptocytidine into Oligonucleotides via Phosphoramidite Chemistry. *J. Org. Chem.* **1997**, *62*, 3415–3420. [CrossRef] [PubMed]

Sample Availability: *Sample Availability*: Samples are not available from authors.

![molecules logo]

Article

Construction of an Isonucleoside on a 2,6-Dioxobicyclo[3.2.0]-heptane Skeleton

Yuichi Yoshimura [1,*], Satoshi Kobayashi [1], Hitomi Kaneko [1], Takeshi Suzuki [1] and Tomozumi Imamichi [2]

[1] Faculty of Pharmaceutical Sciences, Tohoku Pharmaceutical University, 4-4-1 Komatsushima, Aoba-ku, Sendai 981-8558, Japan; s-kobayashi@fujichemical.co.jp (S.K.); hitomi.f528@gmail.com (H.K.); suzutake19851109@coral.plala.or.jp (T.S.)

[2] Laboratory of Human Retrovirology, Leidos Biochemical Research Inc., Frederick National Laboratory for Cancer Research, Frederick, MD 21702, USA; timamichi@mail.nih.gov

* Correspondence: yoshimura@tohoku-pharm.ac.jp; Tel./Fax: +81-22-727-0144

Academic Editors: Mahesh K. Lakshman and Fumi Nagatsugi
Received: 2 February 2015; Accepted: 4 March 2015; Published: 12 March 2015

Abstract: We have built a new isonucleoside derivative on a 2,6-dioxobicyclo[3.2.0]heptane skeleton as a potential anti-HIV agent. To synthesize the target compound, an acetal-protected dihydroxyacetone was first converted to a 2,3-epoxy-tetrahydrofuran derivative. Introduction of an azide group, followed by the formation of an oxetane ring, gave a pseudosugar derivative with a 2,6-dioxobicyclo[3.2.0]heptane skeleton. The desired isonucleoside was obtained by constructing a purine base moiety on the scaffold, followed by amination.

Keywords: nucleoside; bicyclo; oxetane ring; conformation

1. Introduction

Since the discovery of 3'-azidothymidine (AZT), much attention has been paid to the development of effective chemotherapeutic agents against the human immunodeficiency virus (HIV), a causative agent for AIDS [1,2]. More than 20 anti-HIV drugs have now been approved and are clinically used for the treatment of AIDS. Among them, nucleoside reverse transcriptase inhibitors (NRTIs) play a critical role in the treatment of AIDS patients. In the most successful regimen for AIDS referred to as ART (Anti-Retroviral Therapy), a cocktail of anti-HIV drugs, including NRTIs, non-nucleoside reverse transcriptase inhibitors (NNRTIs), and protease inhibitors (PIs) [3], is used. Although ART greatly contributes to increasing the lifespan of patients, drug-resistant strains of the virus are still a serious problem [4,5]. Therefore, new drugs that are effective against the resistant virus strains are constantly needed.

Most NRTIs belong to a category of dideoxynucleosides, e.g., zalcitabine (ddC) [6] and didanosine (ddI) [6]. AZT [7] and lamivudine [8] are 3'-substituted dideoxynucleoside derivatives, and abacavir is a carbocyclic analogue of dideoxynucleoside. Only tenofovir [9], which is a nucleoside phosphonate (Figures 1 and 2), is different. From the viewpoint of designing new anti-HIV agents, nucleosides constructed on a novel scaffold are expected to have antiviral activity against the resistant virus strains and may avoid cross-resistance to the known NRTIs.

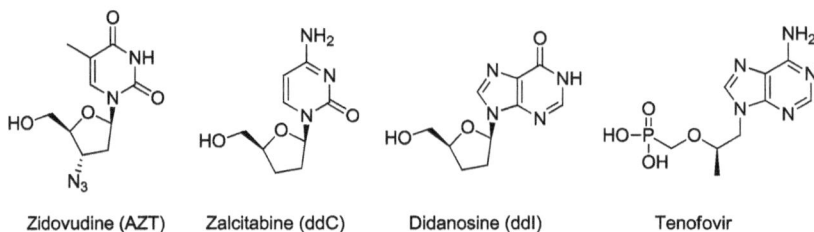

Figure 1. Approved NRTIs.

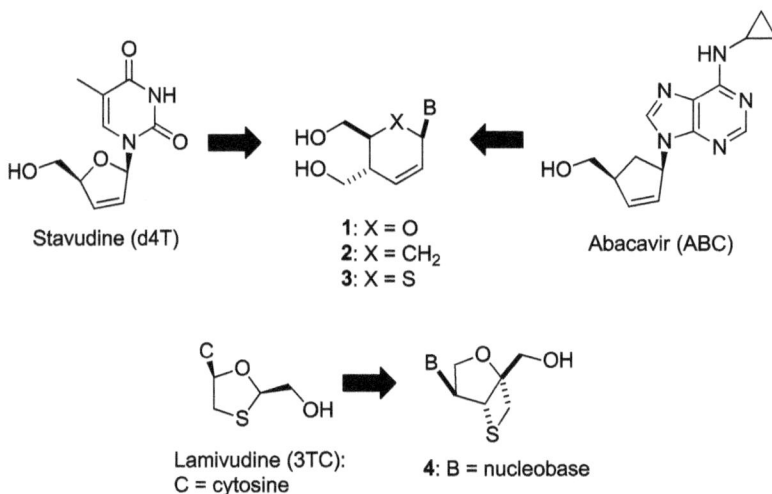

Figure 2. Our previous works searching for anti-HIV nucleosides built on a new scaffold.

Thus, we have been focusing on the design and synthesis of nucleoside derivatives attached to a pseudosugar scaffold [10–17]. Among them, nucleosides with cyclohexenyl [13], dihydrothiophenyl [15], and dihydropyranyl [17] groups in place of a furanose ring have been synthesized as "ring-expanded" analogues of stavudine and abacavir. Dihydropyranyl derivative **1** did not show any activity, whereas cyclohexenyl derivative **2** showed weak anti-HIV activity [13,17]. On the other hand, dihydrothiophenyl derivative **3** showed significant anti-HIV activity [15]. In addition, we have applied the "ring-expanding" concept to lamivudine and synthesized isonucleosides **4** constructed on 2-oxa-6-thiabicyclo[3.2.0]heptane [14]. The isonucleoside **4** was also considered as conformationally-restricted analogue of lamivudine by introducing a fused thietane ring (*vide infra*). However, **4** showed no anti-HIV activity (Figure 2). In this study, we planned to build isonucleoside **6** on a 2,6-dioxobicyclo[3.2.0]heptane skeleton, an analogue of dioxolane nucleoside **5** which exhibited potent anti-HIV activity [18–20]. The similar conformationally-restricted analogue of d4T was known: cyclopropane-fused carbocyclic d4T (*N*-MCd4T), fixed in north conformation, was originally reported by Marquez and his colleagues and had significant anti-HIV activity with lesser cytotoxicity [21]. In addition, D-enantiomer of **5** was known to have potent cytotoxicity [18–20]. Thus, isonucleoside **6** should be promising although the thietane-fused derivative **4** was inactive (Figure 3).

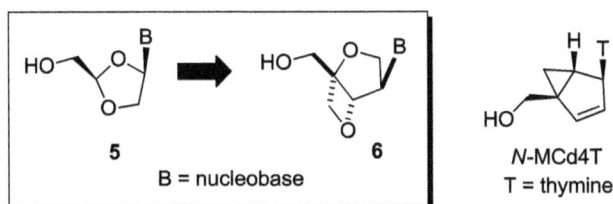

Figure 3. Design of nucleoside derivative built on a 2,6-bicyclo[3.2.0]heptane skeleton.

2. Results and Discussion

Following our previous reports [11,14], epoxide **7** was synthesized. We first attempted to introduce an adenine onto **7** by treating it with DBU [22]. However, the reaction did not give the desired product **9** (Scheme 1).

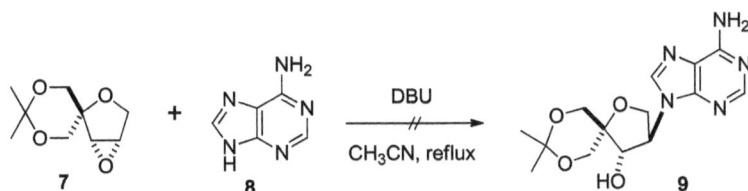

Scheme 1. Attempt to introduce adenine moiety.

In addition, Lewis acid-catalyzed reactions did not afford **9** either (data not shown). Since the low reactivity of **7** might be due to its rigid structure, we next tried nucleophilic substitution using a more reactive cyclic sulfate derivative [23]. *Cis*-allyl alcohol **10**, a precursor of epoxide **7** [11,14], was cyclized under Mitsunobu conditions, as in the case of epoxide **7** [11,14], to give dihydrofuran **11** in 71% yield. Treatment of dihydrofuran **11** with potassium osmate in the presence of N-methylmorpholine N-oxide afforded *cis*-diol **12**. The desired cyclic sulfate **13** was obtained by treatment of **12** with thionyl chloride, followed by oxidation. However, the nucleophilic substitution of **13** with adenine did not afford the desired isonucleoside **14** (Scheme 2).

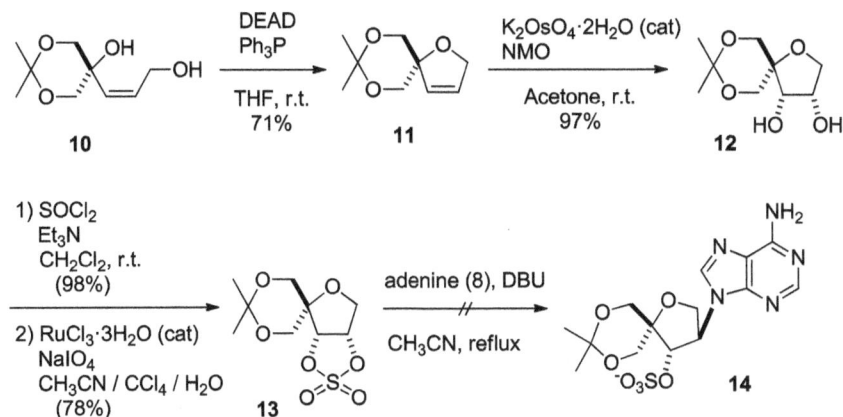

Scheme 2. Second attempt to introduce adenine using a cyclic sulfate.

Therefore, we revised our plan to synthesize an isoadenosine constructed on a 2,6-dioxobicyclo[3.2.0]heptane scaffold, and the revised scheme is shown in Scheme 3 in a retrosynthetic manner. Instead of the direct introduction of adenine, we decided to build the adenine ring on the 2,6-dioxobicyclo[3.2.0]heptane pseudosugar skeleton in a stepwise manner. According to this plan, compound **16** was thought to be a suitable intermediate for preparing **15** since it can be transformed to **6** by the formation of an imidazole ring, followed by amination. Fused oxetane derivative **16** can be obtained from dimesylate **17**. Finally, epoxide **7**, described above, was selected as the starting compound because it can be converted to **17** by the selective cleavage of the oxirane ring with an azide anion (Scheme 2).

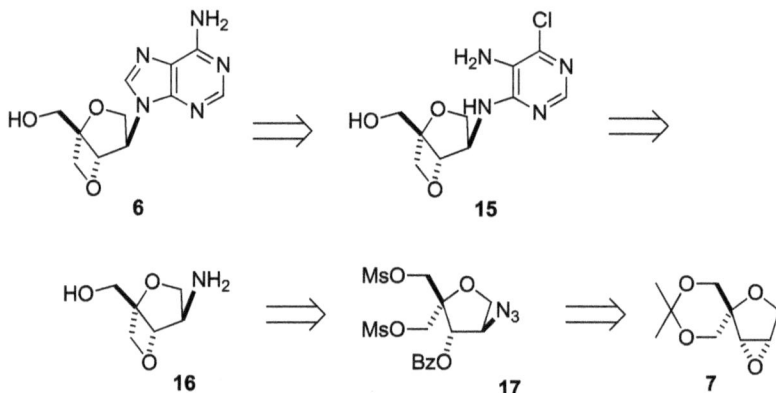

Scheme 3. Revised retrosynthesis of isoadenosine **6** built on a 2,6-dioxobicyclo[3.2.0]heptane skeleton.

First, regioselective cleavage of the oxirane ring of **7** with sodium azide in 2-methoxyethanol under reflux conditions gave the desired azide-alcohol **18** as a single regioisomer in 66% yield. It is obvious that the nucleophilic azide anion attacked from the less hindered side since similar regioselective epoxide opening was observed in our previous report [11,14]. After benzoylation of the hydroxyl group, the acetal group of **19** was removed by using acidic hydrolysis, and the resulting diol was mesylated to give dimesylate **20** in good yield. Deprotection of the benzoyl group and the subsequent formation of an oxetane ring were achieved by treating **20** with sodium methoxide under reflux conditions to give mesylate **21** in 72% yield. The structure of **21** was unambiguously determined by comparison of 1D NMR spectrum with that of 2-oxa-6-thiabicyclo[3.2.0]heptane skeleton [14] after converting it to benzoate **22** by treatment with benzoic acid in the presence of cesium fluoride. In ^{1}H-NMR spectra of **22**, the peaks corresponding to the methyl groups of the dimesylate were absent, and only the peaks corresponding to the benzoyl group in the range of 8.1–7.4 ppm were present. In addition, one of the methylene protons at the 2-position was observed as a doublet at 4.42 ppm, meaning that the coupling with H-3 disappeared. This indicates that the conformation around the tetrahydrofuran ring changes and becomes fixed, which causes a loss of coupling between one pair of H-2 and H-3 protons. A similar correlation between conformation and couplings in ^{1}H-NMR spectra has been reported for the 2-oxa-6-thiabicyclo[3.2.0]heptane skeleton [14]. Moreover, in the mass spectrum of the compound, a molecular ion peak was observed at m/z = 276, further supporting the assignment of the structure. Finally, **22** was deprotected to afford azido-alcohol **23** in 88% yield (Scheme 4).

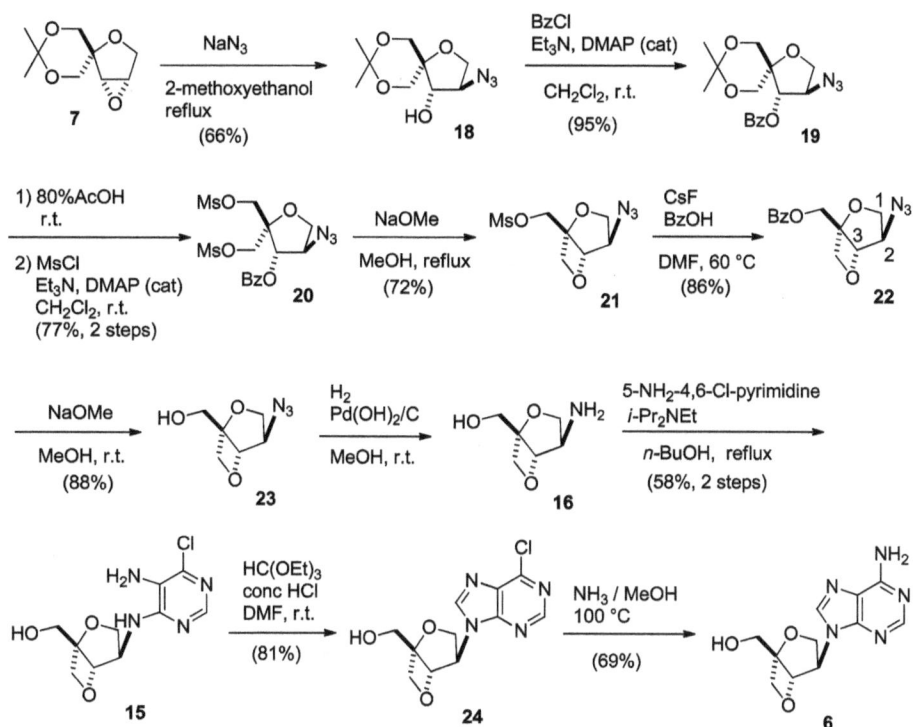

Scheme 4. Synthesis of isoadenosine 6.

Azido-alcohol **23** was reduced by catalytic hydrogenation to give key intermediate **16**, which was treated with 5-amino-4,6-dichloropyrimidine and diisopropylethylamine in refluxing *n*-butanol [23] to give diaminopyrimidine derivative **15** in 58% yield from **23**. Formation of the imidazole ring of **15** was accomplished by treatment with orthoethyl formate under acidic conditions [24] to give 6-chloropurine nucleoside **24**. Finally, the isoadenosine was built on the 2,6-dioxobicyclo[3.2.0]heptane scaffold **6** by heating **24** with methanolic ammonia in a sealed tube in 69% yield (Scheme 4). Isoadenosine **6** did not show any significant activity against HIV even at a concentration of 100 μM.

3. Experimental Section

General Information

Melting points are uncorrected. NMR spectra were recorded at 400 MHz (^{1}H), 100 MHz (^{13}C) using CDCl$_3$ as a solvent. As an internal standard, tetramethylsilane was used for CDCl$_3$. Mass spectra were obtained by EI or FAB mode. Silica gel for chromatography was Silica Gel 60N (spherical, neutral, 100–210 μm, Kanto Chemical Co. Inc., Tokyo, Japan). When the reagents sensitive to moisture were used, the reaction was performed under argon atmosphere.

8,8-Dimethyl-1,7,9-trioxaspiro[4,5]dec-3-ene (**11**). To a solution of PPh$_3$ (2.49 g, 9.48 mmol) in THF (10 mL) was added DEAD (4.31 mL, 9.48 mmol) and the mixture was stirred at room temperature for 5 min. To this mixture, a solution of **10** [11,14] (1.04 g, 5.58 mmol) in THF (10 mL) was added. The mixture was stirred at room temperature for 1 h. After the solvent was removed under reduced pressure, the residue was purified by silica gel column chromatography (hexane–ethyl acetate = 19:1) to give **11** (677 mg, 71%) as a white crystal. mp 47–49 °C; ^{1}H-NMR (CDCl$_3$) δ 1.45 (3H, s), 1.48 (3H, s), 3.76 (2H, d,

J = 11.6 Hz), 3.81 (2H, d, *J* = 11.6 Hz), 4.71 (2H, t, *J* = 1.9 Hz), 5.84-5.87 (1H, m), 6.02 (1H, d, *J* = 6.3 Hz); ^{13}C-NMR (CDCl$_3$) δ 22.6, 24.6, 66.7, 74.9, 98.0, 128.1, 129.1; IR (neat) 2924.2, 2853.1, 1724.1, 1215.9, 758.3 cm^{-1}; FAB-MS (*m/z*) 155 [M−15]$^+$; Anal. Calcd for C$_9$H$_{14}$O$_3$; C, 62.84; H, 8.32. Found; C, 62.78; H, 8.44.

(3S*,4S*)-8,8-Dimethyl-1,7,9,-trioxaspiro[4.5]decane-3,4-diol (**12**). To a solution of **11** (73 mg, 0.43 mmol) and NMO (0.22 mL) in acetone (4 mL), was added a solution of K$_2$OsO$_4$·2H$_2$O (1 mg, 0.004 mmol) in H$_2$O (0.4mL) at 0 °C. After stirred at room temperature for 60 h, Na$_2$S$_2$O$_3$·5H$_2$O (125 mg) was added and the mixture was stirred at room temperature for 30 min. After the whole mixture was dried over Na$_2$SO$_4$, the solid materials were removed by suction and washed with ethyl acetate. The combined filtrate was concentrated under reduced pressure. The residue was purified by silica gel column chromatography (CHCl$_3$–MeOH = 19:1) to give **12** (84 mg, 97%). ^{1}H-NMR (CDCl$_3$) δ 1.43 (3H, s), 1.49 (3H, s), 3.59 (1H, dd, *J* = 11.6, 1.9 Hz), 3.80 (1H, d, *J* = 9.7 Hz), 3.82–3.87 (2H, m), 3.94 (1H, dd, *J* = 9.7, 4.9 Hz), 4.14 (1H, dd, *J* = 11.6, 1.9 Hz), 4.22 (1H, d, *J* = 5.3 Hz) 4.34 (1H, q, *J* = 5.0 Hz); ^{13}C-NMR (CDCl$_3$) δ 21.0, 25.8, 63.3, 66.5, 71.0, 71.4, 74.9, 76.7, 98.4; IR (KBr) 3306.4, 2953.6, 2741.6, 1452.2, 524.19 cm^{-1}; EI-MS (*m/z*): 204 [M+1]$^+$; HRMS Calcd for C$_9$H$_{15}$N$_3$O$_4$: 204.0998, Found: 204.0992.

(3S*,4S*)-8,8-Dimethyl-1,7,9,-trioxaspiro[4.5]decane-3,4-cyclicsulfate (**13**). To a solution of **12** (410 mg, 2.01 mmol) and Et$_3$N (672 μL, 4.82 mmol) in CH$_2$Cl$_2$ (10 mL), was added dropwise a solution of SOCl$_2$ (113 μL, 1.55 mmol) in CH$_2$Cl$_2$ (10 mL) at 0 °C. After stirred at room temperature for 15 min, the mixture was washed with water. The water layer was extracted with CHCl$_3$ twice and the combined organic layer was washed with brine, then dried over Na$_2$SO$_4$. After filtration, the residue was passed through a short silica gel column (eluate: hexane–ethyl acetate = 1:1). After the solvents were removed under reduced pressure, the residue was dissolved in CCl$_4$–CH$_3$CN–H$_2$O (2:2:3, 3 mL). To this solution, were added RuCl$_3$·3H$_2$O (2.7 mg) and NaIO$_4$ (73 mg, 0.34 mmol) at 0 °C. The mixture was stirred at the same temperature for 1.5 h. After diluted with ether, the mixture was washed with water, sat.NaHCO$_3$ and brine, then dried over Na$_2$SO$_4$. After filtration, the solvents were removed under reduce pressure, the residue was purified by silica gel column chromatography (hexane-ethyl acetate = 6:1) to give **13** (82 mg, 77%). ^{1}H-NMR (CDCl$_3$) δ1.43 (3H, s), 1.50 (3H, s), 3.56 (1H, dd, *J* = 12.1, 2.4 Hz), 3.82 (1H, d, *J* = 12.1 Hz), 3.91 (1H, d, *J* = 12.1 Hz), 3.97 (1H, dd, *J* = 12.6, 4.4 Hz), 4.06 (1H, dd, *J* = 12.1, 2.4 Hz), 4.28 (1H, d, *J* = 12.6 Hz), 5.42 (1H, d, *J* = 6.3 Hz), 5.48 (1H, t, *J* = 5.1 Hz); ^{13}C-NMR (CDCl$_3$) δ 19.6, 27.1, 61.0, 62.4, 69.7, 79.0, 83.3, 84.5, 99.2; IR (KBr) 3000.8, 2892.3, 1699.7, 1380.9, 1089.9 cm^{-1}; EI-MS (*m/z*): 267 [M+1]$^+$; HRMS Calcd for C$_9$H$_{15}$N$_3$O$_4$: 266.0460, Found: 266.0467.

(3R*,4S*)-3-Azido-8,8-dimethyl-1,7,9,-trioxaspiro[4.5]decan-4-ol (**18**). A mixture of **7** [11,14] (433 mg, 2.33 mmol) and NaN$_3$ (752 mg 11.6 mmol) in 2-methoxyethanol (26 mL) was kept at 100 °C for 5 h. After the solvent was removed under reduced pressure, the residue was dissolved in ethyl acetate. After washed with water and brine, the organic layer was dried over Na$_2$SO$_4$. After filtration, the solvents were removed under reduce pressure, the residue was purified by silica gel column chromatography (hexane–ethyl acetate = 5:1) to give **18** (351 mg, 66%). ^{1}H-NMR (CDCl$_3$) δ 1.40 (3H, s), 1.49 (3H, s), 3.69-3.80 (3H, m), 3.84 (1H, d, *J* = 11.6 Hz), 4.01–4.08 (3H, m), 4.38 (1H, t, *J* = 1.9, 2.4 Hz); ^{13}C-NMR (CDCl$_3$) δ 19.5, 27.4, 62.3, 66.0, 67.6, 69.4, 78.2, 79.1, 98.5; IR (neat) 3419.0, 2104.8, 1086.6, 831.7 cm^{-1}; EI-MS (*m/z*): 229 [M+1]$^+$; HRMS Calcd for C$_9$H$_{15}$N$_3$O$_4$: 229.1063, Found: 229.1062.

(3R*,4S*)-3-Azido-8,8-dimethyl-1,7,9,-trioxaspiro[4.5]decan-4-yl benzoate (**19**). To a solution of **18** (432 mg, 1.88 mmol), Et$_3$N (0.59 mL, 4.24 mmol), and DMAP (23 mg, 0.19 mmol) in CH$_2$Cl$_2$ (15 mL) was added benzoyl chloride (0.40 mL, 3.39 mmol) and the mixture was stirred at room temperature for 6.5 h. The reaction was quenched by addition of MeOH, and the whole was stirred at room temperature for 10 min. The mixture was diluted with CH$_2$Cl$_2$ and washed with water and brine, then dried over Na$_2$SO$_4$. After filtration, the solvents were removed under reduce pressure, the residue was purified by silica gel column chromatography (hexane–ethyl acetate = 4:1) to give **19** (598 mg, 95%). ^{1}H-NMR (CDCl$_3$) δ 1.37 (3H, s), 1.47 (3H, s), 3.84-3.91 (3H, m), 3.99 (1H, dd, *J* = 1.4, 10.6 Hz), 4.07 (1H, dd, *J* = 1.4, 10.6 Hz), 4.18-4.25 (2H, m) 5.47 (1H, d, *J* = 1.0 Hz), 7.48 (2H, t, *J* = 7.2 Hz), 7.62 (1H, *J* = 7.2 Hz), 8.03

(2H, *J* = 7.2 Hz); ^{13}C-NMR (CDCl$_3$) δ 22.4, 24.2, 62.1, 65.0, 66.2, 69.9, 78.9, 79.2, 98.4, 128.6, 129.0, 129.6, 133.6, 165.3; IR (neat) 2993.1, 2107.2, 1725.4, 1267.4, 1091.3, 711.5 cm^{-1}; EI-MS (*m/z*): 333 [M]$^+$; HRMS Calcd for C$_{16}$H$_{19}$N$_3$O$_5$: 333.1325, Found: 333.1336.

(3S*,4R*)-4-Azido-2,2-bis((methylsulfonyloxy)methyl)tetrahydrohuran-3-yl benzoate (**20**). A mixture of **19** (1.01 g, 3.04 mmol) in 80% AcOH (80 mL) was stirred at room temperature for 5 h. After the solvent was removed under reduced pressure, the residue was co-evaporated with EtOH five times to remove residual AcOH. The resulting crude product was dissolved in CH$_2$Cl$_2$ (40 mL). To this mixture, were added, MsCl (1.19 mL, 15.18 mmol), Et$_3$N (2.14 mL, 15.18 mmol), and DMAP (38 mg, 0.30 mmol). After stirred at room temperature for 1 h, the mixture was diluted with CH$_2$Cl$_2$ washed with 5%HCl, sat.NaHCO$_3$ and brine. The separated organic layer was dried over Na$_2$SO$_4$. After filtration, the solvents were removed under reduce pressure, the residue was purified by silica gel column chromatography (hexane–ethyl acetate = 2:1) to give **20** (1.05 g, 77%). ^{1}H-NMR (CDCl$_3$) δ 3.00 (3H, s), 3.12 (3H, s), 3.94 (1H, dd, *J* = 5.80, 4.37 Hz), 4.33~4.48 (6H, m), 5.48 (1H, d, *J* = 3.4 Hz), 7.50 (2H, t, *J* = 7.7 Hz), 7.64 (1H, t, *J* = 8.0 Hz), 8.04 (2H, d, *J* = 7.3 Hz); ^{13}C-NMR (CDCl$_3$) δ 37.7, 65.5, 65.8, 67.4, 70.3, 77.2, 78.9, 82.8, 128.1, 128.8, 129.9, 134.2, 165.2; IR (near) 2110.7, 1728.8, 1360.0, 1267.0 cm^{-1}; FAB-MS (*m/z*): 450 [M+1]$^+$; HRMS Calcd for C$_{15}$H$_{20}$N$_3$O$_9$S$_2$: 450.0641, Found: 450.0631.

((1R*,4R*,5S*)-4-Azido-2,6-dioxabicyclo[3.2.0]heptan-1-yl)methyl methanesulfonate (**21**). A mixture of **20** (36 mg, 0.08 mmol) and NaOCH$_3$ (4.6 mg, 0.08 mmol) in MeOH (2 mL) was kept at 75 °C overnight. After the solvent was removed under reduced pressure, the residue was dissolved in CHCl$_3$ and washed with water and brine, then dried over Na$_2$SO$_4$. After filtration, the solvents were removed under reduce pressure, the residue was purified by silica gel column chromatography (hexane–ethyl acetate = 2:1) to give **21** (14 mg, 72%). ^{1}H-NMR (CDCl$_3$) δ 3.09 (3H, s), 4.05 (1H, d, *J* = 3.4 Hz), 3.40 (1H, d, *J* = 11.1 Hz), 4.29~4.77 (4H, m), 4.76 (1H, d, *J* = 8.2 Hz), 5.06 (1H, s); ^{13}C-NMR (CDCl$_3$) δ 37.8, 63.6, 67.4, 71.9, 77.1, 84.4, 89.9; IR (neat) 2098.2, 1360.4, 1175.0, 960.2, 815.5 cm^{-1}; FAB-MS (*m/z*): 250 [M+1]$^+$; HRMS Calcd for C$_7$H$_{12}$N$_3$O$_5$S: 250.0498, Found: 250.0490.

((1R*,4R*,5S*)-4-Azido-2,6-dioxabicyclo[3.2.0]heptan-1-yl)methyl benzoate (**22**). A mixture of CsF (332 mg, 2.19 mmol) and PhCOOH (267 mg, 2.19 mmol) in DMF (40 mL) was stirred at room temperature for 20 min. To this mixture was added a solution of **21** (182 mg, 0.73 mmol) in DMF (20 mL). After stirred at 60 °C overnight, the mixture was partitioned between EtOAc and H$_2$O. The separated water layer was extracted with EtOAc, and the organic layer was washed with *sat*.NaHCO$_3$, brine, then dried over Na$_2$SO$_4$. After filtration, the filtrated was concentrated under reduced pressure. The residue was purified by silica gel column chromatography (hexane-ethyl acetate = 5:1) to give **22** (170 mg, 86%). ^{1}H-NMR (CDCl$_3$) δ 4.08 (1H, d, *J* = 3.4 Hz), 4.42 (1H, d, *J* = 10.6 Hz), 4.52 (2H, dd, *J* = 10.6, 3.4 Hz), 4.60 (2H, s), 4.83 (1H, d, *J* = 7.7 Hz), 5.15 (1H, s), 7.61 (2H, t, *J* = 15.5 Hz) 7.59 (1H, t, *J* = 17.4 Hz), 8.10 (2H, d, *J* = 9.7 Hz); ^{13}C-NMR (CDCl$_3$) δ 63.1, 63.8, 71.5, 77.8, 85.1, 90.4, 128.4, 129.4, 129.7, 133.3, 166.1; IR (neat) 2099.7, 1723.1, 1284.3, 713.4 cm^{-1}; FAB-MS (*m/z*): 276 [M+1]$^+$; HRMS Calcd for C$_{13}$H$_{14}$N$_3$O$_4$: 276.0984, Found: 276.0977.

((1S*,4R*,5S*)-4-Azido-2,6-dioxa-bicyclo[3.2.0]heptan-1-yl)methanol (**23**). A mixture of **22** (184 mg, 0.67 mmol) and NaOCH$_3$ (19 mg, 0.33 mmol) in MeOH (15 mL) was stirred at room temperature. After the mixture was neutralized with AcOH (19 μL), the solvents were removed under reduced pressure and the residue was purified by silica gel column chromatography (hexane–ethyl acetate = 4:1) to give **23** (90 mg, 79%). ^{1}H-NMR (CDCl$_3$) δ 3.91 (2H, d, *J* = 6.3 Hz), 4.04 (1H, d, *J* = 3.9 Hz), 4.36 (1H, d, *J* = 10.6 Hz), 4.48 (2H, dd, *J* = 10.6, 3.4 Hz), 4.48 (1H, d, *J* = 8.2 Hz), 4.68 (1H, d, *J* = 8.2 Hz), 5.05 (1H, s); ^{13}C-NMR (CDCl$_3$) δ 62.3, 64.0, 71.5, 77.6, 86.9, 90.2; IR (neat) 3431.5, 2101.6, 1248.2, 870.88 cm^{-1}; EI-MS (*m/z*): 171 [M]$^+$; HRMS Calcd for C$_6$H$_9$N$_3$O$_3$: 171.0644, Found: 171.0643.

((1S*,4R*,5S*)-4-(5-Amino-6-chloropyrimidin-4-ylamino)-2,6-dioxa-bicyclo[3.2.0]heptan-1-yl)methanol (**15**). A mixture of **23** (69 mg, 0.40 mmol) and Pd(OH)$_2$ (6.2 mg, 0.04 mmol) in MeOH (5 mL) was stirred at room temperature overnight under H$_2$ atmosphere. After insoluble materials were removed by

filtration, the solvents were removed under reduced pressure. The resulting crude product was dissolved in *n*-BuOH (3 mL). To this mixture, were added 5-amino-4,6-dichloropyrimidine (140.1 mg, 0.86 mmol) and *i*-Pr$_2$NEt (298 µL, 1.71 mmol). The mixture was kept under reflux overnight. After the solvents were removed under reduced pressure, the residue was purified by silica gel column chromatography (chloroform–methanol = 19:1) to give **15** (64 mg, 58%). ^{1}H-NMR (CD$_3$OD) δ3.67 (2H, q, *J* = 12.6 Hz), 4.24 (1H, d, *J* = 10.1 Hz), 4.38 (1H, d, *J* = 7.7 Hz), 4.45 (1H, d, *J* = 4.4 Hz), 4.50 (1H, d, *J* = 10.1, 4.4 Hz), 4.63 (1H, d, *J* = 7.2 Hz) 4.90 (1H, s), 7.72 (1H, s); ^{13}C NMR (CD$_3$OD) δ 58.1, 58.3, 62.7, 73.1, 88.3, 92.1, 125.5, 138.9, 147.3, 153.3; IR (KBr) 3381.3, 2926.7, 1578.6, 1056.3 cm^{-1}; EI-MS (*m/z*): 272 [M]$^+$; HRMS Calcd for C$_{10}$H$_{13}$ClN$_4$O$_3$: 272.0676, Found: 272.0673.

((1S*,4R*,5S*)-4-(6-Chloro-9H-purin-9-yl)-2,6-dioxa-bicyclo[3.2.0]heptan-1-yl)methanol (**24**). To a solution of **15** (18 mg, 0.07 mmol) in DMF (0.5mL), were added orthoethyl formate (0.7 mL, 4.21 mmol) and *conc* HCl (2 µL, 0.024 mmol) at 0 °C. After the mixture was stirred at room temperature, the solvents were removed under reduced pressure. The residue was dissolved in 0.5 M *aq*HCl (1 mL) and the mixture was stirred at room temperature for 1 h. The mixture was neutralized with 1M aqNaOH (0.5 mL) and concentrated under reduced pressure. The residue was extracted with a solution of chloroform–methanol = 1:1. After the insoluble materials were removed by filtration, the solvents were removed under reduced pressure. The residue was purified by *p*TLC (developed by chloroform-methanol = 5:1) to give **24** (14.9 mg, 81%). ^{1}H-NMR (CDCl$_3$) δ 4.12 (2H, q, *J* = 12.6 Hz), 4.59 (1H, d, *J* = 11.6 Hz), 4.66 (1H, d, *J* = 9.7 Hz), 4.68 (1H, d, *J* = 9.7 Hz), 4.66 (1H, d, *J* = 9.7 Hz), 4.90 (1H, dd, *J* = 11.1, 4.8 Hz), 5.24 (1H, s), 5.38 (1H, d, *J* = 4.4 Hz), 8.68 (1H, s), 8.76 (1H, s); ^{13}C-NMR (CDCl$_3$-CD$_3$OD = 19 : 1) δ 58.6, 61.2, 71.9, 77.2, 87.8, 90.9, 130.7, 144.6, 150.7, 151.2, 151.8; IR (KBr) 3401.4, 2931.1, 1597.4, 1567.4, 1056.6 cm^{-1}; EI-MS (*m/z*): 282 [M]$^+$; HRMS Calcd for C$_{11}$H$_{11}$ClN$_4$O$_3$: 282.0520, Found: 282.0506.

((1S*,4R*,5S*)-4-(6-Amino-9H-purin-9-yl)-2,6-dioxa-bicyclo[3.2.0]heptan-1-yl)methanol (**6**). Compound **24** (29.4 mg, 0.10 mmol) was dissolved in *sat.* methanolic ammonia (7 mL) and the mixture was kept at 100 °C for 21 h in a glass sealed tube. After the solvents were removed under reduced pressure, the residue was purified by *p*TLC (developed by chloroform–methanol = 5:1) to give **6** (18.8 mg, 69%). ^{1}H-NMR (CDCl$_3$-CD$_3$OD = 17:3) δ 3.91 (1H, d, *J* = 12.6 Hz), 4.00 (1H, d, *J* = 12.1 Hz), 4.57 (1H, d, *J* = 10.6 Hz), 4.64 (1H, d, *J* = 7.7 Hz), 4.70 (1H, d, *J* = 7.7 Hz), 4.87 (1H, dd, *J* = 11.1, 4.8 Hz), 5.19 (1H, s), 5.25 (1H, d, *J* = 4.8 Hz), 8.25 (1H, s), 8.28 (1H, s); ^{13}C-NMR (CDCl$_3$–CD$_3$OD = 17:3) δ 29.5, 58.3, 61.3, 72.1, 87.6, 91.0, 118.2, 139.1, 148.9, 152.6, 155.3; IR (KBr) 3192.2, 2409.9, 1660.5, 1615.0, 1054.9 cm^{-1}; EI-MS (*m/z*): 263 [M]$^+$; HRMS Calcd for C$_{11}$H$_{13}$N$_5$O$_3$: 263.1018, Found: 263.1021.

4. Conclusions

We constructed an isoadenosine derivative on a 2,6-dioxobicyclo[3.2.0]heptane scaffold. Since our initial attempt to synthesize **6** by directly introducing the adenine moiety was not successful, we synthesized it by the de novo synthesis of an adenine ring on a pseudosugar moiety. However, this unique adenosine analogue showed no activity against HIV. Previously, we have reported that neither thymine nor adenine analogues **4** built on a 2-oxa-6-thiabicyclo[3.2.0]heptane skeleton inhibit HIV [14]. The structural rigidities of these analogues and isoadenosine **6** due to the introduction of fused thietane and oxetane rings, respectively, appear to inhibit anti-HIV activity. In particular, phosphorylation at the 5'-hydroxyl group would be inhibited since deoxynucleoside kinase recognizes the puckering of sugars [25]. Thus, we are currently preparing new substituted nucleoside derivatives based on **4** and **6**, and the results will be reported elsewhere.

Acknowledgments: This work was supported in part by a Grant-in-Aid for Scientific Research (No. 24590143, Y.Y.) from JSPS and by a grant of Strategic Research Foundation Grant-aided Project for Private Universities from Ministry of Education, Culture, Sport, Science, and Technology, Japan (MEXT), 2010–2014.

Author Contributions: YY and TI designed research; SK, HK and TS performed the synthesis of compounds and TI assayed anti-HIV activity. YY wrote the paper. All authors read and approved the final manuscript.

Conflicts of Interest: The authors declare no conflict of interest.

References

1. Cihlar, T.; Ray, A.S. Nucleoside and nucleotide HIV reverse transcriptase inhibitors: 25 years after zidovudine. *Antivir. Res.* **2010**, *85*, 39–58. [CrossRef] [PubMed]
2. Mehellou, Y.; de Clercq, E. Twenty-six years of anti-HIV drug discovery: Where do we stand and where do we go? *J. Med. Chem.* **2010**, *53*, 521–538. [CrossRef] [PubMed]
3. WHO. Consolidated Guidelines on the Use of Antiretroviral Drugs for Treating and Preventing HIV Infection. Available online: http://www.who.int/hiv/pub/guidelines/arv2013/download/en/ (accessed on 30 June 2013).
4. Meadows, D.C.; Gervey-Hague, J. Current developments in HIV chemotherapy. *ChemMedChem* **2006**, *1*, 16–29. [CrossRef] [PubMed]
5. Imamichi, T. Action of anti-HIV drugs and resistance: Reverse transcriptase inhibitors and protease inhibitors. *Curr. Pharm. Des.* **2004**, *10*, 4039–4053. [CrossRef] [PubMed]
6. Mitsuya, H.; Broder, S. Inhibition of the *in vitro* infectivity and cytopathic effect of human T-lymphotrophic virus type III/lymphadenopathy-associated virus (HTLV-III/LAV) by 2′,3′-dideoxynucleosides. *Proc. Natl. Acad. Sci. USA* **1986**, *83*, 1911–1915. [CrossRef] [PubMed]
7. Mitsuya, H.; Weinhold, K.J.; Furman, P.A.; St Clair, M.H.; Lehrman, S.N.; Gallo, R.C.; Bolognesi, D.; Barry, D.W.; Broder, S. 3′-Azido-3′-deoxythymidine (BW A509U): An antiviral agent that inhibits the infectivity and cytopathic effect of human T-lymphotropic virus type III/lymphadenopathy-associated virus *in vitro*. *Proc. Natl. Acad. Sci. USA* **1985**, *82*, 7096–7100. [CrossRef] [PubMed]
8. Schinazi, R.F.; Chu, C.K.; Peck, A.; McMillan, A.; Mathis, R.; Cannon, D.; Jeong, L.S.; Beach, J.W.; Choi, W.B.; Yeola, S.; *et al.* Activities of the four optical isomers of 2′,3′-dideoxy-3′-thiacytidine (BCH-189) against human immunodeficiency virus type 1 in human lymphocytes. *Antimicrob. Agents Chemother.* **1992**, *36*, 672–676. [CrossRef]
9. Balzarini, J.; Holy, A.; Jindrich, J.; Naesens, L.; Snoeck, R.; Schols, D.; de Clercq, E. Differential antiherpesvirus and antiretrovirus effects of the (*S*) and (*R*) enantiomers of acyclic nucleoside phosphonates: Potent and selective *in vitro* and *in vivo* antiretrovirus activities of (*R*)-9-(2-phosphonomethoxypropyl)-2,6-diaminopurine. *Antimicrob. Agents Chemother.* **1993**, *37*, 332–338. [CrossRef] [PubMed]
10. Yoshimura, Y.; Yamazaki, Y.; Kawahata, M.; Yamaguchi, K.; Takahata, H. Design and synthesis of a novel ring-expanded 4′-Thio-*apio*-nucleoside derivatives. *Tetrahedron Lett.* **2007**, *48*, 4519–4522. [CrossRef]
11. Yoshimura, Y.; Asami, K.; Matsui, H.; Tanaka, H.; Takahata, H. New synthesis of (±)-isonucleosides. *Org. Lett.* **2006**, *8*, 6015–6018. [CrossRef] [PubMed]
12. Yoshimura, Y.; Yamazaki, Y.; Saito, Y.; Takahata, H. Synthesis of 1-(5,6-dihydro-2H-thiopyran-2-yl)uracil by a Pummerer-type thioglycosylation reaction: The regioselectivity of allylic substitution. *Tetrahedron* **2009**, *65*, 9091–9102. [CrossRef]
13. Yoshimura, Y.; Ohta, M.; Imahori, T.; Imamichi, T.; Takahata, H. A new entry to carbocyclic nucleosides: Oxidative coupling reaction of cycloalkenylsilanes with a nucleobase mediated by hypervalent iodine reagent. *Org. Lett.* **2008**, *10*, 3449–3452. [CrossRef] [PubMed]
14. Yoshimura, Y.; Asami, K.; Imamichi, T.; Okuda, T.; Shiraki, K.; Takahata, H. Design and synthesis of isonucleosides constructed on a 2-oxa-6-thiabicyclo[3.2.0]heptane scaffold. *J. Org. Chem.* **2010**, *75*, 4161–4171. [CrossRef] [PubMed]
15. Yoshimura, Y.; Yamazaki, Y.; Saito, Y.; Natori, Y.; Imamichi, T.; Takahata, H. Synthesis of 5-thiodidehydropyranylcytosine derivatives as potential anti-HIV agents. *Bioorg. Med. Chem. Lett.* **2011**, *21*, 3313–3316. [CrossRef] [PubMed]
16. Kiran, Y.B.; Wakamatsu, H.; Natori, Y.; Takahata, H.; Yoshimura, Y. Design and synthesis of a nucleoside and a phosphonate analogue constructed on a branched-*threo*-tetrofuranose skeleton. *Tetrahedron Lett.* **2013**, *54*, 3949–3952. [CrossRef]
17. Kan-no, H.; Saito, Y.; Omoto, S.; Minato, S.; Wakamatsu, H.; Natori, Y.; Imamichi, T.; Takahata, H.; Yoshimura, Y. Synthesis of a dihydropyranonucleoside using an oxidative glycosylation reaction mediated by hypervalent iodine. *Synthesis* **2014**, *46*, 879–886. [CrossRef]

18. Chu, C.K.; Ahn, S.K.; Kim, H.O.; Beach, J.W.; Alves, A.J.; Jeong, L.S.; Islam, Q.; van Roey, P.; Schinazi, R.F. Asymmetric synthesis of enantiomerically pure (−)-(1'R,4'R)-dioxolane-thymine and its anti-HIV activity. *Tetrahedron Lett.* **1991**, *32*, 3791–3794. [CrossRef]

19. Kim, H.O.; Schinazi, R.F.; Nampalli, S.; Shanmuganathan, K.; Cannon, D.L.; Alves, A.J.; Jeong, L.S.; Beach, J.W.; Chu, C.K. 1,3-dioxolanylpurine nucleosides (2R,4R) and (2R,4S) with selective anti-HIV-1 activity in human lymphocytes. *J. Med. Chem.* **1993**, *36*, 30–37. [CrossRef] [PubMed]

20. Kim, H.O.; Schinazi, R.F.; Shanmuganathan, K.; Jeong, L.S.; Beach, J.W.; Nampalli, S.; Cannon, D.L.; Chu, C.K. L-beta-(2S,4S)- and L-alpha-(2S,4R)-dioxolanyl nucleosides as potential anti-HIV agents: Asymmetric synthesis and structure-activity relationships. *J. Med. Chem.* **1993**, *36*, 519–528. [CrossRef]

21. Choi, Y.; George, C.; Comin, M.J.; Brachi, J.J.; Kim, H.S.; Jacobsen, K.A.; Balzarini, J.; Mitsuya, H.; Boyer, P.L.; Hughes, S.H.; *et al.* A conformationally locked analogue of the anti-HIV agent stavudine. An important correlation between pseudorotation and maximum amplitude. *J. Med. Chem.* **2003**, *46*, 3292–3299. [CrossRef] [PubMed]

22. D'Alonzo, D.; van Aerschot, A.; Guaragna, A.; Palumbo, G.; Schepers, G.; Capone, S.; Rozenski, J.; Herdewijn, P. Synthesis and base pairing properties of 1',5'-anhydro-L-hexitol nucleic acids (L-HNA). *Chem. Eur. J.* **2009**, *15*, 10121–10131. [CrossRef] [PubMed]

23. Bera, S.; Nair, V. A new general synthesis of isomeric nucleosides. *Tetrahedron Lett.* **2001**, *42*, 5813–5815. [CrossRef]

24. Quadrelli, P.; Piccanello, A.; Mella, M.; Corsaro, A.; Pistarà, V. From cyclopentadiene to isoxazoline-carbocyclic nucleosides: A rapid access to biological molecules through aza-Dielse Alder reactions. *Tetrahedron* **2008**, *64*, 3541–3547. [CrossRef]

25. Comin, M.J.; Vu, B.C.; Boyer, P.L.; Liao, C.; Hughes, S.H.; Marquez, V.E. D-(+)-iso-methanocarbathymidine: A high-affinity substrate for herpes simplex virus 1 thymidine kinase. *ChemMedChem* **2008**, *3*, 1129–1134. [CrossRef]

Sample Availability: *Sample Availability*: Sample of the final compound is available from the authors. About the other compounds, please contact the authors.

![molecules logo] *molecules*

MDPI

Article

Synthesis of Peptide Nucleic Acids Containing a Crosslinking Agent and Evaluation of Their Reactivities

Takuya Akisawa, Yuki Ishizawa and Fumi Nagatsugi *

Institute of Multidisciplinary Research for Advanced Materials, Tohoku University, 2-1-1 Katahira, Aoba-ku, Sendai-shi, Miyagi 980-8577, Japan; akisawat@mail.tagen.tohoku.ac.jp (T.A.); yi87653@yahoo.co.jp (Y.I.)
* Correspondence should be addressed; nagatugi@tagen.tohoku.ac.jp; Tel./Fax: +81-22-217-5633.

Academic Editor: Mahesh K. Lakshman
Received: 26 January 2015; Accepted: 9 March 2015; Published: 13 March 2015

Abstract: Peptide nucleic acids (PNAs) are structural mimics of nucleic acids that form stable hybrids with DNA and RNA. In addition, PNAs can invade double-stranded DNA. Due to these characteristics, PNAs are widely used as biochemical tools, for example, in antisense/antigene therapy. Interstrand crosslink formation in nucleic acids is one of the strategies for preparing a stable duplex by covalent bond formation. In this study, we have synthesized PNAs incorporating 4-amino-6-oxo-2-vinylpyrimidine (AOVP) as a crosslinking agent and evaluated their reactivities for targeting DNA and RNA.

Keywords: peptide nucleic acid (PNA); crosslink; antisense; invasion

1. Introduction

Peptide nucleic acids (PNAs) are synthetic nucleic acid analogs, in which the sugar phosphate backbone is replaced with N-(2-aminoethyl)glycine [1] (Figure 1). The hybrid complexes between PNA and DNA or RNA exhibit an extraordinary thermal stability due to the lack of a negatively-charged backbone. Additionally, PNAs are chemically stable in comparison to DNA over a wide range of temperatures and pH values and are resistant to nucleases and proteases [2]. These characteristics of the PNAs provide a variety of applications in therapeutic approaches [3–5], including use as biochemical tools [6]. The PNA was used as an antisense to interfere with dimerization of the HIV-1 RNA transcript and to inhibit the template switching process in HIV-1 [7]. Many modifications have been made to the PNA backbone [8] and the nucleobases [9–11] for improving PNA's properties, such as their cellular uptake and solubility. GPNA is a backbone-modified PNA constructed of guanidine and is cell permeable. The GPNA designed to target the epidermal growth factor receptor (EGFR) induced potent antitumor effects due to its antisense effect [12]. A higher cellular uptake of the PNA was also achieved by conjugation with cysteine and lysine at the N-terminal position, which exhibited the anti-miRNA effect for miR-122 [13]. The most remarkable property of PNAs is their ability to invade the secondary structure of nucleic acids, for example, the DNA duplex [14] and the G-quadruplex [15]. This invasion ability of the PNAs raises the possibility their use to control gene expression at the DNA level with high efficiency. Strand invasion requires a high stability of the PNA-DNA complex due to competition from the displaced DNA strand. One strategy for increasing the stability of the PNA-DNA complex is by forming a covalent bond between the PNA and the target DNA.

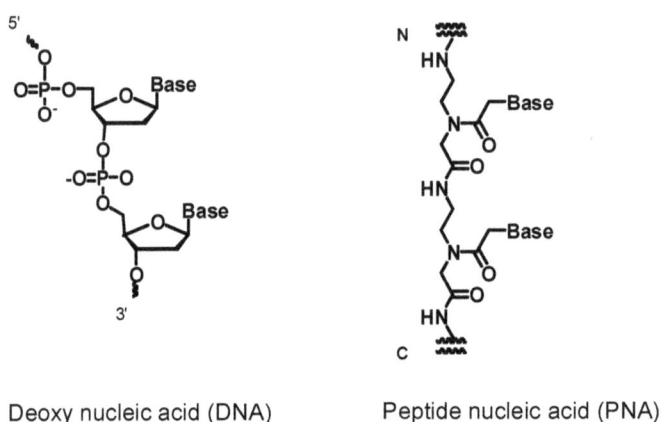

Deoxy nucleic acid (DNA)   Peptide nucleic acid (PNA)

Figure 1. Chemical structures of DNA and PNA.

Interstrand crosslinking (ICL) in nucleic acids is one strategy for preparing a stable duplex by covalent bond formation [16]. Several crosslinking oligonucleotides (ON) have been reported to react with DNA or RNA when activated by photo-irradiation [17] or a chemical reaction [18]. The PNA containing a modified thymine derivative is reported to form PNA/DNA ICL by photolysis or under oxidative conditions [19]. The PNA conjugated with a quinone methide precursor caused an inducible alkylation of the target DNA [20]. The bis-PNA conjugated with nitrogen mustard is reported to suppress transcription in a model system by forming a strand invasion complex and covalent bond with the target duplex DNA [21]. Thus a PNA bearing an alkylating group provides an efficient strategy for gene targeting via the inhibition of transcription with high sequence selectivity.

We have previously developed a 4-amino-6-oxo-2-vinyl pyrimidine (AOVP) derivative and demonstrated that this nucleoside derivative **1** exhibited a highly selective and very fast ICL reaction with thymine [22] (Figure 2). The high selectivity and reactivity of **1** could be attributed to the close proximity effect between the vinyl group of AOVP and thymine in the hybridized complex. In this paper, we describe the synthesis of PNAs containing AOVP and the evaluation of their crosslinking reactivities.

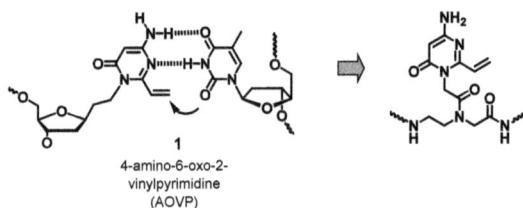

Figure 2. Design of the crosslink-forming PNA.

2. Results and Discussion

2.1. Synthesis of PNAs Containing 4-Amino-6-oxo-2-vinyl Pyrimidine (AOVP)

The synthesis of the PNA monomer (**8**) as a stable precursor of AOVP is summarized in Scheme 1. The base portion (**2**) was synthesized as previously described [22].

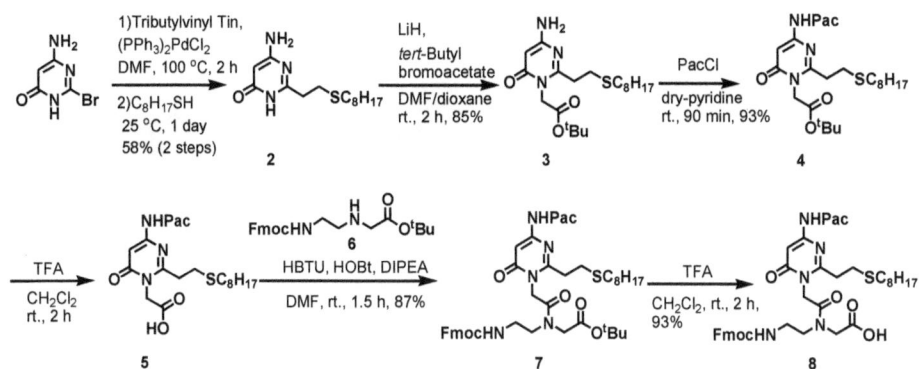

Scheme 1. Synthesis of the AOVP PNA monomer.

The N1 position of **2** was alkylated with *t*-butyl bromoacetate to yield the *t*-butyl ester **3** in 85% yield. The 4-amino group of **3** was protected with a phenoxyacetyl group, and subsequent deprotection of the *t*-butyl group with trifluoroacetic acid (TFA) afforded the acid **5**, which was coupled with the Fmoc/*t*-Bu PNA backbone **6** using HBTU/HOBT/DIPEA to form the monomer **7**, which was then deprotected of the *t*-butyl ester by treatment with TFA to produce the PNA monomer **8**.

We designed the PNA sequence in order to target the template RNA included in the telomerase ribonucleoprotein based on a previous report, in which a quinone methide cross-linking agent was successfully incorporated [20]. It was also reported that non-covalent binding of the PNA to the template RNA of the telomerase efficiently inhibited the telomerase activity in cell cultures [23]. Therefore, we expected that our crosslinking agent in PNA would further improve the inhibitory activity for the telomerase. In this study, we synthesized two sequences that contained AOVP in the middle position or in the N-terminal position.

Generally, the C- or N-terminal position of the PNA is conjugated with lysine (Lys) to increase its solubility in water. Our previous study showed that the vinyl group of AOVP in the PNA caused an intramolecular reaction with the amino group of Lys conjugated at the C terminal position. In this study, glycine (Gly) and two arginines (Arg) were conjugated with the PNA at the C-terminal position. The solid-phase synthesis of the PNA oligomer was manually performed on the NovaSin TGR resin using Fmoc strategies. The N-terminal position of the PNA probe was modified with an acetyl group to avoid side reactions between the vinyl group in AOVP and the amino group at the N-terminal position. The resin was treated with potassium carbonate (K_2CO_3) in MeOH for the deprotection of the phenoxy acetyl group. Each synthesized PNA was cleaved from the resin using water/triisopropylsilane/TFA (2.5%/2.5%/95%) and purified by reverse phase HPLC to give the desired PNA1 (Scheme 2).

The sulfide-protected PNA1 was smoothly converted to PNA2 by oxidation with magnesium monoperoxyphthalate (MMPP), followed by treatment with an aqueous NaOH solution to generate the vinyl group of PNA3, as shown in Figure 3.

The structure of PNA3 was confirmed by MALDI-TOF mass spectra measurements. The presence of the vinyl group in PNA3 was further supported by the fact that the treatment of PNA3 with an aqueous NaSMe solution generated the sulfide derivatives PNA4.

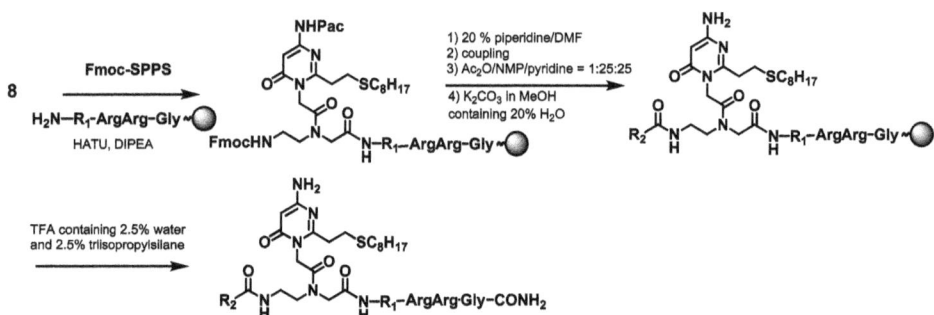

Scheme 2. Synthesis of the PNAs containing the stable precursors of AOVP.

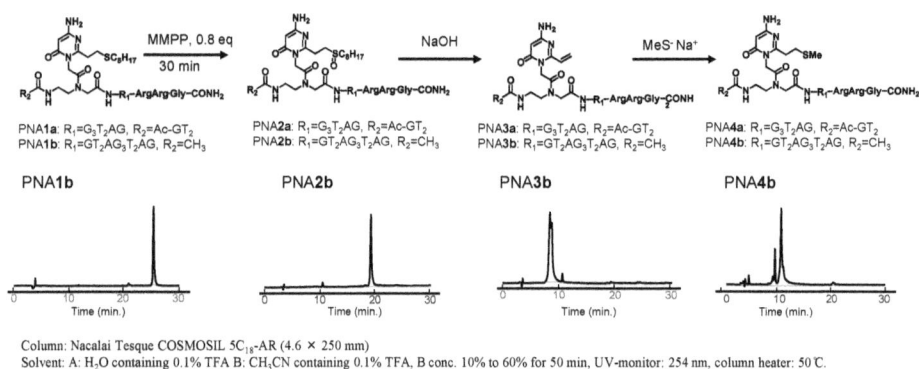

Column: Nacalai Tesque COSMOSIL 5C$_{18}$-AR (4.6 × 250 mm)
Solvent: A: H$_2$O containing 0.1% TFA B: CH$_3$CN containing 0.1% TFA, B conc. 10% to 60% for 50 min, UV-monitor: 254 nm, column heater: 50 ℃.

Figure 3. Synthesis of the PNAs containing AOVP and analysis of these reactions by HPLC.

2.2. Evaluation of the Crosslinking Reactivity of the PNA Containing AOVP

The crosslinking reaction was investigated under neutral conditions using the reactive PNA3 and the target DNA (Y = dG, dA, dC, dT) or RNA (Y = rG, rA, rC, U) labelled with fluorescein at the 5' end. After 24 h, the reaction mixtures were analyzed by gel electrophoresis with 20% denaturing gel. The crosslinked products were identified as the less mobile bands (Figure 4). The crosslink yields were calculated from the ratio of the crosslinked product to the total amount of the remaining single stranded DNA1/RNA1 and crosslinked product. PNA3a, which contained AOVP in the middle position, did not produced significant adducts to any targets (Figure 4A,B). Conversely the crosslink product was observed in the reaction between PNA3b containing AOVP at the terminal position and DNA2 (Y=T) (Figure 4C). The crosslinking reaction of PNA3b did not efficiently occur with an RNA target (Figure 4D). The lower mobility bands in the reaction between PNA3b and DNA2 (Y=T) was extracted from the denaturing gel and proved to be the crosslinked adduct by MALDI-TOF MS measurements (calcd.10,071.8 found 10,072.3).

(A) PNA3a N (Ac)GTT**X**GGGTTAG *RRG* C
    DNA1   3' GAC <u>CAA**Y**CCCAATC</u> TGGTT -FAM 5'

X = AOVP   Y = A, G, C, T

Y	:	A		G		C		T	
PNA3a	:	−	+	−	+	−	+	−	+
Yield (%)	:	0	3	0	3	0	3	0	15

(B) PNA3a N (Ac)GTT**X**GGGTTAG *RRG* C
    RNA1   3' GAC <u>CAA**Y**CCCAAUC</u> UGGUU -FAM 5'

X = AOVP   Y = A, G, C, U

Y	:	A		G		C		U	
PNA3a	:	−	+	−	+	−	+	−	+
Yield (%)	:	0	0	0	2	0	0	0	2

(C) PNA3b N (Ac)**X**GTTAGGGTTAG *RRG* C
    DNA2   3' GAC <u>**Y**CAATCCCAATC</u> TGGTT -FAM 5'

X = AOVP   Y = A, G, C, T

Y	:	A		G		C		T	
PNA3b	:	−	+	−	+	−	+	−	+
Yield (%)	:	0	7	0	5	0	13	0	42

(D) PNA3b N (Ac)**X**GTTAGGGTTAG *RRG* C
    RNA2   3' GAC <u>**Y**CAAUCCCAAUC</u> UGGUU -FAM 5'

X = AOVP   Y = A, G, C, U

Y	:	A		G		C		U	
PNA3b	:	−	+	−	+	−	+	−	+
Yield (%)	:	0	2	0	2	0	5	0	17

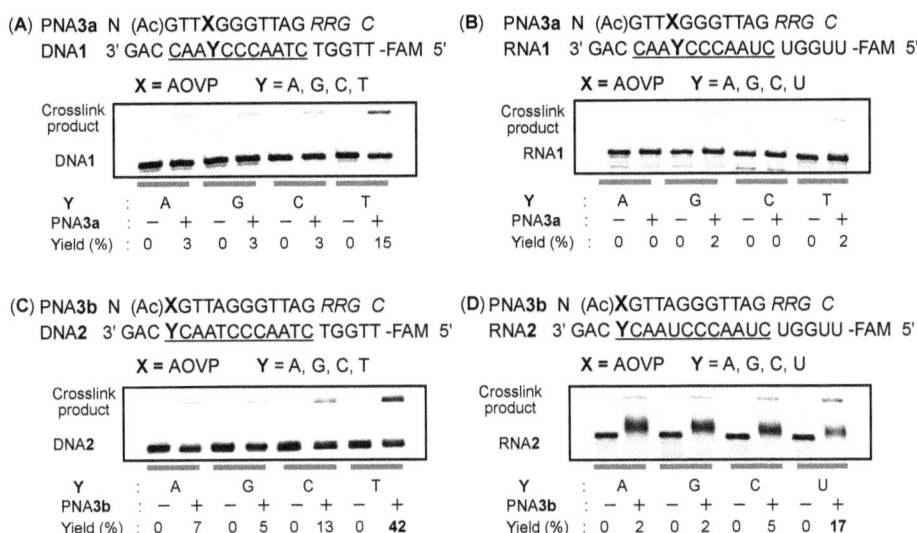

The reactions were performed with 4 µM modified PNA and 1 µM target DNA or RNA in 50 mM MES, (pH 7.0) at 37 °C.

Figure 4. Evaluation of the crosslink reactivity of PNA3 to DNA and RNA. (**A**) PNA3a and target DNA1; (**B**) PNA3a and target RNA1; (**C**) PNA3b and target DNA2; (**D**) PNA3b and target RNA2.

The time course of the crosslink yields for DNA2 and RNA2 with PNA3b is summarized in Figure 5. PNA3b exhibited the highest yield to thymine compared to the other target bases in DNA. In contrast, lower yields with any of the target bases in RNA were observed, although the selectivity was retained to some extent during the reaction with uridine. The crosslink reactions with DNA and RNA with PNA3b were accelerated at 50 °C (Figure S6). The DNA oligonucleotide with the same sequence containing the AOVP derivative (**1**) did not form the duplex at 50 °C, resulting in no crosslink formation with the target DNA or RNA at this temperature. In our previous report, the AOVP derivative (**1**) at the middle position in DNA exhibited high reactivity and selectivity to thymine in DNA (~70% after 2 h) and to uridine in RNA (~80% after 2 h) [22]. The PNA CONTAINING AOVP showed lower yields than that of the DNA derivative.

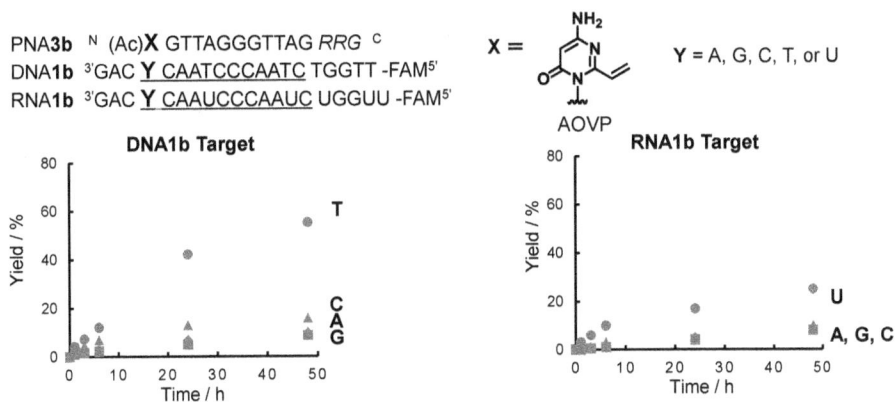

PNA3b N (Ac)**X** GTTAGGGTTAG *RRG* C
DNA1b 3'GAC **Y** <u>CAATCCCAATC</u> TGGTT -FAM5'
RNA1b 3'GAC **Y** <u>CAAUCCCAAUC</u> UGGUU -FAM5'

X =

AOVP

Y = A, G, C, T, or U

Figure 5. Time course of the crosslink yields with PNA3b.

Table 1. Melting temperatures of PNA4 with DNA and RNA.

	PNA4a			PNA4b	
Target	Tm (°C)	ΔTm/°C	Target	Tm (°C)	ΔTm/°C
DNA1 (Y=G)	69.7	−7.6	DNA2 (Y=G)	79.2	−1.1
DNA1 (Y=A)	68.0	−9.3	DNA2 (Y=A)	79.2	−1.1
DNA1 (Y=C)	66.1	−11.2	DNA2 (Y=C)	78.6	−1.7
DNA1 (Y=T)	68.1	−9.2	DNA2 (Y=T)	77.9	−2.4
RNA1 (Y=G)	71.3	−11.7	RNA2 (Y=G)	83.2	−1.3
RNA1 (Y=A)	72.9	−10.1	RNA2 (Y=A)	84.1	−0.4
RNA1 (Y=C)	69.1	−13.9	RNA2 (Y=C)	83.2	−1.3
RNA1 (Y=U)	69.1	−13.9	RNA2 (Y=U)	82.9	−1.6

UV melting profiles measured using 1.5 μM each of the strands in 50 mM MES buffer, pH 7.0; PNA5a: GTTAGGGTTAG-RRG, PNA5b AGTTAGGGTTAG-RRG; Tm (°C) for PNA5a/DNA1 (Y=T): 77.3, PNA5a/RNA1 (Y=U): 83.0, PNA5b/DNA2 (Y=T): 80.3, PNA5a/DNA2 (Y=U): 84.5.

To explain the difference in the crosslink yields, the thermal stability was estimated by the melting temperature (Tm) of the duplex between PNA4 containing the stable precursor of AOVP and the target DNA or RNA. The Tm values for PNA4a/DNA1 or RNA1 were 8–14 °C lower than those of PNA5 containing adenosine instead of AOVP (Table 1). Conversely, no significant difference in the Tm values was observed between PNA4b/DNA, RNA and PNA5b/DNA, RNA. These results indicated that the incorporation of AOVP in the middle site may destabilize the duplex and that the incorporation of AOVP at the terminal position does not affect the thermal stability of the PNA/DNA, RNA hetero-duplex. It may be assumed that the lower reactivity of PNA3a might be attributed to the lower stability of the duplex. However, despite a similar stability of the duplexes formed with PNA3b or RNA, the reactivity of PNA3b with DNA was higher than that with RNA. Accordingly, the lower thermal stability of the duplex might not be the major cause of the lower reactivity of PNA containing AOVP. As the crosslink reactivity of AOVP is highly dependent on the close proximity between the reactants, a small difference in the local structure around AOVP and the target base might become a major factor of the reactivity (Figure S7). Further study about the local structure of the PNA containing AOVP is also needed for a rational explanation of the higher reactivity of PNA3b with DNA than that with RNA.

3. Experimental Section

3.1. General

All air-sensitive reactions were carried out under argon in oven-dried glassware using standard syringe and septa techniques, unless otherwise noted. The ^{1}H- and ^{13}C-NMR spectra (500 MHz for ^{1}H and 125 MHz for ^{13}C) were recorded on a Bruker spectrometer (Bruker BioSpin K.K., Kanagawa, Japan). Chemical shifts (δ) are reported in parts per million (ppm) and are referenced to the solvent CDCl$_3$ (7.26 ppm), DMSO (2.50 ppm), for ^{1}H-NMR and CDCl$_3$ (77.16 ppm), DMSO (39.52 ppm) for ^{13}C-NMR. Multiplicity and qualifier abbreviations are as follows: s = singlet, d = doublet, t = triplet, q = quartet, quint. = quintet, m = multiplet, br = broad. Coupling constants (J) are reported in hertz (Hz). High resolution mass spectra were obtained using electrospray ionization (ESI). MALDI-TOF mass spectra were measured by using an Autoflex Speed mass spectrometer (Bruker Daltonics, Kanagawa, Japan) with the laser at 337 nm in negative or positive mode using 3-hydroxypicolinic acid or gentisic acid as the matrix. Thin-layer chromatography was performed on Merck 60 F254 pre-coated silica gel plates (EMD Millipore Co., Darmstadt, Germany). Merck 60 F254 pre-coated silica gel on glass in a thickness of 0.9 mm was used for preparative TLC. Column chromatography was performed on silica gel (Silica Gel 60 N; 40–100 μm, or 100–210 μm, Kanto Chemical Co., Inc, Tokyo, Japan). The ultraviolet-visible (UV-vis) absorption spectra were recorded on a Beckman Coulter DU800 (Beckman Coulter K.K., Tokyo, Japan). PNA synthesis was carried out by Fmoc-SPPS. High performance liquid chromatography

(HPLC) was performed using Nacalai Tesque (Kyoto, Japan) Cosmosil 5C18AR (4.6 or 10 × 250 mm) as the column, a JASCO PU-2089 as the pump, JASCO 2075 for the UV monitoring, and JASCO 2067 as the column oven (all Jasco Co., Tokyo, Japan). pH measurements were measured performed with a Mettler Toledo (Schwerzenbaha, switzerland) Seven Easy pH meter using a 8220BNWP electrode (Thermo Scientific, Waltham, MA, USA). UV-melting experiments were performed on a Beckman Coulter DU800. Densitometric analysis of the gel was carried out on 20% denaturing polyacrylamide gel plates, and visualized and quantified using a FLA-5100 Fluor Imager (Fujifilm Co., Tokyo, Japan). Commercially available reagents were obtained from Wako Pure Chemical Industries Ltd. (Osaka, Japan) or Kanto Chemical Co., Inc. and used without further purification. DNA and RNA oligomers were purchased from Japan Bio Services Co., Ltd. (Saitama, Japan), buffers and salts were from Nacalai Tesque.

3.2. Chemistry

3.2.1. Synthesis of 1-(*tert*-Butoxycarbonylmethyl)-4-amino-6-oxo-2-(2-octylthioethyl)pyrimidine (3)

Compound **2** (300 mg, 1.1 mmol) and lithium hydride (84 mg, 11 mmol) were dissolved in dry dioxane (20 mL) and dry DMF (10 mL), and the reaction mixture was stirred at room temperature. After 1.5 h, *tert*-butyl bromoacetate (233 µL, 1.6 mmol) was added and stirred at room temperature. After 1.5 h, the reaction mixture was diluted with 50 mL EtOAc and washed with sat. NH_4Cl aq. (50 mL), and brine (50 mL). The organic layer was dried over Na_2SO_4, filtered, and concentrated under reduced pressure. The residue was purified by column chromatography (hexane/EtOAc = 1:1) to give **3** as a colorless oil (362 mg, 0.91 mmol 87%). ^{1}H-NMR (CDCl$_3$): δ 5.35 (s, 1H), 4.69 (s, 2H), 4.67 (s, 2H), 2.92–2.89 (m, 2H), 2.80–2.76 (m, 2H), 2.53 (t, *J* = 7.5 Hz, 2H), 1.59 (quin, *J* = 7.5Hz, 2H), 1.48 (s, 9H), 1.38–1.27 (m, 10H), 0.88 (t, *J* = 7.5 Hz, 3H). ^{13}C-NMR (CDCl$_3$): δ 167.1, 162.8, 161.3, 160.3, 85.5, 83.1, 44.5, 35.2, 32.7, 31.9, 29.8, 29.3, 29.0, 28.5, 28.1, 22.7, 14.2. HRMS (ESI-MS) *m/z* calcd for $C_{20}H_{35}N_3O_3S$ [M+H]$^+$ 398.24719, found 398.24711.

3.2.2. Synthesis of 1-(*tert*-Butoxycarbonylmethyl)-4-phenoxyacetylamino-6-oxo-2-(2-octylthioethyl) pyrimidine (4)

To a solution of **3** (130 mg, 0.33 mmol) in dry pyridine (5 mL) was added phenoxyacetyl chloride (68 µL, 4.9 mmol), and the reaction mixture was stirred at room temperature. After 1.5 h, the reaction mixture was diluted with 50 mL EtOAc and washed with water (50 mL), and brine (50 mL). The organic layer was dried over Na_2SO_4, filtered, and concentrated under reduced pressure. The residue was purified by column chromatography (CH$_2$Cl$_2$/EtOAc=9:1 to 4:1) to give **4** as a white solid (169 mg, 0.32 mmol, 93%). ^{1}H-NMR (CDCl$_3$): δ 8.49 (s, 1H), 7.37–7.33 (m, 2H), 7.25 (s, 1H), 7.08–7.05 (m, 1H), 7.00–6.98 (m, 2H), 4.70 (s, 2H), 4.59 (s, 2H), 2.93–2.90 (m, 2H), 2.85–2.82(m, 2H), 2.54 (t, *J* = 7.5 Hz, 2H), 1.59 (quin, *J* = 7.5Hz, 2H), 1.48 (s, 9H), 1.39–1.26 (m, 10H), 0.87 (t, *J* = 7.0 Hz, 3H). ^{13}C-NMR (CDCl$_3$): δ 167.3, 166.4, 162.9, 160.5, 157.0, 152.8, 130.1, 122.8, 115.1, 97.0, 83.5, 67.7, 45.0, 35.1, 32.8, 31.9, 29.8, 29.3, 29.0, 28.5, 28.1, 22.7, 14.2. HRMS (ESI-MS) *m/z* calcd for $C_{28}H_{41}N_3O_5S$ [M+H]$^+$ 532.28397, found 532.28398.

3.2.3. Synthesis of 1-(Carbonylmethyl)-4-phenoxyacetylamino-6-oxo-2-(2-octylthioethyl)pyrimidine (5)

To a solution of **4** (200 mg, 0.38 mmol) in CH$_2$Cl$_2$ (1 mL) was added TFA (9 mL) and the reaction mixture was stirred for 1 h at room temperature. The solvent was removed under reduced pressure and the residue was diluted with 5 mL diethyl ether. The solution was added to hexane (30 mL) to cause precipitation. The resulting precipitates were collected by filtration and washed with hexane/diethyl ether (9:1) to give a pure sample of **5** as a white powder (131 mg, 0.28 mmol, 89%). ^{1}H-NMR (DMSO): δ 10.4 (s, 1H), 7.30 (t, *J* = 8.0 Hz, 2H), 6.98–6.92 (m, 3H), 6.86 (s, 1H), 4.81 (s, 2H), 4.75 (s, 2H), 2.96 (t, *J* = 7.0 Hz, 2H), 2.88 (t, *J* = 7.0 Hz, 2H), 2.54 (t, *J* = 7.5 Hz, 2H), 1.52 (quin, *J* = 7.5 Hz, 2H), 1.34–1.24 (m,

10H), 0.85 (t, J = 7.0 Hz, 3H). ^{13}C-NMR (DMSO): δ 169.1, 168.3, 162.1, 161.0, 157.7, 153.8, 129.5, 121.1, 114.4, 94.7, 66.6, 44.3, 34.2, 31.2, 31.1, 29.0, 28.6, 28.5, 28.2, 27.3, 22.0, 13.9. HRMS (ESI-MS) *m/z* calcd for $C_{24}H_{33}N_3O_5S$ [M+H]$^+$ 476.22137, found 476.22136.

3.2.4. Synthesis of *tert*-Butyl-*N*-[2-(*N*-9-fluorenylmethoxycarbonyl)aminoethyl]-*N*-[[4-phenoxyacetyl-amino-6-oxo-2-(2-octylthioethyl)pyrimidin-1-yl]acetyl]glycinate (**7**)

Compound **5** (100 mg, 0.21 mmol) was dissolved in dry DMF (5 mL). HBTU (159 mg, 0.42 mmol), HOBt (57 mg, 0.42 mmol), and DIPEA (146 μL, 0.84 mmol) were added and stirred for 5 min at room temperature. The reaction mixture was added to a solution of **6** (82 mg, 0.21 mmol) in dry DMF (5 mL) and stirred at room temperature. After 1.5 h, the reaction mixture was diluted with 50 mL EtOAc, and washed with sat.NaHCO$_3$ (2 × 50 mL) and brine (50 mL). The organic layer was dried over Na$_2$SO$_4$, filtered, and concentrated under reduced pressure. The residue was purified by column chromatography (hexane/EtOAc=2:1) to give **7** as a white foam (140 mg, 0.16 mmol, 87%). ^{1}H-NMR (DMSO): δ 10.42 (s, 1H), 7.81 (d, J = 7.5 Hz, 2H), 7.68 (t, J = 7.0 Hz, 2H), 7.44–7.39 (m, 2H), 7.33–7.29 (m, 4H), 6.96–6.92 (m, 3H), 6.85 (s, 1H), 5.07 and 4.86 (br s, 2H), 4.81 (s, 2H), 4.35–4.20 (m, 4H), 3.95 (s, 1H), 3.51–3.12 (m, 4H), 2.86 (br s, 4H), 1.48–1.39 (m, 11H), 1.29–1.20 (m, 10H), 0.84–0.82 (m, 3H). ^{13}C-NMR (DMSO): δ 168.6, 168.3, 168.2, 167.9, 166.9, 166.6, 162.0, 161.50, 161.45, 157.7, 156.4, 156.1, 153.7, 143.8, 142.6, 140.7, 139.4, 137.4, 129.5, 128.9, 127.6, 127.2, 127.0, 125.1, 121.3, 121.1, 120.1, 120.0, 114.4, 109.6, 94.6, 82.0, 80.9, 66.6, 65.6, 65.4, 50.0, 49.0, 47.3, 46.7, 43.7, 43.3, 38.0, 34.0, 31.2, 31.09, 31.05, 29.0, 28.6, 28.5, 28.24, 28.21, 27.6, 27.4, 27.3, 22.0, 13.9. HRMS (ESI-MS) *m/z* calcd for $C_{47}H_{59}N_5O_8S$ [M+Na]$^+$ 876.39766, found 876.39760.

3.2.5. Synthesis of *N*-[2-(*N*-9-Fluorenylmethoxycarbonyl)aminoethyl]-*N*-[[4-phenoxyacetyl-amino-6-oxo-2-(2-octylthioethyl)pyrimidin-1-yl]acetyl]glycine (**8**)

To a solution of **7** (40 mg, 47 μmol) in CH$_2$Cl$_2$ (1 mL) was added TFA (9 mL) and the resulting mixture was stirred at room temperature for 1.5 h. The solvent was removed and the residue was diluted with 5 mL diethyl ether. The solution was added to hexane (30 mL) to cause the precipitation. The resulting precipitates were collected by filtration and washed with hexane/diethyl ether (9/:1) to give a pure sample of **8** as a white powder (34 mg, 41 μmol, 93%). ^{1}H-NMR (DMSO): δ 10.42 and 10.41 (s, 1H), 7.88 (d, J = 7.5 Hz, 2H), 7.68 (t, J = 7.5 Hz, 2H), 7.46–7.39 (m, 2H), 7.33–7.39 (m, 4H), 6.98–6.92 (m, 3 H), 6.85 (s, 1H), 5.06 and 4.89 (br s, 2H), 4.81 (s, 2H), 4.34–4.21 (m, 4H), 4.00 (s, 1H), 3.52–3.13 (m, 4H), 2.85 (br s, 4H), 1.46–1.45 (m, 2H), 1.27–1.19 (m, 10H), 0.84–0.80 (m, 3H). ^{13}C-NMR (DMSO): δ 170.9, 170.2, 168.3, 167.0, 166.6, 162.0, 161.6, 161.5, 157.7, 156.4, 156.1, 153.7, 143.9, 140.7, 129.5, 127.6, 127.0, 125.1, 121.1, 120.1, 114.4, 94.6, 66.6, 65.6, 65.5, 49.2, 47.9, 47.2, 46.7, 43.7, 43.4, 38.0, 34.04, 33.96, 31.2, 31.12, 31.06, 29.0, 28.60, 28.57, 28.2, 27.44, 27.35, 22.0, 13.9. HRMS (ESI-MS) *m/z* calcd for $C_{43}H_{51}N_5O_8S$ [M+Na]$^+$ 820.33506, found 820.33508.

3.2.6. Synthesis of PNA**1**

Modified PNAs were synthesized on Novasyn TGR resin by the Fmoc-SPPS procedure. After acetylation of the N-terminal amino group, the resin was treated with 0.1 M K$_2$CO$_3$ in MeOH containing 20% H$_2$O for 3 h for deprotection of the Pac group in the precursor of AOVP. The resulting oligomer was cleaved from the resin by treatment with a cleaving cocktail containing water/triisopropylsilane/TFA (12.5:12.5:475 μL for 20 mg of resin) for 90 min. The crude mixture was eluted and precipitated in diethyl ether, dissolved in water containing 0.1% TFA, purified by reversed-phase HPLC, and characterized by MALDI-TOF mass spectrometry. The concentration of each oligomer was determined by UV absorption at 260 nm in water using the following extinction coefficients: 115,800 M^{-1} cm^{-1} (PNA**1a**), 131,000 M^{-1} cm^{-1} (PNA**1b**). MALDI-TOF MS (*m/z*) PNA**1a**: calcd 3647.5, found 3,647.4, PNA**1b**: calcd 3922.7, found 3922.3.

3.2.7. Synthesis of PNA3

To a solution of PNA1 (0.1 mM, 5 μL, 0.5 nmol) was added a solution of magnesium monoperoxyphthalate (MMPP) (0.5 mM, 0.6 μL, 0.3 nmol) in carbonate buffer adjusted to pH 10 at room temperature. After 30 min, a solution of NaOH (0.1 M, 0.5 μL, 50 nmol) was added, and the mixture was left for an additional 1 h to give PNA3. MALDI-TOF MS (*m/z*) PNA**3a**: calcd 3500.4, found 3500.4, PNA**3b**: calcd 3776.6, found 3776.0.

3.2.8. Synthesis of PNA4

A solution of sodium thiomethoxide (1 M, 0.5 μL, 0.5 μmol) was added to the above mentioned mixture of PNA3. The mixture was left for 1h and purified by HPLC to give PNA4. MALDI-TOF MS (*m/z*) PNA**4a**: calcd 3549.4, found 3549.9, PNA**4b**: calcd 3824.6, found 3824.5.

3.3. General Procedure for the Crosslinking Reaction

The crosslinking reaction was performed with 4 μM of **PNA3** and 1 μM of the target DNA or RNA labeled by fluorescein at the 5'-end in a buffer of 50 mM MES (pH 7.0). The reaction mixture was incubated at 37 °C. An aliquot of the reaction mixture was collected at each indicated time and quenched by the addition of loading dye (formamide [~100% v/v], 0.5 M EDTA [20 μL: 1 mM], xylene cyanol [0.05% w/v], and bromophenol blue [0.05% w/v]). The cross-linked products were analyzed by denaturing 20% polyacrylamide gel electrophoresis containing urea (7 M) with TBE buffer [PNA**3a**] or denaturing 20% polyacrylamide gel electrophoresis containing urea (7 M) and 10% v/v formamide at 50 °C with TBE buffer [PNA**3b**]. The labeled bands were visualized and quantified using a FLA-5100 Fluor Imager.

3.4. Melting Temperature (T_m) Measurements

UV-melting experiments were performed on a Beckman Coulter DU800. For the hetero-duplex formation study, equimolar amounts of each oligonucleotide and PNA were dissolved in 50 mM MES buffer (pH 7.0) to provide final concentrations of 1.5 μM. The solutions were heated to 90 °C for 5 min and allowed to cool to room temperature slowly. The melting profiles were recorded at 260 nm from 20 to 90 °C at a scan rate of 1 °C/min.

4. Conclusions

In this study, we have synthesized PNAs containing AOVP as a crosslinking agent and evaluated their reactivities toward the targets DNA and RNA. The PNA incorporating AOVP at the terminal position exhibited a highly selective crosslinking reactivity to the thymine in DNA and lower reactivity to any bases in RNA. This PNA is expected to provide a new probe for sequence specific PNA-DNA interactions for a regulation of the gene expression. The application of the PNA-incorporated AOVP for crosslinking reactions with the duplex DNA by forming an invasion complex is currently under investigation.

Supplementary Materials: Supplementary materials can be accessed at: http://www.mdpi.com/1420-3049/20/03/4708/s1.

Acknowledgments: This work was supported by a Grant-in-Aid for Scientific Research on Innovative Areas ("Chemical Biology of Natural Products: Target ID and Regulation of Bioactivity") and a Grant-in-Aid for Scientific Research (B) from the Japan Society for the Promotion of Science (JSPS). This work was also supported in part by the Management Expenses Grants National Universities Corporations from the Ministry of Education, Science, Sports and Culture of Japan (MEXT).

Author Contributions: F. Nagatsugi was responsible for supervision of the students, T. Akisawa and Y. Ishizawa. T. Akisawa and Y. Ishizawa performed the synthesis of the PNAs and the evaluation of their reactivities. F. Nagatsugi and T. Akisawa wrote the paper.

Conflicts of Interest: The authors declare no conflict of interest.

References

1. Nielsen, P.E.; Egholm, M.; Berg, R.H.; Buchardt, O. Sequence-Selective Recognition of DNA by Strand Dispacement with a Thymine-Substituted Polyamide. *Science* **1991**, *254*, 1497–1500. [CrossRef] [PubMed]
2. Demidov, V.V.; Potaman, V.N.; Frankkamenetskii, M.D.; Egholm, M.; Buchard, O.; Sonnichsen, S.H.; Nielsen, P.E. Stability of Peptide Nucleic-Acids in Human Serum and Cellular-Extracts. *Biochem. Pharmacol.* **1994**, *48*, 1310–1313. [CrossRef] [PubMed]
3. Nielsen, P.E. Peptide Nucleic Acids (PNA) in Chemical Biology and Drug Discovery. *Chem. Biodivers.* **2010**, *7*, 786–804. [CrossRef] [PubMed]
4. Nielsen, P.E. Targeted Gene Repair Facilitated by Peptide Nucleic Acids (PNA). *Chembiochem* **2010**, *11*, 2073–2076. [CrossRef] [PubMed]
5. Nielsen, P.E. Gene Targeting and Expression Modulation by Peptide Nucleic Acids (PNA). *Curr. Pharm. Des.* **2010**, *16*, 3118–3123. [CrossRef] [PubMed]
6. Shakeel, S.; Karim, S.; Ali, A. Peptide nucleic acid (PNA)—A review. *J. Chem. Technol. Biotechnol.* **2006**, *81*, 892–899. [CrossRef]
7. Parkash, B.; Ranjan, A.; Tiwari, V.; Gupta, S.K.; Kaur, N.; Tandon, V. Inhibition of 5′-UTR RNA Conformational Switching in HIV-1 Using Antisense PNAs. *PLoS One* **2012**, *7*, e49310. [CrossRef] [PubMed]
8. Sugiyama, T.; Kittaka, A. Chiral Peptide Nucleic Acids with a Substituent in the N-(2-Aminoethy)glycine Backbone. *Molecules* **2013**, *18*, 287–310. [CrossRef]
9. Xia, X.; Piao, X.; Bong, D. Bifacial Peptide Nucleic Acid as an Allosteric Switch for Aptamer and Ribozyme Function. *J. Am. Chem. Soc.* **2014**, *136*, 7265–7268. [CrossRef] [PubMed]
10. Devi, G.; Yuan, Z.; Lu, Y.; Zhao, Y.; Chen, G. Incorporation of thio-pseudoisocytosine into triplex-forming peptide nucleic acids for enhanced recognition of RNA duplexes. *Nucl. Acids Res.* **2014**, *42*, 4008–4018. [CrossRef] [PubMed]
11. Chenna, V.; Rapireddy, S.; Sahu, B.; Ausin, C.; Pedroso, E.; Ly, D.H. A Simple Cytosine to G-Clamp Nucleobase Substitution Enables Chiral gamma-PNAs to Invade Mixed-Sequence Double-Helical B-form DNA. *Chembiochem* **2008**, *9*, 2388–2391. [CrossRef] [PubMed]
12. Thomas, S.M.; Sahu, B.; Rapireddy, S.; Bahal, R.; Wheeler, S.E.; Procopio, E.M.; Kim, J.; Joyce, S.C.; Contrucci, S.; Wang, Y.; *et al.* Antitumor Effects of EGFR Antisense Guanidine-Based Peptide Nucleic Acids in Cancer Models. *ACS Chem. Biol.* **2013**, *8*, 345–352. [CrossRef] [PubMed]
13. Torres, A.G.; Fabani, M.M.; Vigorito, E.; Williams, D.; Al-Obaidi, N.; Wojciechowski, F.; Hudson, R.H.E.; Seitz, O.; Gait, M.J. Chemical structure requirements and cellular targeting of microRNA-122 by peptide nucleic acids anti-miRs. *Nucleic Acids Res.* **2012**, *40*, 2152–2167. [CrossRef] [PubMed]
14. Nielsen, P.E. Peptide nucleic acid. A molecule with two identities. *Acc. Chem. Res.* **1999**, *32*, 624–630. [CrossRef]
15. Gupta, A.; Lee, L.-L.; Roy, S.; Tanious, F.A.; Wilson, W.D.; Ly, D.H.; Armitage, B.A. Strand Invasion of DNA Quadruplexes by PNA: Comparison of Homologous and Complementary Hybridization. *Chembiochem* **2013**, *14*, 1476–1484. [CrossRef] [PubMed]
16. Nagatsugi, F.; Sasaki, S. Synthesis of Reactive Oligonucleotides for Gene Targeting and Their Application to Gene Expression Regulation. *Bull.Chem. Soc. Jpn.* **2010**, *83*, 744–755. [CrossRef]
17. Fujimoto, K.; Yamada, A.; Yoshimura, Y.; Tsukaguch, T.; Sakamoto, T. Details of the Ultrafast DNA Photo-Cross-Linking Reaction of 3-Cyanovinylcarbazole Nucleoside: Cis-Trans Isomeric Effect and the Application for SNP-Based Genotyping. *J. Am. Chem. Soc.* **2013**, *135*, 16161–16167. [CrossRef] [PubMed]
18. Op de Beeck, Madder, A. Unprecedented C-Selective Interstrand Cross-Linking through in Situ Oxidation of Furan-Modified Oligodeoxynucleotides. *J. Am. Chem. Soc.* **2011**, *133*, 796–807.
19. Kim, Y.; Hong, I.S. PNA/DNA interstrand cross-links from a modified PNA base upon photolysis or oxidative conditions. *Bioorg.Med. Chem. Lett.* **2008**, *18*, 5054–5057. [CrossRef] [PubMed]
20. Liu, Y.; Rokita, S.E. Inducible Alkylation of DNA by a Quinone Methide-Peptide Nucleic Acid Conjugate. *Biochemistry* **2012**, *51*, 1020–1027. [CrossRef] [PubMed]
21. Zhilina, Z.V.; Ziemba, A.J.; Nielsen, P.E.; Ebbinghaus, S.W. PNA-nitrogen mustard conjugates are effective suppressors of HER-2/neu and biological tools for recognition of PNA/DNA interactions. *Bioconjugate Chem.* **2006**, *17*, 214–222. [CrossRef]

22. Hattori, K.; Hirohama, T.; Imoto, S.; Kusano, S.; Nagatsugi, F. Formation of highly selective and efficient interstrand cross-linking to thymine without photo-irradiation. *Chem. Commun.* **2009**, *45*, 6463–6465. [CrossRef]

23. Hamilton, S.E.; Pitts, A.E.; Katipally, R.R.; Jia, X.Y.; Rutter, J.P.; Davies, B.A.; Shay, J.W.; Wright, W.E.; Corey, D.R. Identification of determinants for inhibitor binding within the RNA active site of human telomerase using PNA scanning. *Biochemistry* **1997**, *36*, 11873–11880. [CrossRef] [PubMed]

Sample Availability: *Sample Availability*: Samples of the compounds (**2**–**4**) are available from the authors. On the other compounds, please contact with the authors.

Article

Synthesis and Biological Properties of 5-(1H-1,2,3-Triazol-4-yl)isoxazolidines: A New Class of C-Nucleosides

Salvatore V. Giofrè [1,*], Roberto Romeo [1,*], Caterina Carnovale [1,†], Raffaella Mancuso [2,†], Santa Cirmi [1,†], Michele Navarra [1,†], Adriana Garozzo [3,†] and Maria A. Chiacchio [4,†]

[1] Dipartimento Scienze del Farmaco e Prodotti per la Salute, Università di Messina, Via S.S. Annunziata, Messina 98168, Italy; ccarnovale@unime.it (C.C.); scirmi@unime.it (S.C.); mnavarra@unime.it (M.N.)

[2] Dipartimento di Chimica e Tecnologie Chimiche, Università della Calabria, Via P. Bucci, 12/C, Arcavacata di Rende (CS) 87036, Italy; raffaella.mancuso@unical.it

[3] Dipartimento di Scienze Bio-mediche, Università di Catania, Via Androne 81, Catania 95124, Italy; agar@unict.it

[4] Dipartimento Scienze del Farmaco, Università di Catania, Viale A. Doria 6, Catania 95125, Italy; ma.chiacchio@unict.it

* Correspondence: sgiofre@unime.it (S.V.G.); robromeo@unime.it (R.R.); Tel.: +39-090-356230 (S.V.G. & R.R.); Fax: +39-090-676-6562 (S.V.G. & R.R)

† These authors contributed equally to this work.

Academic Editors: Mahesh K. Lakshman and Fumi Nagatsugi

Received: 4 February 2015; Accepted: 13 March 2015; Published: 24 March 2015

Abstract: A novel series of C-nucleosides, featuring the presence of a 1,2,3-triazole ring linked to an isoxazolidine system, has been designed as mimetics of the pyrimidine nucleobases. An antiproliferative effect was observed for compounds **17a** and **17b**: the growth inhibitory effect reaches the 50% in HepG2 and HT-29 cells and increases up to 56% in the SH-SY5Y cell line after 72 h of incubation at a 100 µM concentration.

Keywords: vinyl triazoles; C-Nucleosides; 1,3-dipolar cycloaddition; antiproliferative activity; microwave

1. Introduction

Structural modification of natural N-nucleosides on either the sugar unit or the nucleobase has led to the discovery of a variety of new therapeutic agents, which includes antiviral and anticancer agents [1–9]. In this context, the family of C-nucleosides constitutes a group of molecules characterized by the link of the sugar unit to a carbon atom of the pyrimidine/purine nucleobase: some of these compounds have been reported to exhibit significant antibacterial, antiviral and antitumoral activities [10].

While natural and synthetic N-nucleosides are vulnerable to enzymatic and acid-catalyzed hydrolysis of nucleosidic bond, the C-analogues are much more stable. Many natural and synthetic C-nucleosides, also containing modified heterocyclic bases, biologically active, have been described in literature. Natural antibiotics formycins [11] (in particular **FA**, **FB**, and oxoformycin B **OFB**) have been known since the early 1960s to possess antibiotic and cytotoxic properties; their antiparasitic activity (antimalarial, antischisostoma, *etc.*) was unraveled later. Pseudouridine is the most abundant natural C-nucleoside present in most tRNA and rRNA structures [12], where it has been shown to stabilize RNA duplex [13,14]. Showdomycin [15] possesses well known antibiotic and cytotoxic properties, involving inhibition of nucleoside transport into the cell [16]. Pyrazofurin [17] and tiazofurin [18] have been shown to possess a wide range of medicinal properties, including antibiotic, antiviral, and antitumor activity (Figure 1).

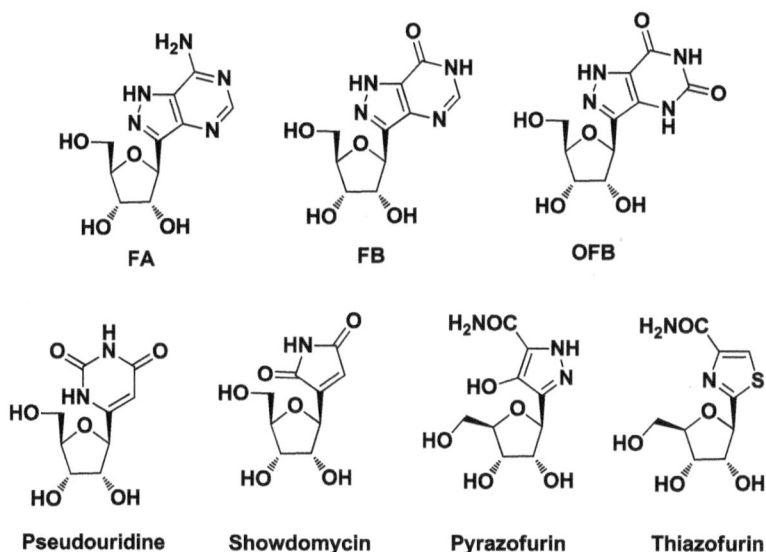

Figure 1. Examples of *C*-Nucleosides.

As part of ongoing efforts to identify new antiviral/antitumor agents, we were interested in the investigation of *N,O*-nucleosides, where the ribose moiety has been replaced by an isoxazolidine system, as mimetic of the sugar unit [19–23]. Phosphonated carbocyclic 2'-oxa-3'-azanucleosides **1** have shown to be potent inhibitors of RT of different retroviruses, following incubation with human PBMCs crude extract [24–26]; truncated phosphonated azanucleosides **2** are able to inhibit HIV and HTLV-1 viruses at concentration in the nanomolar range [27]; truncated phosphonated *N,O*-psiconucleosides **3** inhibit HIV *in vitro* infection with low or absent cytotoxicity (Figure 2) [28].

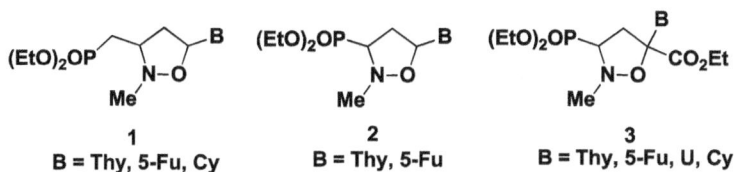

Figure 2. Phosphonated *N,O*-nucleosides.

New base-modified *N,O*-nucleosides have been also synthesized. *N,O*-pseudouridine **4** [29] and *N,O*-tiazofurin **5** [30] are characterized by a C-C bond between the isoxazolidine and the heterocyclic systems.

Recently, 1,2,3-triazolyl *N,O*-nucleosides have been designed: 3-hydroxymethyl-5-(1H-1,2,3-triazol)isoxazolidine **6** are able to inhibit proliferation of follicular and anaplastic human thyroid cancer cell lines, with IC$_{50}$ values ranging from 3.87 to 8.76 mM (Figure 3) [31]. In the same context, novel 1,2,3-triazole-appended *N,O*-nucleoside analogs **7** were developed [32]: some of these compounds show a good anticancer activity against the follicular (FTC-133), the anaplastic (8305C) human thyroid cancer cell lines, and especially on the U87MG human primary glioblastoma cell line (Figure 3).

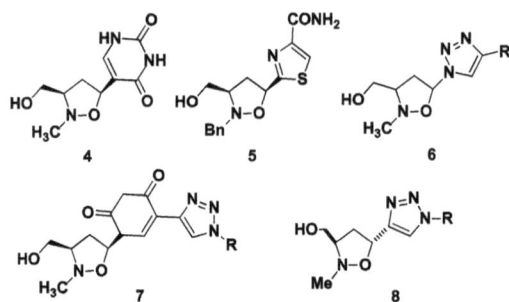

Figure 3. Examples of *N,O*-Nucleosides.

According to these promising results, we have developed a novel series of *C*-nucleosides featured by the presence of a 1,2,3-triazole ring, linked to an isoxazolidine system, as mimetic of the pyrimidine nucleobases. We report in this paper the synthesis of 5-(1H-1,2,3-triazol-4-yl)isoxazolidines **8** and their biological activities as antiviral and/or antitumoral agents.

2. Results and Discussion

C-Vinyl triazoles **12a–g** have been prepared according to Scheme 1 [33]. Thus, 3-butyn-1-ol **10** was reacted with a variety of azides **9a–g**, by click chemistry reactions, performed in H$_2$O/*tert*-BuOH (1:1) in the presence of sodium ascorbate, copper(II) sulfate and TEA at room temperature. The obtained triazole derivatives have been tosylated and then converted into vinyl triazoles **12a–g** by reaction with potassium *tert*-butoxide in *tert*-butanol (78%–90% yields).

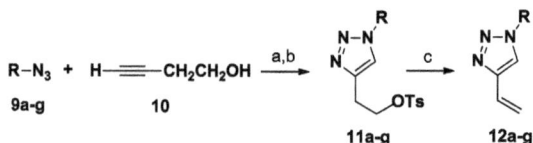

Scheme 1. Synthesis of *C*-Vinyl triazoles **12a–g**. *Reagents and conditions*: (a) CuSO$_4$·5H$_2$O, sodium ascorbate, TEA, 4 h, rt; (b) tosyl chloride, TEA, CH$_2$Cl$_2$, rt, 12 h; (c) *t*-BuOK, *t*-BuOH, 40 °C, 12 h.

The 1,3-dipolar cycloaddition of 1-substituted-4-vinyl-1,2,3-triazoles **12a–g** with *C*-[(*tert*-butyldiphenylsilyl)oxy]-*N*-methylnitrone **13** [34,35], at 150 W, 80 °C for 2 h in CHCl$_3$, proceeded with a good yield and a complete regioselectivity to give a mixture of *trans*/*cis* isoxazolidines **14a–g** and **15a–g** in a 1:1.3 relative ratio (global yield 80%–85%). Removal of the TBDPS protecting group was accomplished under standard conditions, by treating the diastereomeric mixture with TBAF in THF, to afford the triazolyl nucleosides **16a–f** and **17a–f**, which were separated by silica gel chromatography (Scheme 2, Table 1).

The diastereomeric ratio of the products was determined by ^{1}H NMR spectroscopy of the crude reaction mixture, whereas the relative configuration was assigned by NOEDS spectra. In particular, in the *cis* derivative **17a**, chosen as model compound, a positive NOE effect observed for H-4' and H-5'b (the downfield resonance of protons at C-5', 2.91 ppm) upon irradiation of H-1'(δ = 5.25 ppm), is clearly indicative of their *cis* relationship. Analogously, irradiation of H-4'(δ = 3.25 ppm) in the same compound gives rise to an enhancement in the signals corresponding to H-1' and H-5'b (δ = 5.25 and 2.91 ppm, respectively). On the contrary, no NOE effect was detected between H-4' and H-1' in compound **16a**.

The absence of *cis*/*trans* diastereoselectivity can be rationalized by assuming that the *E*-endo attack of the dipolarophile on the nitrone, which leads to *cis* adducts, competes efficiently with the *E*-exo attack, the preferred reaction pathway (steric control) leading to *trans* adducts, because of secondary orbital interactions exerted by the triazole ring. This behavior is also in agreement with literature data [36].

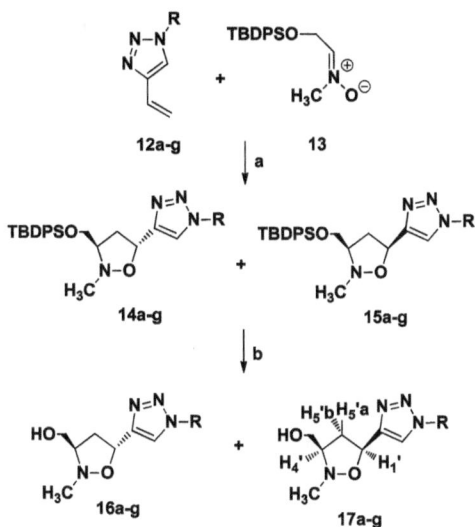

Scheme 2. Synthesis of isoxazolidinyl-triazoles **16a–g** and **17a–g** by 1,3-dipolar cycloaddition. *Reagents and conditions*: (a) CHCl3, 150 W, 80 °C, 2 h, 85%; (b) TBAF, THF, r.t., 4–5 h, global yield 93%–96% ratio *trans/cis* 1:1.3 or 1:1.

Table 1. Vinyl triazoles **12a–g** and isoxazolidinyl-triazoles **16a–g** and **17a–g** produced via Schemes 1 and 2.

R	Vinyl-Triazole	Yield %	Product	Ratio *trans:cis*	Yield % [a]
	12a	88	16a 17a	1:1.3	92
	12b	85	16b 17b	1:1.3	94
	12c	78	16c 17c	1:1.3	93
	12d	82	16d 17d	1:1	93
	12e	85	16e 17e	1:1.3	93
	12f	81	16f 17f	1:1.3	95
	12g	90	16g 17g	1:1	96

[a] Combined yield.

3. Experimental Section

3.1. General Information

Solvents and reagents were used as received from commercial sources. Melting points were determined with a Kofler apparatus (Fisher Scientific, Loughborough, UK). Elemental analyses were performed with a Perkin–Elmer elemental analyzer (PerkinElmer, Waltham, MA, USA). NMR spectra (^{1}H-NMR recorded at 500 MHz, ^{13}C-NMR recorded at 125 MHz) were obtained with a Varian instrument (Agilent Technologies, Palo Alto, CA, USA), and data are reported in ppm relative to tetramethylsilane. Thin-layer chromatographic separations were carried out on Merck silica gel 60-F254 precoated aluminum plates (Merk, Darmstadt, Germany). Flash chromatography was carried out using Merck silica gel (200–400 mesh). Preparative separations were carried out using an Büchi C-601 MPLC instrument (BUCHI Italia S.r.l., Milano, Italy) using Merck silica gel 0.040–0.063 mm, and the eluting solvents were delivered by a pump at the flow rate of 3.5–7.0 mL/min. C-[(*tert*-Butyldiphenylsilyl)oxy]-*N*-methyl nitrone was prepared according to described procedures [34,35]. Benzyl/alkyl and aromatic azides were synthesized according to literature procedures [32].

3.2. General 1,3-Dipolar Cycloaddition Procedure

A solution of **12a** (0.50 g, 2.92 mmol) and nitrone **13** (1.10 g, 3.50 mmol) in CHCl$_3$ (5 mL) was put in a sealed tube and irradiated under microwave conditions at 150 W, 80 °C, for 2 h. The removal of the solvent *in vacuo* afforded a crude material which, after flash chromatography purification by using as eluent a mixture of cyclohexane/ethyl acetate 7:3, gave the unseparable mixture (*trans/cis*) of compound **14a** and **15a**, as yellow oil, that was used for the next reaction, yield 1.24 g (85%). The ^{1}H-NMR spectrum of the crude reaction mixture shows the presence of *trans* and *cis* isomers in 1:1.3 ratio, respectively. Compounds **14b–g** and **15b–g** were prepared by the 1,3-dipolar cycloaddition procedure in 80%–85% yield as yellow oil and then used for the next reaction.

3.3. General Desilylation of the Hydroxymethyl Group Procedure: Synthesis of 16 and 17

A solution of compounds **14a** and **15a** (1.24 g, 2.49 mmol) and TBAF (0.90 mL, 3.73 mmol) in freshly distilled THF (30 mL) was stirred until desilylation was completed (TLC, 4–5 h). Volatiles were flash evaporated, and the residue was purified by MPLC (CH$_2$Cl$_2$/MeOH, 98:2) to afford **17a–g** first eluted isomer (*trans*) and **16a–g** second eluted isomer (*cis*) in 92% global yield.

((3RS,5RS)-2-Methyl-5-(1-phenyl-1H-1,2,3-triazol-4-yl)isoxazolidin-3-yl) methanol (**16a**): White solid, 52% yield, mp = 57–59 °C. ^{1}H-NMR (CDCl$_3$): δ = 8.00 (s, 1H), 7.70 (d, *J* = 7.7 Hz, 2H), 7.50 (t, *J* = 7.8 Hz, 2H), 7.42 (t, *J* = 7.4 Hz, 1H), 5.52 (dd, *J* = 8.5, 6.6 Hz, 1H), 3.66–3.52 (m, 2H), 3.31–3.19 (m, 1H), 2.94 (dt, *J* = 12.8, 8.5 Hz, 2H), 2.81 (s, 3H), 2.36 (ddd, *J* = 12.8, 6.6, 4.8, 1H). ^{13}C-NMR (CDCl$_3$): δ = 148.50, 137.07, 129.86, 128.95, 120.66, 119.84, 71.06, 69.53, 63.42, 44.94, 36.62; Anal. Calcd for C$_{13}$H$_{16}$N$_4$O$_2$: C, 59.99; H, 6.20; N, 21.52; found C, 60.04; H, 6.27; N, 21.60.

((3RS,5SR)-2-Methyl-5-(1-phenyl-1H-1,2,3-triazol-4-yl)isoxazolidin-3-yl) methanol (**17a**): White solid, 40% yield, mp = 100–102 °C. ^{1}H-NMR (CDCl$_3$): δ = 7.97 (s, 1H), 7.71 (d, *J* = 7.8 Hz, 2H), 7.51 (t, *J* = 7.8 Hz, 2H), 7.43 (t, *J* = 7.4 Hz, 1H), 5.25 (t, *J* = 8.0 Hz, 1H), 3.68 (ddd, *J* = 17.2, 11.4, 4.8 Hz, 2H), 3.25–3.18 (m, 1H), 2.98–2.85 (m, 1H), 2.83 (s, 3H), 2.62 (ddd, *J* = 12.7, 8.0, 5.0 Hz, 1H). ^{13}C-NMR (CDCl$_3$): δ = 148.15, 137.06, 129.89, 128.99, 120.72, 120.49, 72.39, 69.36, 62.35, 45.70, 36.78; Anal. Calcd for C$_{13}$H$_{16}$N$_4$O$_2$: C, 59.99; H, 6.20; N, 21.52; found C, 60.08; H, 6.26; N, 21.48.

((3RS,5RS)-5-(1-Benzyl-1H-1,2,3-triazol-4-yl)-2-methylisoxazolidin-3-yl) methanol (**16b**): Compound **16b** was prepared by the general desilylation procedure in 53.1% yield as yellow oil. ^{1}H-NMR (CDCl$_3$): δ = 7.46 (s, 1H), 7.38–7.30 (m, 3H), 7.26–7.23 (m, 2H), 5.51–5.43 (m, 2H), 5.40 (dd, *J* = 8.2, 7.0 Hz, 1H), 3.55–3.46 (m, 2H), 3.23–3.14 (m, 1H), 2.84 (dt, *J* = 12.8, 8.4 Hz, 1H), 2.74 (s, 3H), 2.26 (ddd, *J* = 12.8, 6.7,

4.7, 1H), 2.15 (bs, 1H). ^{13}C-NMR (CDCl$_3$): δ = 147.87, 134.48, 129.21, 128.88, 128.24, 121.55, 71.00, 69.48, 63.37, 54.32, 44.90, 36.44; Anal. Calcd for C$_{14}$H$_{18}$N$_4$O$_2$: C, 61.30; H, 6.61; N, 20.42; found C, 61.36; H, 6.67; N, 20.45.

((3RS,5SR)-5-(1-Benzyl-1H-1,2,3-triazol-4-yl)-2-methylisoxazolidin-3-yl) methanol (**17b**): Compound **17b** was prepared by the general desilylation procedure in 40.9% yield as white solid, mp = 117–119 °C. ^{1}H-NMR (CDCl$_3$): δ = 7.43 (s, 1H), 7.39–7.33 (m, 3H), 7.28–7.24 (m, 2H), 5.54–5.44 (m, 2H), 5.12 (t, J = 7.9 Hz, 1H), 3.63 (ddd, J = 17.2, 11.4, 4.8, 2H), 3.10–3.18 (m, 1H), 2.82 (dt, J = 12.7, 8.3, 1H), 2.76 (s, 3H), 2.52 (ddd, J = 12.7, 7.9, 5.0, 1H). ^{13}C-NMR (CDCl$_3$): δ = 147.72, 134.52, 129.24, 128.92, 128.26, 122.10, 72.35, 69.28, 62.31, 54.31, 45.55, 36.77; Anal. Calcd for C$_{14}$H$_{18}$N$_4$O$_2$: C, 61.30; H, 6.61; N, 20.42; found C, 61.38; H, 6.68; N, 20.47.

((3RS,5RS)-2-Methyl-5-(1-(pyridin-2-ylmethyl)-1H-1,2,3-triazol-4-yl) isoxazolidin-3-yl)methanol (**16c**): Compound **16c** was prepared by the general desilylation procedure in 52.6% yield as yellow oil. ^{1}H-NMR (CDCl$_3$): δ = 7.71 (s, 1H), 7.65 (td, J = 7.8, 1.7 Hz, 1H), 7.23 (dd, J = 7.0, 5.1 Hz, 1H), 7.16 (d, J = 7.8 Hz, 1H), 5.65–5.54 (m, 2H), 5.41 (dd, J = 8.5, 6.8, 1H), 3.58–3.48 (m, 2H), 3.23–3.15 (m, 1H), 2.84 (dt, J = 12.8, 8.5, 1H), 2.74 (s, 3H), 2.39 (bs, 1H), 2.28 (ddd, J = 12.8, 6.5, 4.8 Hz, 1H). ^{13}C-NMR (CDCl$_3$): δ = 154.31, 149.82, 147.92, 137.49, 123.55, 122.58, 122.37, 70.93, 69.50, 63.32, 55.68, 44.86, 36.42; Anal. Calcd for C$_{13}$H$_{17}$N$_5$O$_2$: C, 56.71; H, 6.22; N, 25.44; found C, 56.76; H, 6.27; N, 25.48.

((3RS,5SR)-2-Methyl-5-(1-(pyridin-2-ylmethyl)-1H-1,2,3-triazol-4-yl) isoxazolidin-3-yl)methanol (**17c**): Compound **17c** was prepared by the general desilylation procedure in 40.4% yield as yellow solid, mp = 76–78 °C. ^{1}H-NMR (CDCl$_3$): δ = 8.66–8.56 (m, 1H), 7.72 (s, 1H), 7.70 (dt, J = 7.7, 1.7 Hz, 1H), 7.33–7.26 (m, 1H), 7.23 (d, J = 7.8 Hz, 1H), 5.73–5.59 (m, 2H), 5.18 (t, J = 7.8 Hz, 1H), 3.66 (ddd, J = 17.2, 11.4, 4.8 Hz, 2H), 3.17 (s, 1H), 2.93–2.81 (m, 1H), 2.80 (s, 3H), 2.62–2.52 (m, 1H). ^{13}C-NMR (CDCl$_3$): δ = 154.31, 149.82, 137.49, 123.55, 122.58, 122.37, 70.93, 69.50, 63.32, 55.68, 44.86, 36.42; Anal. Calcd for C$_{13}$H$_{17}$N$_5$O$_2$: C, 56.71; H, 6.22; N, 25.44; found C, 56.78; H, 6.30; N, 25.50.

((3RS,5RS)-5-(1-(4-Chloro-3-(trifluoromethyl)phenyl)-1H-1,2,3-triazol-4-yl)-2-methyl isoxazolidin-3-yl)methanol (**16d**): Compound **16d** was prepared by the general desilylation procedure in 46.5% yield as yellow solid, mp = 94–96 °C. ^{1}H-NMR (CDCl$_3$): δ = 8.07 (s, 1H), 8.06 (d, J = 2.6 Hz, 1H), 7.86 (dd, J = 8.7, 2.6 Hz, 1H), 7.65 (d, J = 8.7 Hz, 1H), 5.49 (dd, J=8.7, 6.3 Hz, 1H), 3.58 (d, J = 5.7 Hz, 2H), 3.26–3.18 (m, 1H), 2.94 (dt, J = 12.8, 8.7, 1H), 2.79 (s, 3H), 2.35 (ddd, J = 12.8, 6.3, 4.9, 1H). ^{13}C-NMR (CDCl$_3$): δ = 149.50, 135.51, 133.11, 132.57, 130.06 (q, J = 32.4 Hz), 124.38, 123.24, 121.07, 119.73, 119.64 (q, J = 5.5 Hz), 70.94, 69.55, 63.25, 44.92, 36.78, 29.78; Anal. Calcd for C$_{14}$H$_{14}$ClF$_3$N$_4$O$_2$: C, 46.36; H, 3.89; N, 15.45; found C, 46.41; H, 3.93; N, 15.48.

((3RS,5SR)-5-(1-(4-Chloro-3-(trifluoromethyl)phenyl)-1H-1,2,3-triazol-4-yl)-2-methyl isoxazolidin-3-yl)methanol (**17d**): Compound **17d** was prepared by the general desilylation procedure in 46.5% yield as yellow thick oil. ^{1}H-NMR (CDCl$_3$): δ = 8.08 (d, J = 2.5 Hz, 1H), 8.00 (s, 1H), 7.89 (dd, J = 8.6, 2.5 Hz, 1H), 7.69 (d, J = 8.6 Hz, 1H), 5.26 (t, J = 7.9, 1H), 3.69 (ddd, J=17.1, 11.4, 4.7, 2H), 3.25–3.18 (s, 1H), 2.90 (dt, J = 12.8, 8.0, 1H), 2.84 (s, 3H), 2.66 (ddd, J = 12.8, 8.0, 5.0, 1H). ^{13}C-NMR (CDCl$_3$): δ = 149.02, 135.55, 134.94, 133.22, 132.77, 130.11, 127.85, 124.51, 119.77 (q, J = 5.3), 72.29, 69.31, 62.28, 45.62, 36.84; Anal. Calcd for C$_{14}$H$_{14}$ClF$_3$N$_4$O$_2$: C, 46.36; H, 3.89; N, 15.45; found C, 46.40; H, 3.95; N, 15.50.

((3RS,5RS)-5-(1-(4-Methoxyphenyl)-1H-1,2,3-triazol-4-yl)-2-methylisoxazolidin -3-yl)methanol (**16e**): Compound **16e** was prepared by the general desilylation procedure in 52.6% yield as white solid, mp = 57–59 °C. ^{1}H-NMR (CDCl$_3$): δ = 7.91 (s, 1H), 7.56 (dd, J =8.9, 1.3 Hz, 2H), 6.96 (dd, J = 8.9, 1.3 Hz, 2H), 5.48 (dd, J = 8.3, 6.8 Hz, 1H), 3.82 (s, H), 3.62–3.53 (m, 2H), 3.25–3.18 (m, 1H), 2.94–2.85 (m, 1H), 2.78 (s, 3H), 2.37–2.29 (m, 1H). ^{13}C-NMR (CDCl$_3$): δ = 159.89, 148.21, 130.45, 122.21, 120.01, 114.81, 70.99, 69.53, 63.38, 55.67, 44.91, 36.64; Anal. Calcd for C$_{14}$H$_{18}$N$_4$O$_3$: C, 57.92; H, 6.25; N, 19.30; found C, 57.96; H, 6.28; N, 19.27.

((3RS,5SR)-5-(1-(4-Methoxyphenyl)-1H-1,2,3-triazol-4-yl)-2-methyl isoxazolidin-3-yl)methanol (**17e**): Compound **17e** was prepared by the general desilylation procedure in 40.4% yield as white solid, mp = 103–105 °C. ^{1}H-NMR (CDCl$_3$): δ = 7.88 (s, 1H), 7.60 (d, *J* = 8.9 Hz, 2H), 7.00 (d, *J* = 8.9 Hz, 2H), 5.24 (t, *J* = 7.8 Hz, 1H), 3.85 (s, 3H), 3.67 (ddd, *J*=17.1, 11.4, 4.8 Hz, 2H), 3.22–3.15 (m, 1H), 2.97–2.86 (m, 1H), 2.82 (s, 3H), 2.76 (bs, 1H), 2.60 (ddd, *J* = 12.7, 7.9, 5.0 Hz, 1H). ^{13}C-NMR (CDCl$_3$): δ = 160.00, 147.89, 130.51, 122.36, 120.66, 114.89, 72.39, 69.35, 55.74, 45.68, 36.79; Anal. Calcd for C$_{14}$H$_{18}$N$_4$O$_3$: C, 57.92; H, 6.25; N, 19.30; found C, 57.99; H, 6.30; N, 19.33.

((3RS,5RS)-5-(1-(4-Fluorophenyl)-1H-1,2,3-triazol-4-yl)-2-methylisoxazolidin-3-yl)methanol (**16f**): Compound **16f** was prepared by the general desilylation procedure in 53.7% yield as white solid, mp = 68–70 °C. ^{1}H-NMR (CDCl$_3$): δ = 7.96 (s, 1H), 7.71–7.64 (m, 2H), 7.20 (t, *J* = 8.5 Hz, 2H), 5.51 (dd, *J* = 8.5, 6.6 Hz, 1H), 3.58 (d, *J* = 5.7 Hz, 2H), 3.31–3.18 (m, 1H), 2.94 (dt, *J* = 12.8, 8.5 Hz, 1H), 2.80 (s, 3H), 2.36 (ddd, *J* = 12.8, 6.2, 4.9, 1H). ^{13}C-NMR (CDCl$_3$): δ = 163.56, 161.57, 148.73, 133.37, 122.65 (d, *J* = 8.7 Hz), 120.00, 116.84 (d, *J* = 23.2 Hz), 71.04, 69.51, 63.38, 44.93, 36.65; Anal. Calcd for C$_{13}$H$_{15}$FN$_4$O$_2$: C, 56.11; H, 5.43; N, 20.13; found C, 56.14; H, 5.49; N, 20.17.

((3RS,5SR)-5-(1-(4-Fluorophenyl)-1H-1,2,3-triazol-4-yl)-2-methylisoxazolidin-3-yl)methanol (**17f**): Compound **17f** was prepared by the general desilylation procedure in 41.3% yield as white solid, mp = 102–104 °C. ^{1}H-NMR (CDCl$_3$): δ = 7.93 (s, *J* = 1.0, 1H), 7.73–7.63 (m, 2H), 7.24–7.14 (m, 2H), 5.24 (t, *J* = 7.8 Hz, 1H), 3.68 (ddd, *J* = 17.2, 11.5, 4.8 Hz, 2H), 3.24–3.17 (m, 1H), 2.97–2.86 (m, 1H), 2.82 (s, 3H), 2.66–2.56 (m, 1H). ^{13}C-NMR (CDCl$_3$): δ = 163.58, 161.59, 148.26, 133.34, 122.70 (d, *J* = 8.5 Hz), 120.67, 116.86 (d, *J* = 23.2 Hz), 72.30, 69.38, 62.32, 45.68, 36.79; Anal. Calcd for C$_{13}$H$_{15}$FN$_4$O$_2$: C, 56.11; H, 5.43; N, 20.13; found C, 56.15; H, 5.47; N, 20.19.

((3RS,5RS)-2-Methyl-5-(1-(naphthalen-2-ylmethyl)-1H-1,2,3-triazol-4-yl) isoxazolidin-3-yl)methanol (**16g**): Compound **16g** was prepared by the general desilylation procedure in 48.0% yield as white solid, mp = 95–97 °C. ^{1}H-NMR (CDCl$_3$): δ = 7.83–7.80 (m, 3H), 7.74 (s, 1H), 7.51 (dd, *J* = 6.2, 3.3 Hz, 2H), 7.48 (s, 1H), 7.33 (dd, *J* = 8.4, 1.4 Hz, 1H), 5.70–5.56 (m, 2H), 5.41 (dd, *J* = 8.3, 7.0 Hz, 1H), 3.51 (d, *J* = 6.2 Hz, 2H), 3.21–3.17 (m, 1H), 2.84 (dt, *J* = 12.8, 8.3 Hz, 1H), 2.74 (s, 3H), 2.31–2.22 (m, 1H), 1.91 (bs, 1H). ^{13}C-NMR (CDCl$_3$): δ = 148.00, 133.32, 131.83, 129.32, 128.07, 127.90, 127.67, 126.87, 125.51, 71.06, 69.44, 63.40, 54.57, 44.89, 36.40; Anal. Calcd for C$_{18}$H$_{20}$N$_4$O$_2$: C, 66.65; H, 6.21; N, 17.27; found C, 66.68; H, 6.25; N, 17.30.

((3RS,5RS)-2-Methyl-5-(1-(naphthalen-2-ylmethyl)-1H-1,2,3-triazol-4-yl) isoxazolidin-3-yl)methanol (**17g**): Compound **17g** was prepared by the general desilylation procedure in 48.0% yield as white solid, mp = 115–117 °C. ^{1}H-NMR (CDCl$_3$): δ = 7.88–7.79 (m, 3H), 7.74 (s, *J* = 6.8 Hz, 1H), 7.55–7.48 (m, 2H), 7.46 (s, 1H), 7.34 (d, *J* = 8.3 Hz, 1H), 5.72–5.60 (m, 2H), 5.12 (t, *J* = 7.9 Hz, 1H), 3.62 (ddd, *J* = 17.2, 11.5, 4.8, 2H), 3.08–3.16 (m, 1H), 2.90–2.76 (m, 1H), 2.75 (s, 3H), 2.56–2.47 (m, 1H). ^{13}C-NMR (CDCl$_3$): δ = 147.79, 133.29, 131.85, 129.29, 128.03, 127.89, 127.62, 126.87, 126.84, 125.49, 122.15, 72.35, 69.27, 62.28, 54.50, 45.54, 36.80; Anal. Calcd for C$_{18}$H$_{20}$N$_4$O$_2$: C, 66.65; H, 6.21; N, 17.27; found C, 66.70; H, 6.27; N, 17.32.

3.4. Biological Tests

3.4.1. Antiviral Activity

Compounds **16–17** were evaluated for their ability to inhibit the replication of a variety of DNA and RNA viruses, using the following cell-based assays: (a) Vero cell cultures: poliovirus 1, human echovirus 9, coxsackievirus B4, adenovirus type 2, herpes simplex type 1 (HSV-1), herpes simplex type 2 (HSV-2); (b) human embryonic lung fibroblast cells (MRC-5): cytomegalovirus (CMV: VR-538); (c) African green monkey kidney cells (BS-C-1): varicella-zoster virus (VZV). Acyclovir was used as the reference compounds. Unfortunately, no inhibitory activity against any virus was detected for the evaluated compounds (data not shown).

Molecules **2015**, *20*, 5260–5275

3.4.2. Antiproliferative Activity

The antiproliferative activity of all the synthesized C-Nucleosides was evaluated. An antiproliferative effect was observed for compounds **17a** and **17b**, while other derivatives show a IC_{50} in the range 150–200 µM. In particular, as shown in Figures 4A and 5A cell culture incubation with increasing concentration of **17a** and **17b**, ranging from 1 µM to 100 µM for 24, 48 and 72 h, reduced the proliferation in all the cancer cell lines with a similar trend.

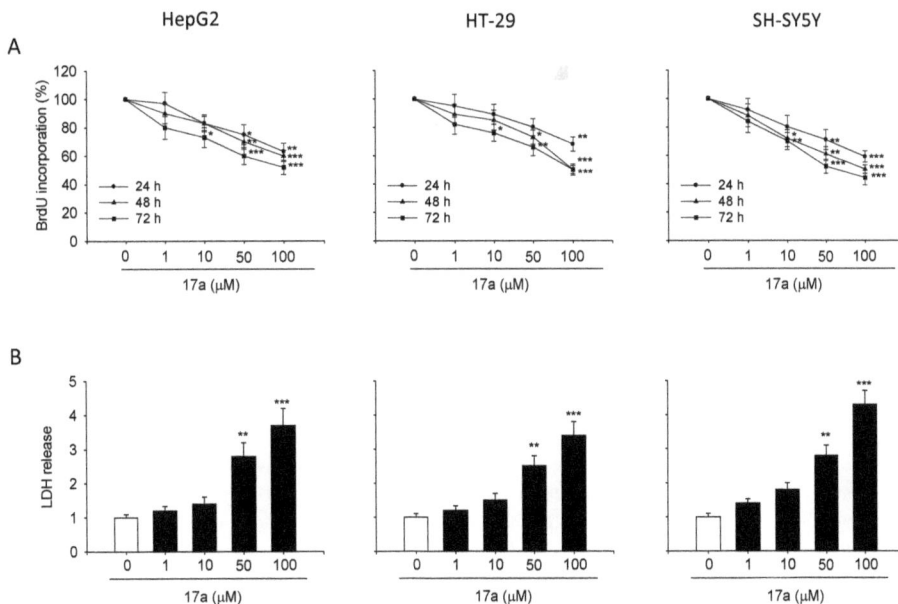

Figure 4. Compound **17a** reduces cell proliferation and induce LDH release. HepG2, HT-29 and SH-SY5Y cells were exposed to increased concentration of **17a** compound for 24, 48 and 72 h. (**A**) Proliferation rate assessed by BrdU assay. (**B**) Cytotoxic effect assessed in terms of LDH release after 24 h of exposure. Each value is the mean ± S.E.M. of three experiments performed eight times (BrdU) or in triplicate (LDH) and repeated three different times. * $p < 0.05$ *vs.* ctrl, ** $p < 0.01$ *vs.* ctrl, *** $p < 0.001$ *vs.* ctrl.

In particular, data of the BrdU assay shows that the growth inhibitory effect induced by **17a** reaches the 50% in both HepG2 and HT-29 cells and increases up to 56% in the SH-SY5Y cell line ($p < 0.001$ *vs.* control) after 72 h of incubation at a 100 µM concentration. Reduction of cell proliferation was also observed when the cells were exposed for 48 ($p < 0.001$ *vs.* control) and 24 hours ($p < 0.01$ *vs.* control in HepG2 and HT-29 cells and $p < 0.001$ in SH-SY5Y cell line; Figure 4A) to 100 µM concentration of the compound **17a**.

Lesser antiproliferative effect was observed treating the cells with compound **17a** at concentrations of 50 and 10 µM (expecially for longer time of exposure). Similar results were obtained by the compound **17b** (56% of cell growth inhibition for the HepG2, 62% for the HT-29 and the SH-SY5Y cell lines at the 100 µM concentration for 72 h; Figure 5A). Results of the BrdU assay show very close alignment with those of MTT test performed after 24, 48 and 72 h of incubation with both **17a** and **17b** (see Supplementary Materials Figures S1A and S2A).

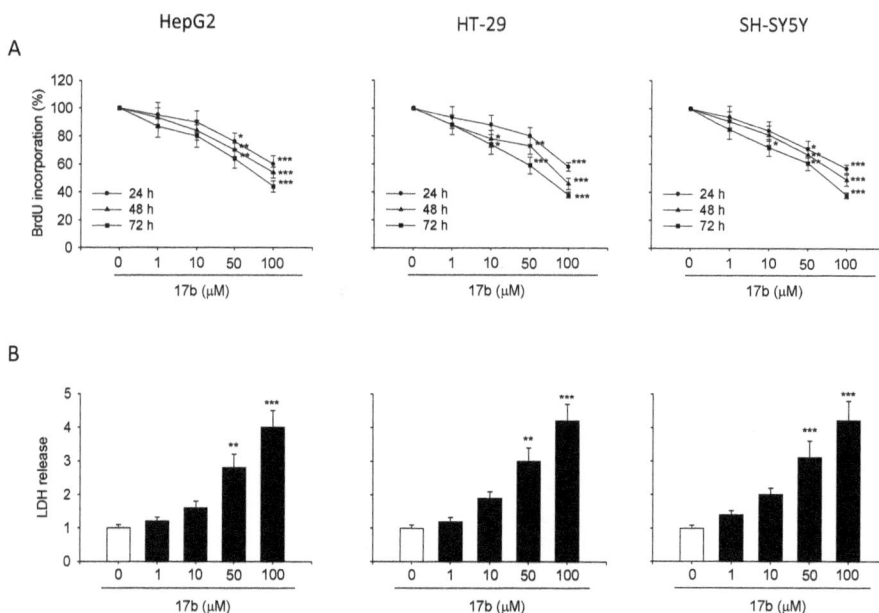

Figure 5. Compound **17b** inhibit cell growth and determine LDH release. (**A**) Proliferation rate assessed by BrdU assay after 24–72 h of treatment. (**B**) Cytotoxic activities evaluated by the test of LDH after 24 h of incubation. Each value is the mean ± S.E.M. of three experiments performed in eight times (BrdU) or in triplicate (LDH) and repeated three different times. * $p < 0.05$ *vs.* ctrl, ** $p < 0.01$ *vs.* ctrl, *** $p < 0.001$ *vs.* ctrl.

3.4.3. Cytoxicity Evaluation

The cytotoxic effect induced by **17a** and **17b** was evaluated through either LDH and trypan blue dye exclusion assays. Figures 4B and 5B show that both **17a** and **17b** caused increase of the LDH release at 50 and 100 μM concentration in all the investigated cell lines. Furthermore, LDH release was accompanied by a significant increase of cell death, as detected by the trypan blue test (see Supplementary Materials Figures S1B and S2B), underlying the cytotoxic activity of the compounds **17a** and **17b**.

3.4.4. Cell Culture and Treatments

HepG2 (hepatocellular carcinoma), HT-29 (colorectal adenocarcinoma) and SH-SY5Y (neuroblastoma) cells were obtained originally from ATCC (Rockville, MD, USA). These cell lines, maintained in RPMI supplemented with 10% heat-inactivated fetal bovine serum, 1 mM sodium pyruvate, 2 mM glutamine, penicillin/streptomycin (100 units/mL and 100 μg/mL, respectively), were grown at 37 °C and 5% CO_2 conditions. All the reagents for cell cultures were from Gibco (Life Technologies, Monza, Italy). For biological investigations, 100 mM stock solutions were prepared dissolving the tested compounds in dimethyl sulfoxide (DMSO). Small aliquots were stored at −20 °C and were diluted in culture media to the final concentration, ranging from 1 to 100 μM, just prior the use. The highest DMSO concentration used in this study (0.1%) did not have any appreciable effect on cell proliferation or cytotoxicity.

3.4.5. Antiproliferative Activity

The antiproliferative activity of all the synthesized compounds was evaluated by either BrdU and MTT assays. For the first assay, the cells were seeded in 96-well plates at a density of 10×10^4 cell/well (HT-29 cells and SH-SY5Y) and 12×10^4 (HepG2) and allowed to stand overnight. Then, the culture medium was replaced with clean media containing the tested compounds at a concentrations ranging from 1 to 100 µM or with media with equivalent dilutions of DMSO (control cultures). After 24, 48 or 72 h incubation, BrdU assay was performed as described [37]. BrdU is a uridine derivative, structural analog of thymidine, which can be used as a marker for proliferation, because it is incorporated into DNA during the synthesis-phase of the cell cycle as a substitute for thymidine. Cells marked by BrdU incorporation may be detected after addition of goat anti-mouse IgG-peroxidase conjugated secondary antibody. Results are expressed as percentages of the absorbance measured in control cells.

Cell growth was also detected by the MTT assay as reported with modification [38]. The cells were seeded onto 96-well plates at a density of 5×10^3 cells/well (HT-29 cells and SH-SY5Y) or 6×10^3 cells/well (HepG2). The following day, cells were treated with the test compounds as described above. At the end of the exposure time, the plates were centrifuged at 1200 rpm for 10 min, the supernatants were replaced with clean medium without phenol red containing 0.5 mg/mL of 3-(4,5-dimethylthiazole-2-yl)-2,5-diphenyltetrazolium bromide (MTT; Sigma-Aldrich, Milan, Italy), and the plates were returned in the incubator for 4 h. Then, the solutions were removed from the wells and crystals of formazan (MTT metabolic product) were solubilised by 100 µL HCl/isopropanol 0.1 N lysis buffer. The latter were spectrophotometrically quantified at a wavelength of 595 nm (iMark™ microplate reader, Bio-Rad Laboratories, Milan, Italy). Results of the cell proliferation assays are expressed as percentages in untreated cultures. The tests were performed in triplicate (BrdU) or eight times (MTT) and repeated three different times.

3.4.6. Detection of Cell Viability

Lactate dehydrogenase is a stable cytosolic enzyme which is released into the surrounding culture medium when the plasma membrane is damaged, and can be considered a marker of cytotoxicity. The released LDH can be quantified by a coupled enzymatic reaction. First, LDH catalyzes the conversion of lactate to pyruvate via reduction of NAD^+ to NADH. Second, diaphorase uses NADH to reduce a tetrazolium salt to a red formazan product. Therefore, the level of formazan formation is directly proportional to the amount of released LDH in the medium. HepG2, HT-29 and SH-SY5Y cells were seeded in 96-well plates in a number of 15×10^3 cells/well. The following day, cells were incubated with the synthesized C-Nucleosides for 24 h. Then the LDH concentration was measured as reported [39] by using commercial LDH kit (CytoTox 96® Non-Radioactive Cytotoxicity Assay, Promega, Milan, Italy), according to the manufacturer's protocol. LDH levels are extrapolated as the values detected in control cells, which are arbitrarily expressed as 1.

Cell viability in presence of the tested compounds was assessed also by the trypan blue dye (0.4% w/v) exclusion test [40] and cell death was reported as the percentage of stained (non-viable) *vs.* total cells counted. All the experiments were carried out in triplicate and repeated three times.

3.4.7. Statistical Analysis

Data were expressed as mean $\pm$ S.E.M. and statistically evaluated for differences using one-way analysis of variance (ANOVA), followed by Turkey-Kramer multiple comparisons test (GraphPAD Software for Science, GraphPad Software, Inc., La Jolla, CA, USA).

4. Conclusions

We report in this paper the synthesis of 5-(1*H*-1,2,3-triazol-4-yl)isoxazolidines, a novel series of C-nucleosides, featured by the presence of a 1,2,3-triazole ring linked to an isoxazolidine system, as mimetic of the pyrimidine nucleobases. The synthesized compounds have been evaluated for their ability to inhibit the replication of a variety of DNA and RNA viruses: unfortunately, no inhibitory activity against any virus was detected for the evaluated compounds. The antiproliferative activity of all the synthesized *C*-Nucleosides was also tested. An antiproliferative effect was observed for compounds **17a** and **17b**: the induced growth inhibitory effect reaches the 50% in HepG2 and HT-29 cells and increases up to 56% in the SH-SY5Y cell line ($p < 0.001$ *vs.* control) after 72 h of incubation at a 100 µM concentration.

Supplementary Materials: Supplementary materials can be accessed at: http://www.mdpi.com/1420-3049/20/04/5260/s1.

Acknowledgments: We gratefully acknowledge the Italian Ministry of Education, Universities, and Research (MIUR), the Universities of Messina and Catania, and the Interuniversity Consortium for Innovative Methodologies and Processes for Synthesis (CINMPIS) for partial financial support.

Author Contributions: SVG and RR, designed research; MN, SC, AG performed biological data; SVG, RR, CC, RM and MAC performed research and analyzed the data; SVG and RR wrote the paper. All authors read and approved the final manuscript.

Conflicts of Interest: The authors declare no conflict of interest.

References

1. De Clercq, E. The history of antiretrovirals: Key discoveries over the past 25 years. *Rev. Med. Virol* **2009**, *19*, 287–299.

2. De Clercq, E. *Antiviral Drug Strategies*; Wiley-VCH: Weinheim, Germany, 2011.

3. Parker, W.B. Enzymology of Purine and Pyrimidine Antimetabolites Used in the Treatment of Cancer. *Chem. Rev.* **2009**, *109*, 2880–2893. [CrossRef] [PubMed]

4. Vanek, V.; Budesinsky, M.; Rinnova, M.; Rosenberg, I. Prolinol-based nucleoside phosphonic acids: New isosteric conformationally flexible nucleotide analogues. *Tetrahedron* **2009**, *65*, 862–876. [CrossRef]

5. Ishitsuka, H.; Shimma, N. Capecitabine Preclinical Studies: From Discovery to Translational Research. In *Modified Nucleosides in Biochemistry, Biotechnology and Medicine*; Herdewijn, P., Ed.; Wiley-VCH: Weinheim, Germany, 2008; pp. 587–600.

6. Kumamoto, H.; Topalis, D.; Broggi, J.; Pradere, U.; Roi, V.; Berteina-Raboin, S.; Nolan, S.P.; Deville-Bonne, D.; Andrei, G.; Snoeck, R.; *et al.* Preparation of acyclo nucleoside phosphonate analogues based on cross-metathesis. *Tetrahedron* **2008**, *64*, 3517–3526.

7. Jung, K.-H.; Marx, A. Synthesis of 4'-C-modified 2'-Deoxyribonucleoside Analogues and Oligonucleotides. *Curr. Org. Chem.* **2008**, *12*, 343–354. [CrossRef]

8. Littler, E.; Zhou, X.-X. Deoxyribonucleic Acid Viruses: Antivirals for Herpesviruses and Hepatitis B Virus. In *Comprehensive Medicinal Chemistry II*; Taylor, J.B., Triggle, D.J., Eds.; Elsevier: Oxford UK, 2006; Volume 7, pp. 295–327.

9. Thottassery, J.V.; Westbrook, L.; Someya, H.; Parker, W.B. c-Abl-independent p73 stabilization during gemcitabine- or 4'-thio-beta-D-arabinofuranosylcytosine-induced apoptosis in wild-type and p53-null colorectal cancer cells. *Mol. Cancer Ther.* **2006**, *5*, 400–410. [CrossRef] [PubMed]

10. Stambasky, J.; Hocek, M.; Kocovsky, P. C-Nucleosides: Synthetic Strategies and Biological Applications. *Chem. Rev.* **2009**, *109*, 6729–6764. [CrossRef] [PubMed]

11. Bzowska, A. Formycins and their Analogues: Purine Nucleoside Phosphorylase Inhibitors and Their Potential Application in Immunosuppression and Cancer. In *Modified Nucleosides: in Biochemistry, Biotechnology and Medicine*; Herdewijn, P., Ed.; Wiley-VCH: Weinheim, Germany, 2008; pp. 473–510.

12. Desaulniers, J.P.; Chang, Y.C.; Aduri, R.; Abeysirigunawardena, S.C.; Santa Lucia, J.; Chow, C.S. Pseudouridines in rRNA helix 69 play a role in loop stacking interactions. *Org. Biomol. Chem.* **2008**, *6*, 3892–3895. [CrossRef] [PubMed]

13. Matsuura, S.; Shiratori, O.; Katagiri, K. Antitumor activity of showdomycin. *J. Antibiot.* **1964**, *17*, 234–237. [PubMed]

14. Nakagawa, Y.; Kano, H.; Tsukuda, Y.; Koyama, H. Structure of a new class of C-nucleoside antibiotic, showdomycin. *Tetrahedron Lett.* **1967**, *42*, 4105–4109. [CrossRef] [PubMed]

15. Uehara, Y.I.; Rabinowicz, J. Showdomycin and its reactive moiety, maleimide. A comparison in selective toxicity and mechanism of action *in vitro. Biochem. Pharmacol.* **1980**, *29*, 2199–2204.

16. Shaban, M.A., E; Nasr, A.Z. The Chemistry of C-Nucleosides and Their Analogs I: C-Nucleosides of Hetero Monocyclic Bases. *Adv. Heterocycl. Chem.* **1997**, *68*, 223–432.

17. Sallam, M.A.E.; Luis, F.F.; Cassady, J.M. Studies on Epimeric-D-arabino-and-D-ribo-tetritol-1-yl-2-phenyl-2 *H*-1,2,3-triazoles. Synthesis and Anomeric Configuration of 4-(α- and β-D-Erythrofuranosyl)-2-phenyl-2 *H*-1,2,3-Triazole C-Nucleoside Analogs. *Nucleos. Nucleot.* **1998**, *17*, 769–783.

18. Romeo, G.; Chiacchio, U.; Corsaro, A.; Merino, P. Chemical Synthesis of Heterocyclic–Sugar Nucleoside Analogues. *Chem. Rev.* **2010**, *110*, 3337–3370. [CrossRef] [PubMed]

19. Romeo, G.; Giofré, S.V.; Piperno, A.; Romeo, R.; Chiacchio, M.A. Synthesis of N,O-homonucleosides with high conformational freedom. *ARKIVOC* **2009**, *viii*, 168–176. [CrossRef]

20. Romeo, R.; Giofrè, S.V.; Iaria, D.; Sciortino, M.T.; Ronsisvalle, S.; Chiacchio, M.A.; Scala, A. Synthesis of 5-Alkynyl Isoxazolidinyl Nucleosides. *Eur. J. Org. Chem.* **2011**, *28*, 5690–5695.

21. Romeo, R.; Giofrè, S.V.; Macchi, B.; Balestrieri, E.; Mastino, A.; Merino, P.; Carnovale, C.; Romeo, G.; Chiacchio, U. Truncated Reverse Isoxazolidinyl Nucleosides: A New Class of Allosteric HIV-RT Inhibitors. *ChemMedChem* **2012**, *7*, 565–569. [CrossRef] [PubMed]

22. Romeo, R.; Giofrè, S.V.; Garozzo, A.; Bisignano, B.; Corsaro, A.; Chiacchio, M.A. Synthesis and biological evaluation of furopyrimidine N,O-nucleosides. *Bioorg. Med. Chem.* **2013**, *21*, 5688–5693. [CrossRef] [PubMed]

23. Chiacchio, U.; Rescifina, A.; Iannazzo, D.; Piperno, A.; Romeo, R.; Borrello, L.; Sciortino, M.T.; Balestrieri, E.; Macchi, B.; Mastino, A.; *et al.* Phosphonated Carbocyclic 2'-Oxa-3'-azanucleosides as New Antiretroviral Agents. *J. Med. Chem.* **2007**, *50*, 3747–3750.

24. Chiacchio, U.; Balestrieri, E.; Macchi, B.; Iannazzo, D.; Piperno, A.; Rescifina, A.; Romeo, R.; Saglimbeni, M.; Sciortino, M.T.; Valveri, V.; *et al.* Synthesis of Phosphonated Carbocyclic 2'-Oxa-3'-aza-nucleosides: Novel Inhibitors of Reverse Transcriptase. *J. Med. Chem.* **2005**, *48*, 1389–1394.

25. Chiacchio, U.; Iannazzo, D.; Piperno, A.; Romeo, R.; Romeo, G.; Rescifina, A.; Saglimbeni, M. Synthesis and biological evaluation of phosphonated carbocyclic 2'-oxa-3'-aza-nucleosides. *Bioorg. Med. Chem.* **2006**, *14*, 955–959. [CrossRef] [PubMed]

26. Romeo, R.; Carnovale, C.; Giofrè, S.V.; Monciino, G.; Chiacchio, M.A.; Sanfilippo, C.; Macchi, B. Enantiomerically Pure Phosphonated Carbocyclic 2'-Oxa-3'-Azanucleosides: Synthesis and Biological Evaluation. *Molecules* **2014**, *19*, 14406–14416. [CrossRef] [PubMed]

27. Piperno, A.; Giofrè, S.V.; Iannazzo, D.; Romeo, R.; Romeo, G.; Chiacchio, U.; Rescifina, A.; Piotrowska, D.G. Synthesis of C-4' Truncated Phosphonated Carbocyclic 2'-Oxa-3'-azanucleosides as Antiviral Agents. *J. Org. Chem.* **2010**, *75*, 2798–2805. [CrossRef] [PubMed]

28. Romeo, R.; Carnovale, C.; Giofrè, S.V.; Romeo, G.; Macchi, B.; Frezza, C.; Merino-Merlo, F.; Pistarà, V.; Chiacchio, U. Truncated phosphonated C-1'-branched N,O-nucleosides: A new class of antiviral agents. *Bioorg. Med. Chem.* **2012**, *20*, 3652–3657. [CrossRef] [PubMed]

29. Chiacchio, U.; Corsaro, A.; Mates, J.; Merino, P.; Piperno, A.; Rescifina, A.; Romeo, G.; Romeo, R.; Tejero, T. Isoxazolidine analogues of pseudouridine: A new class of modified nucleosides. *Tetrahedron* **2003**, *59*, 4733–4738. [CrossRef]

30. Chiacchio, U.; Rescifina, A.; Saita, M.G.; Iannazzo, D.; Romeo, G.; Mates, J.A.; Tejero, T.; Merino, P. Zinc(II) triflate-controlled 1,3-dipolar cycloadditions of C-(2-thiazolyl)nitrones: application to the synthesis of a novel isoxazolidinyl analogue of tiazofurin. *J. Org. Chem.* **2005**, *70*, 8991–9001. [CrossRef] [PubMed]

31. Romeo, R.; Giofrè, S.V.; Carnovale, C.; Campisi, A.; Parenti, R.; Bandini, L.; Chiacchio, M.A. Synthesis and Biological Evaluation of 3-Hydroxymethyl-5-(1H-1,2,3-Triazol) Isoxazolidines. *Bioorg. Med. Chem.* **2013**, *21*, 7929–7937. [CrossRef] [PubMed]

32. Romeo, R.; Giofrè, S.V.; Carnovale, C.; Chiacchio, M.A.; Campisi, A.; Mancuso, R.; Cirmi, S.; Navarra, M. Synthesis and Biological Activity of Triazole-Appended N,O-Nucleosides. *Eur. J. Org. Chem.* **2014**, *25*, 5442–5447. [CrossRef]

33. Nulwala, H.; Takizawa, K.; Odukale, A.; Anzar Khan, A.; Thibault, R.J.; Taft, B.R.; Lipshutz, B.H.; Hawker, C.J. Synthesis and Characterization of Isomeric Vinyl-1,2,3-triazole Materials by Azide–Alkyne Click Chemistry. *Macromolecules* **2009**, *42*, 6068–6074. [CrossRef]

34. Iannazzo, D.; Piperno, A.; Pistarà, V.; Rescifina, A.; Romeo, R. Modified nucleosides. A general and diastereoselective approach to N,O-psiconucleosides. *Tetrahedron* **2002**, *58*, 581–587.

35. Romeo, R.; Carnovale, C.; Giofrè, S.V.; Chiacchio, M.A.; Garozzo, A.; Amata, E.; Romeo, G.; Chiacchio, U. C-5'-Triazolyl-2'-oxa-3'-aza-4'a-carbanucleosides: Synthesis and biological evaluation. *Beilstein J. Org. Chem.* **2015**, *11*, 328–334. [CrossRef]

36. Merino, P.; Revuelta, J.; Tejero, T.; Chiacchio, U.; Rescifina, A.; Romeo, G. A DFT study on the 1,3-dipolar cycloaddition reactions of C-(methoxycarbonyl)-N-methyl nitrone with methyl acrylate and vinyl acetate. *Tetrahedron* **2003**, *59*, 3581–3592. [CrossRef]

37. Visalli, G.; Ferlazzo, N.; Cirmi, S.; Campiglia, P.; Gangemi, S.; Pietro, A.D.; Calapai, G.; Navarra, M. Bergamot juice extract inhibits proliferation by inducing apoptosis in human colon cancer cells. *Anticancer Agents Med. Chem.* **2014**, *14*, 1402–1413. [CrossRef] [PubMed]

38. Navarra, M.; Celano, M.; Maiuolo, J.; Schenone, S.; Botta, M.; Angelucci, A.; Bramanti, P.; Russo, D. Antiproliferative and pro-apoptotic effects afforded by novel Src-kinase inhibitors in human neuroblastoma cells. *BMC Cancer* **2010**, *10*, 602. [CrossRef] [PubMed]

39. Romeo, R.; Navarra, M.; Giofrè, S.V.; Carnovale, C.; Cirmi, S.; Lanza, G.; Chiacchio, M.A. Synthesis and biological activity of new arenediyne-linked isoxazolidines. *Bioorg. Med. Chem.* **2014**, *22*, 3379–3385. [CrossRef] [PubMed]

40. Delle Monache, S.; Sanità, P.; Trapasso, E.; Ursino, M.R.; Dugo, P.; Russo, M.; Ferlazzo, N.; Calapai, G.; Angelucci, A.; Navarra, M. Mechanisms Underlying the Anti-Tumoral Effects of Citrus bergamia Juice. *PLoS ONE* **2013**, *8*, e61484. [CrossRef] [PubMed]

Sample Availability: *Sample Availability:* Samples of the compounds are available from the authors.

molecules

MDPI

Article

Carboxylated Acyclonucleosides: Synthesis and RNase A Inhibition

Kaustav Chakraborty, Swagata Dasgupta * and Tanmaya Pathak *

Department of Chemistry, Indian Institute of Technology Kharagpur, Kharagpur 721302, India;
kaustav@chem.iitkgp.ernet.in
* Correspondence: swagata@chem.iitkgp.ernet.in (S.D.); tpathak@chem.iitkgp.ernet.in (T.P.);
  Tel.: +91-3222-283-306 (S.D.); +91-3222-283-342 (T.P.)

Academic Editor: Mahesh Lakshman
Received: 17 December 2014; Accepted: 20 March 2015; Published: 3 April 2015

Abstract: Strategically designed carboxylated acyclonucleosides have been probed as a new class of RNase A inhibitors. Several experimental and theoretical studies have been performed to compile relevant qualitative and quantitative information regarding the nature and extent of inhibition. The inhibition constant (K_i) values were determined using a UV-based kinetics experiment. The changes in the secondary structure of the enzyme upon binding with the inhibitors were obtained from circular dichroism studies. The binding constants for enzyme-inhibitor interactions were determined with the help of fluorescence spectroscopy. Docking studies were performed to reveal the possible binding sites of the inhibitors within the enzyme. The cytosine analogues were found to possess better inhibitory properties in comparison to the corresponding uracil derivatives. An increment in the number of carboxylic acid groups (-COOH) in the inhibitor backbone was found to result in better inhibition.

Keywords: acyclonucleosides; RNase A; inhibition; kinetics; docking

1. Introduction

Ribonucleases are a family of digestive enzymes that degrade RNA [1]. Effective inhibition of their enzymatic activity has become a topic of growing interest with the realization that the detrimental biological activities manifested by certain members of this family [2–6] are critically dependent upon their ribonucleolytic activity [7,8]. Ribonuclease A (RNase A) is a representative member of this family [9,10] that works at the juncture of the transcription and translation processes, thereby maintaining the cellular RNA levels. Structural homology among the various members of this family [11–13] permits the use of Ribonuclease A (RNase A) as a model system to explore structure–activity relationships.

Recent reports from our laboratory have revealed that modified nucleoside carboxylic acids manifest RNase A inhibitory properties in a competitive, reversible manner [14–21]. These molecules have an added advantage over the reported phosphate- or pyrophosphate-based nucleotide inhibitors [22–31] as the polyionic nature of the latter hampers their migration through the cell membrane [32]. The active site of RNase A consists of several subsites made up of polar amino acid residues for specific recognition [9,33] (Figure 1). The cleavage of the phosphodiester bond of RNA at the P_1 subsite involves the two His residues (His 12 and His 119) participating in a conjugate acid-base mechanism [34,35]. At physiological pH, the carboxylic acid (-COOH) group(s) in the inhibitor remains deprotonated, and interacts electrostatically with the protonated His and Lys residues present at the ribonucleolytic site [36]. This perturbation of the protonating/deprotonating environment of the P_1 subsite results in the inhibition of RNase A.

Following the same hypothesis, it was expected that carboxylated acyclonucleosides may elicit similar inhibitory properties because of their in-built structural features. The absence of the rigid ribose ring further enhances the flexibility of these molecules, and generates additional information on the importance of the furanoside ring. We report the synthesis of several uracil- and cytosine-based modified acyclonucleosides followed by the exploration of their RNase A inhibitory properties. The nucleobases have been carefully selected, as the B_1 subsite of RNase A shows preferential recognition towards pyrimidine bases. The acidity of the molecules has been increased via stepwise incorporation of carboxylic acid groups in the molecular framework to study the resulting effect on their inhibition capacities.

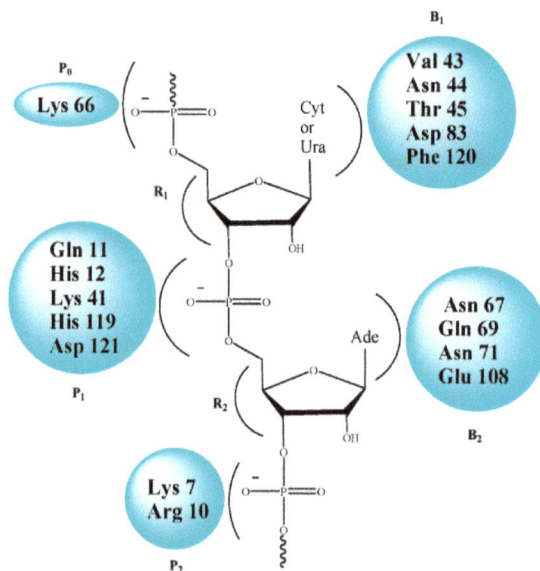

Figure 1. Key residues of the active site of RNase A.

2. Results and Discussion

2.1. Synthesis of Nucleosides

The uracil-glycine conjugate **3** was synthesized according to a literature reported procedure [37] (Scheme 1). Syntheses of other modified uracil derivatives were achieved via coupling of the known uracil-1-acetic acid **1** [37] with suitable secondary amines, followed by hydrolysis and/or deprotection as required (Scheme 1). The synthesis of the uracil-diethanolamine conjugate **5** [38] was achieved by coupling **1** with the *tert*butyldimethyl silyl (TBDMS)-protected diethanolamine, and deprotecting the TBDMS groups of the coupled product **4** with a catalytic amount of CH_3COCl in methanol (Scheme 1). TBDMS-protected N-(2-hydroxyethyl)glycine ethyl ester, on EDC-HOBT-mediated coupling with **1**, generated compound **6**. Deprotection of the TBDMS group produced compound **7**, which was converted to the corresponding acid **8** via hydrolysis (Scheme 1). Again, **1** was coupled with diethyl iminodiacetate to afford the corresponding diester **9**. The diacid derivative **10** was produced from **9** by base-mediated hydrolysis (Scheme 1).

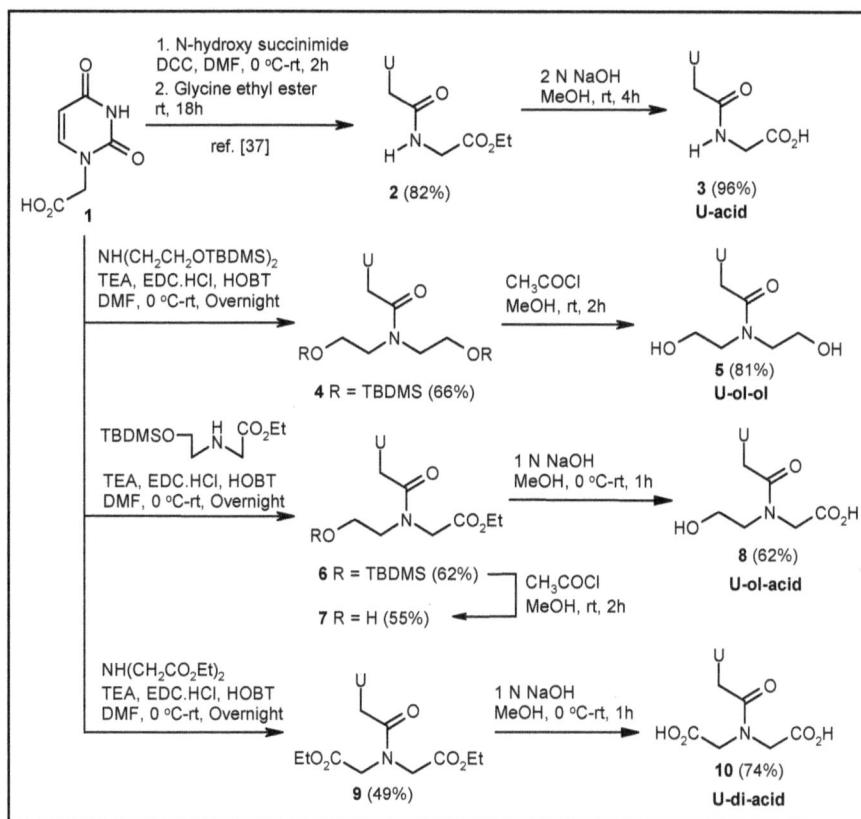

Scheme 1. Synthesis of uracil-based modified acyclonucleosides.

Syntheses of the corresponding cytosine-based molecules were achieved using the reported cytosine derivative **11** [39] as the starting material (Scheme 2). EDC-HOBT-mediated coupling of glycine ethyl ester to compound **11** produced the coupled product **12**. The deprotection of the *tert*butyloxy carbonyl (BOC) groups of **12** by TFA resulted in the formation of compound **13**. The hydrolysis of the ester group provided the desired acid derivative **14** (Scheme 2). Compound **11**, on coupling with TBDMS-protected diethanolamine, generated the coupled derivative **15**. TFA treatment of compound **15** afforded the deprotected cytosine-diethanolamine conjugate **16** (Scheme 2). Again, coupling of the TBDMS-protected N-(2-hydroxyethyl)glycine ethyl ester with compound **11**, followed by hydrolysis and deprotection of the resulting ester **17**, afforded the corresponding hydroxy acid derivative **18** (Scheme 2). Finally, the diacid **21** was obtained from compound **11** following a similar sequence of steps (Scheme 2). Compound **11**, on coupling with diethyl iminodiacetate, produced the coupled product **19**. Compound **19** was transformed to compound **20** by treatment with TFA. Base-mediated hydrolysis of **20** afforded the desired diacid derivative **21**.

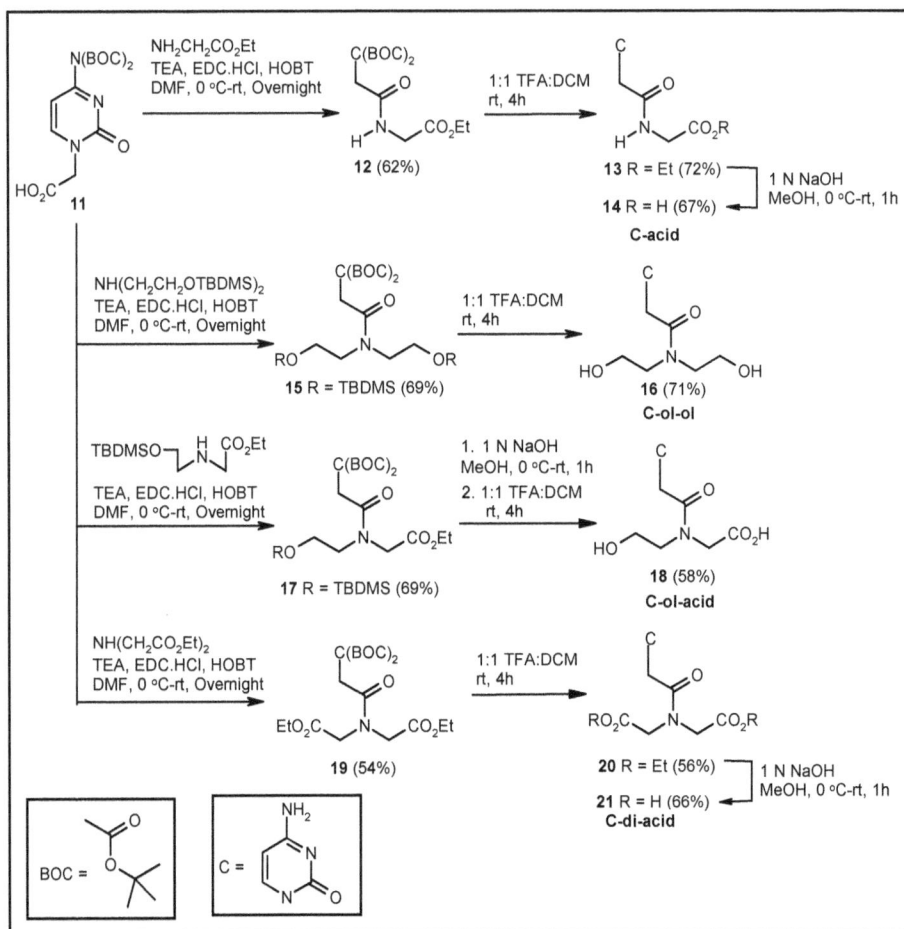

Scheme 2. Synthesis of cytosine-based modified acyclonucleosides.

2.2. RNase A Inhibition

Qualitative indication of RNase A inhibition by the synthetic inhibitors was obtained from a comparative agarose gel-based assay by monitoring the degradation of RNA by RNase A (Figures 2 and 3). The most intense band observed in lane **I** in each gel is due to the presence of only RNA. The faint band in lane **II** is due to the maximum possible degradation of RNA by RNase A. Different intensities of bands from lane **III** to **VI** revealed different extents of inhibition by the inhibitors at a fixed concentration (0.5 mM). The histograms obtained by plotting the relative intensities of the bands revealed that compounds **U-ol-acid (8)**, **C-ol-acid (18)**, **U-di-acid (10)**, and **C-di-acid (21)** were relatively more potent inhibitors in comparison to the others. These experimental observations reaffirmed our assumption that increasing the acidity of the molecules leads to an increase in the inhibitory property of the inhibitor.

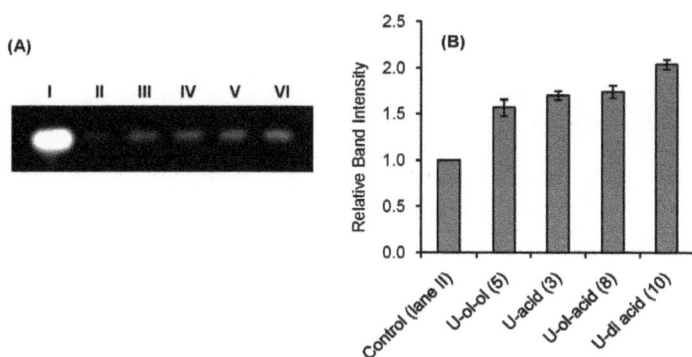

Figure 2. Agarose gel-based assay for RNase A (1 µM) Inhibition. (**A**) lane **I**: RNA (10 mg/mL), lane **II**: RNA + RNase A, lanes **III**, **IV**, **V**, **VI**: RNA + RNase A + **U-ol-ol** (**5**), **U-acid** (**3**), **U-ol-acid** (**8**), and **U-di-acid** (**10**) (0.5 mmol), respectively. (**B**) Histogram showing the relative band intensities of agarose gel assay (the data are the mean ± SD).

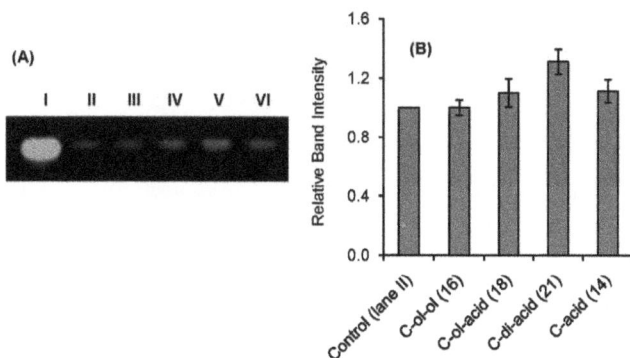

Figure 3. Agarose gel-based assay for RNase A (1 µM) Inhibition. (**A**) lane **I**: RNA (10 mg/mL), lane **II**: RNA + RNase A, lanes **III**, **IV**, **V**, **VI**: RNA + RNase A + **C-ol-ol** (**16**), **C-ol-acid** (**18**), **C-di-acid** (**21**), and **C-acid** (**14**) (0.5 mmol), respectively. (**B**) Histogram showing the relative band intensities of agarose gel assay (the data are the mean ± SD).

In order to determine the type of inhibition and the inhibition constant (K_i) values, kinetic experiments were performed with compounds **U-ol-acid** (**8**), **U-di-acid** (**10**), **C-ol-acid** (**18**), and **C-di-acid** (**21**). The inhibition constant values are given in Table 1. The competitive nature of inhibition in all the cases was apparent from the nature of the Lineweaver–Burk plots obtained from the kinetic experiments (Figure 4 and Figure S1 (SI)). The numerical order of the inhibition constant (K_i) values indicated that **U-di-acid** (**10**) and **C-di-acid** (**21**) are the two most potent inhibitors of the series. The results are in good agreement with the results obtained from agarose gel-based assay, suggesting a correlation of inhibitory efficiency with the number of carboxylic acid groups present in the concerned inhibitor. The cytosine analogue **C-di-acid** (**21**) was found to possess superior inhibitory property in comparison to the corresponding uracil derivative **U-di-acid** (**10**).

Table 1. Inhibition constants (K_i) of the inhibitors.

Inhibitor	K_i * (μM)
U-ol-acid (8)	454 ± 9
C-ol-acid (18)	356 ± 7
U-di-acid (10)	301 ± 15
C-di-acid (21)	235 ± 9

* The data are the mean ± SD.

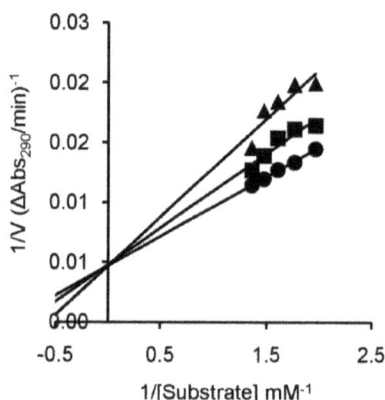

Figure 4. Lineweaver–Burk plot for inhibition of RNase A by **C-di-acid** (**21**) of 0.15 (▲), 0.05 (■), or 0 (●) mM, with 2',3'-cCMP concentrations of 0.75–0.52 mM and RNase A concentration of 9.8 μM.

Inhibitors of RNase A are known to perturb the secondary structure of the enzyme upon binding [40–44]. A model drug–protein interaction study showed that 3'-azido-3'-deoxythymidine increases the α-helix content of RNase A [40]. Similar increments were observed with 3'-*O*-carboxy esters of thymidine, which inhibited RNase A in reversible competitive mode [42]. Therefore, the probable changes in the secondary structure of the enzyme by the inhibitors were monitored by observing the CD spectra of RNase A in the absence or presence of compounds **U-di-acid** (**10**) and **C-di-acid** (**21**) (Figure 5). Both the inhibitors induced moderate changes in the secondary structure of the enzyme, which was reflected in the enhanced α-helix content. The α-helix content in native RNase A was 22.7%, which was found to increase upon binding with **U-di-acid** (**10**) and **C-di-acid** (**21**) to 24.7% and 29.2%, respectively.

The fluorescence emission intensity of RNase A due to the presence of six Tyr residues is found to decrease upon interactions with small molecule inhibitors [40–44]. The emission spectrum of RNase A in presence of **U-di-acid** (**2.10**) and **C-di-acid** (**2.21**) at 25 °C showed quenching of the fluorescence intensity of Tyr residues (Figure S2, (SI)). The fluorescence quenching study was used to calculate the binding parameters for enzyme-inhibitor interactions. Binding constants (K_b) calculated from the experiment were found to be in the order of 10^5 for **C-di-acid** (**21**) and 10^4 for **U-di-acid** (**10**) (Table S1, (SI)), indicating strong binding of the inhibitors with the enzyme.

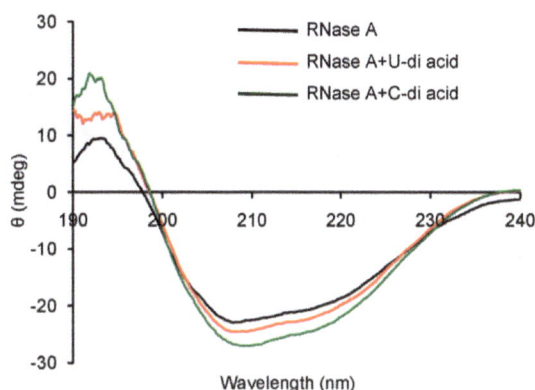

Figure 5. CD spectra of RNase A in absence or presence of **U-di-acid (10)** and **C-di-acid (21)**.

To gain an insight into the probable binding sites, protein-ligand docking studies were undertaken. The docked conformations shown in Figure 6 revealed that the nucleobase-amino acid conjugates were in close proximity to the amino acid residues of the P_1 subsite, resulting in a competitive mode of inhibition, as observed from the kinetic study. The carboxylic acid (-COOH) groups, in each case, were positioned close to His12 and His 119, probably engaged in hydrogen bonding with the residues. For the cytosine-based inhibitors, the N3 and $-NH_2$ groups of the nucleobase were near to several amino acid residues (Arg39, Lys41, Val43), thus increasing the possibility of polar interactions between them. Such interactions were absent for the uracil-derived inhibitors. These extra interactions probably contribute to the experimentally observed better inhibitory potency of the cytosine inhibitors. Apart from the interactions with the active site residues, one of the carboxylic acid (-COOH) groups of **U-di-acid (10)** was found to be within hydrogen bonding distance of Oε1 of Gln11. Similarly, one –COOH group of **C-di-acid (21)** was engaged in hydrogen bonding with Phe120. Such favorable interactions of the –COOH groups may possibly result in the better inhibition capacity of the inhibitors. The docked enzyme-inhibitor complexes also revealed that the nucleobase in the acyclic structure is not in close proximity to the amino acid residues of the pyrimidine binding subsite B_1 (Thr45, Asp83, Phe120). It can, therefore, be assumed that the ribose ring in the nucleosides may have a role in recognition of the inhibitors by the enzyme. A detailed account of all the interactions between the inhibitors and the enzyme has been provided in Table S2 (SI).

A further clarification regarding the possible binding sites of the inhibitors was obtained by calculating the changes in accessible surface area (ΔASA) of the interacting residues between the free and complexed forms of the enzyme (Table 2). The measurements revealed that the accessible surface area of the amino acid residues of the P_1 subsite (His12, His 119 and Lys 41) was largely affected upon binding with the inhibitors. The observed results correlate well with the results of the docking study, suggesting that the inhibitors do bind to the P_1 subsite of RNase A.

Table 2. Changes in accessible surface area (ΔASA) of the interacting residues between the uncomplexed and complexed forms of RNase A.

Amino Acid Residue	ASA (Å^2) in RNase A	ΔASA (Å^2) for Different Inhibitors			
		U-ol-Acid (8)	C-ol-Acid (18)	U-di-Acid (10)	C-di-Acid (21)
Lys 7	88.03	42.88	41.75	29.62	43.82
His 12	12.64	7.09	7.76	9.21	7.53
Arg 39	142.03	25.81	29.72	5.06	30.79
Lys 41	36.39	26.80	31.58	33.00	31.31
His 119	85.05	33.62	28.29	26.69	26.64

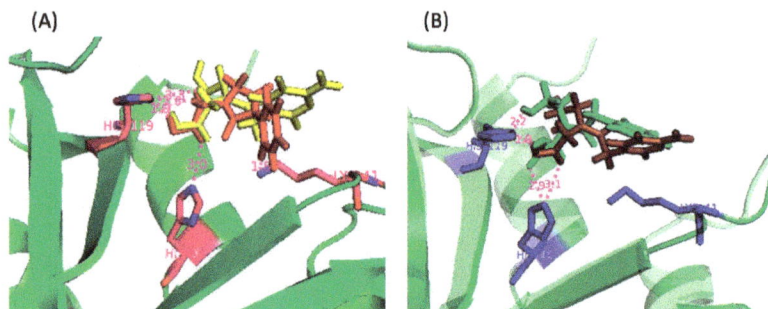

Figure 6. Docked poses of (**A**) **C-di-acid** (**21**) (yellow) and **U-di-acid** (**10**) (red), and (**B**) **U-ol-acid** (**8**) (green) and **C-ol-acid** (**18**) (brown) with RNase A (1FS3).

3. Experimental Section

3.1. General Methods

All reagents and fine chemicals were purchased from commercial suppliers and were used without further purification. Column chromatography was performed with silica gel (230–400 mesh). Solvents were dried and distilled following standard methods. TLC was carried out on precoated plates (Merck silica gel 60, f_{254}). ^{1}H and ^{13}C-NMR for the compounds were recorded at 200 and 50 MHz, respectively, using a Bruker NMR instrument. For ^{1}H and ^{13}C-NMR spectra in D_2O, CH_3CN was used as internal standard. Chemical shifts are reported in parts per million (ppm, δ scale). Methylene carbons have been identified using DEPT spectrum. Melting points were determined in open-end capillary tubes. Bovine pancreatic RNase A, RNA (*Torula utilis*), and 2′,3′-cCMP were purchased from commercial suppliers. UV-vis measurements were made using a UV-vis spectrophotometer (Shimadzu 2450). Concentrations of the solutions were estimated spectrophotometrically using the following data: $\varepsilon_{278.5} = 9800\ M^{-1}\cdot cm^{-1}$ (RNase A) [45] and $\varepsilon_{268} = 8500\ M^{-1}\cdot cm^{-1}$ (2′,3′-cCMP) [22]. CD measurements were carried out on a Jasco-810 automatic recording spectrophotometer. Fluorescence measurements were carried out using a Horiba Jobin Yvon Fluoromax-4 Spectrofluorimeter.

3.2. General Procedure for Carboxylic Acid–Amine Coupling Reaction

To a well-stirred solution of carboxylic acid (**1, 11**) (1 mmol) in DMF (5 mL) at 0 °C was added EDC-HCl (1.2 mmol) followed by triethylamine (1.2 mmol) and the stirring was further continued at this temperature. After 10 min, HOBT (1.2 mmol) was added and the reaction mixture was allowed to warm back to room temperature. Amine (1 mmol) was added to the resulting solution and the stirring was allowed to continue overnight. The reaction mixture was then poured into a brine solution (50 mL) and extracted with EtOAc (3 × 20 mL). Organic extracts were pooled together, dried over anhydrous Na_2SO_4, filtered, and the filtrate was evaporated under reduced pressure. The resulting residue was purified by column chromatography over silica gel (EtOAc/petroleum ether) to obtain pure coupling products (**4, 6, 9, 12, 15, 17, 19**).

3.3. General Procedure for Deprotection of –TBDMS Group

To a well-stirred solution of compound (**4, 6**) (1 mmol) in methanol (10 mL), was added CH_3COCl (cat). The resulting solution was allowed to stir for 2 h at room temperature. The solvent was evaporated under reduced pressure and the resulting residue was purified by column chromatography over silica gel (DCM/MeOH) to obtain pure products (**5, 7**).

3.4. General Procedure for the Deprotection of –BOC Group/Simultaneous Deprotection of –BOC Group and –TBDMS Group

The compound (**12, 15, 19**) (1 mmol) was dissolved in 1:1 TFA/DCM (5 mL) and the resulting solution was stirred at room temperature for 4 h. The solvent was evaporated under reduced pressure and the resulting residue was purified by column chromatography over silica gel (DCM/MeOH) to obtain pure products (**13, 16, 20**).

3.5. General Procedure for Ester Hydrolysis

To a well-stirred solution of the compound (**7, 9, 13, 20**) (1 mmol) in methanol (5 mL) at 0 °C was added 1 N NaOH solution (2 mL) dropwise. The resulting solution was allowed to warm back up to room temperature and stirred for 1 h. After evaporation of methanol under reduced pressure, the resulting residue was dissolved in water (10 mL) and neutralized with acidic amberlyte. The resulting mixture was filtered, and the filtrate was evaporated under reduced pressure. The resulting residue was purified by column chromatography over silica gel (DCM/MeOH) to obtain pure products (**8, 10, 14, 21**).

N,N-Bis-[2-(tert-butyldimethylsilyloxy)-ethyl]-2-(2,4-dioxo-3,4-dihydro-2H-pyrimidin-1-yl)-acetamide (**4**): Compound **1** (0.68 g, 3.99 mmol) was converted to compound **4** (1.28 g, 66%) following the general procedure in Section 3.2. R_f = 0.4 [30% EtOAc in pet ether]. White solid. M.P: 122–124 °C. ^{1}H-NMR (200 MHz, CDCl$_3$): δ 0.02 (s, 6H), 0.06 (s, 6H), 0.86 (s, 9H), 0.88 (s, 9H), 3.47 (t, J = 5.5 Hz, 2H), 3.59 (t, J = 5.1 Hz, 2H), 3.71–3.80 (m, 4H), 4.66 (s, 2H), 5.69 (d, J = 8.0 Hz, 1H), 7.11 (d, J = 7.8 Hz, 1H), 9.28 (bs, 1H). ^{13}C-NMR (50 MHz, CDCl$_3$): δ −5.3, −5.2, 18.3, 18.5, 26.0, 26.1, 48.1 (CH$_2$), 49.2 (CH$_2$), 51.1 (CH$_2$), 61.0(CH$_2$), 61.5 (CH$_2$), 102.1, 145.4, 151.2, 163.9, 167.0. HRMS (ESI$^+$): m/z calcd for C$_{22}$H$_{44}$N$_3$O$_5$Si$_2$ (M+H)$^+$: 486.2820; found: 486.2827.

2-(2,4-Dioxo-3,4-dihydro-2H-pyrimidin-1-yl)-N,N-bis-(2-hydroxy-ethyl) acetamide (**5**): Compound **5** [38] (0.39 g, 81%) was obtained from compound **4** (0.9 g, 1.85 mmol) following the general procedure in Section 3.3. White solid. R_f = 0.4 [15% MeOH in DCM]. ^{1}H-NMR (200 MHz, DMSO-d_6): δ 3.31–3.58 (m, 8H), 4.67 (s, 2H), 4.75–4.80 (m, 1H), 5.01–5.05 (m, 1H), 5.54 (d, J = 7.8 Hz, 1H), 7.47 (d, J = 7.8 Hz, 1H), 11.23 (bs, 1H).

{[2-(tert-Butyldimethylsilyloxy) ethyl]-[2-(2,4-dioxo-3,4-dihydro-2H-pyrimidin-1-yl-acetyl] amino}acetic acid ethyl ester (**6**): Compound **1** (0.93 g, 5.47 mmol) was converted to compound **6** (1.40 g, 62%) following the general procedure in Section 3.2. R_f = 0.3 [40% EtOAc in pet ether]. White solid. M.P: 171–172 °C. The compound was obtained as a mixture of rotamers as indicated by the NMR spectra. ^{1}H-NMR (200 MHz, CDCl$_3$): δ 0.01 (s, 6H), 0.06 (s, 6H), 0.86 (s, 9H), 0.87 (s, 9H), 1.20–1.32 (m, 3H), 3.52–3.56 (m, 2H), 3.70–3.81 (m, 2H), 4.10–4.24 (m, 4H), 4.30 (s, 2H), 4.46 (s, 2H), 4.71 (s, 2H), 5.68–5.73 (m, 1H), 7.13–7.26 (m, 1H), 9.39 (bs, 1H), 9.51 (bs, 1H). ^{13}C-NMR (50 MHz, CDCl$_3$): δ −5.3, 14.3, 18.4, 26.0, 26.1, 47.9 (CH$_2$), 48.6 (CH$_2$), 50.8 (CH$_2$), 61.4 (CH$_2$), 61.6 (CH$_2$), 62.2 (CH$_2$), 102.3, 145.4, 151.1, 163.9, 167.5, 168.9, 172.9. HRMS (ESI$^+$): m/z calcd for C$_{18}$H$_{32}$N$_3$O$_6$Si (M+H)$^+$: 414.2060; found: 414.2088.

[[2-(2,4-Dioxo-3,4-dihydro-2H-pyrimidin-1-yl-acetyl]-(2-hydroxyethyl)-amino] acetic acid ethyl ester (**7**): Compound **7** (0.48 g, 55%) was generated from compound **6** (1.21 g, 2.90 mmol) following the general procedure in Section 3.3. R_f = 0.4 [10% MeOH in DCM]. Light yellow gum. The compound was obtained as a mixture of rotamers as indicated by the NMR spectra. ^{1}H-NMR (200 MHz, DMSO-d_6): δ 1.14–1.26 (m, 3H), 3.34–3.58 (m, 4H), 4.01–4.18 (m, 4H), 4.34 (s, 2H), 4.53 (s, 2H), 4.74 (s, 2H), 5.53–5.57 (m, 1H), 7.34–7.45 (m, 1H), 11.25 (bs, 1H). ^{13}C-NMR (50 MHz, DMSO-d_6): δ 14.0, 48.0 (CH$_2$), 49.9 (CH$_2$), 50.7 (CH$_2$), 58.3 (CH$_2$), 58.9 (CH$_2$), 60.5 (CH$_2$), 61.03(CH$_2$), 100.6, 146.5, 151.0, 163.9, 167.4, 167.5, 169.1, 169.4. HRMS (ESI$^+$): m/z calcd for C$_{12}$H$_{17}$N$_3$O$_6$Na (M+Na)$^+$: 322.1015; found: 322.1003.

[[2-(2,4-Dioxo-3,4-dihydro-2H-pyrimidin-1-yl-acetyl]-(2-hydroxyethyl) amino] acetic acid (**8**): Compound **7** (0.4 g, 1.34 mmol) was converted to compound **8** (0.22 g, 62%) following the general procedure in Section 3.5. R_f = 0.2 [20% MeOH in DCM]. White solid. M.P: 177–179 °C. The compound was obtained

as a mixture of rotamers as indicated by the NMR spectra. ^{1}H-NMR (200 MHz, D$_2$O): δ 3.54–3.82 (m, 4H), 4.19 (s, 2H), 4.37 (s, 2H), 4.57 (s, 2H), 4.90 (s, 2H), 5.80–5.87 (m, 1H), 7.50–7.56 (m, 1H). ^{13}C-NMR (50 MHz, D$_2$O): δ 49.2 (CH$_2$), 50.0 (CH$_2$), 50.6 (CH$_2$), 51.0 (CH$_2$), 59.3 (CH$_2$), 59.4 (CH$_2$), 102.2, 148.1, 152.6, 167.2, 170.0, 173.3. HRMS (ESI$^+$): m/z calcd for C$_{10}$H$_{13}$N$_3$O$_6$Na (M+Na)$^+$: 294.0702; found: 294.0707.

{[2-(2,4-Dioxo-3,4-dihydro-2H-pyrimidin-1-yl-acetyl] ethoxycarbonylmethylamino]acetic acid ethyl ester (9): Compound **1** (0.52 g, 3.05 mmol) was transformed to compound **9** (0.51 g, 49%) following the general procedure in Section 3.2. R$_f$ = 0.5 [70% EtOAc in pet ether]. White solid. M.P: 140–142 °C. ^{1}H-NMR (200 MHz, CDCl$_3$): δ 1.22–1.34 (m, 6H), 4.12–4.31 (m, 8H), 4.61 (s, 2H), 5.73 (d, *J* = 8.0 Hz, 1H), 7.20 (d, *J* = 7.8 Hz, 1H), 9.09 (bs, 1H). ^{13}C-NMR (50 MHz, CDCl$_3$): δ 14.3, 47.7 (CH$_2$), 49.0 (CH$_2$), 50.2 (CH$_2$), 61.8 (CH$_2$), 62.5 (CH$_2$), 102.6, 145.2, 151.0, 163.7, 167.6, 168.6. HRMS (ESI$^+$): m/z calcd for C$_{14}$H$_{19}$N$_3$O$_7$Na (M+Na)$^+$: 364.1121; found: 364.1109.

{Carboxymethyl-[2-(2,4-dioxo-3,4-dihydro-2H-pyrimidin-1-yl) acetyl] amino]acetic acid (10): Compound **9** (0.42 g, 1.23 mmol) was converted to compound **10** (0.26 g, 74%) following the general procedure in Section 3.5. R$_f$ = 0.2 [30% MeOH in DCM]. White solid. M.P: 118–120 °C ^{1}H-NMR (200 MHz, DMSO-d_6): δ 4.00 (s, 2H), 4.25 (s, 2H), 4.62 (s, 2H), 5.55 (d, *J* = 7.2 Hz, 1H), 7.43 (d, *J* = 7.6 Hz, 1H), 11.28 (s, 1H). ^{13}C-NMR (50 MHz, DMSO-d_6): δ 47.7 (CH$_2$), 48.4 (CH$_2$), 49.2 (CH$_2$), 100.7, 146.4, 151.0, 163.8, 167.7, 170.3. HRMS (ESI$^+$): m/z calcd for C$_{10}$H$_{11}$N$_3$O$_7$Na (M+Na)$^+$: 308.0495; found: 308.0475.

{[4-[Bis(1,1-dimethylethoxy)carbonyl]amino-2-oxo-2H-pyrimidin-1-yl] acetylamino]acetic acid ethyl ester (12): Compound **11** (0.39 g, 1.05 mmol) was converted to compound **12** (0.30 g, 62%) following the general procedure in Section 3.2. R$_f$ = 0.5 [60% EtOAc in pet ether]. White solid. M.P: 118–120 °C. ^{1}H-NMR (200 MHz, CDCl$_3$): δ 1.18–1.25 (m, 3H), 1.52 (s, 18H), 3.95 (d, *J* = 5.6 Hz, 2H), 4.12 (q, *J* = 7.2 Hz, 2H), 4.58 (s, 2H), 7.08 (d, *J* = 7.2 Hz, 1H), 7.69–7.76 (m, 1H). ^{13}C-NMR (50 MHz, CDCl$_3$): δ 14.1, 27.7, 41.4 (CH$_2$), 52.8 (CH$_2$), 61.4 (CH$_2$), 85.0, 96.7, 149.0, 149.4, 152.6, 155.5, 162.7, 167.1, 169.4. HRMS (ESI$^+$): m/z calcd for C$_{20}$H$_{30}$N$_4$O$_8$Na (M+Na)$^+$: 477.1961; found: 477.1975.

[2-(4-Amino-2-oxo-2H-pyrimidin-1-yl) acetylamino] acetic acid ethyl ester (13): Compound **12** (0.28 g, 0.62 mmol) was converted to compound **13** (0.11 g, 72%) following the general procedure in Section 3.4. R$_f$ = 0.3 [5% MeOH in DCM]. White solid. M.P: 155–158 °C. ^{1}H-NMR (200 MHz, DMSO-d_6): δ 1.17 (t, *J* = 7.2 Hz, 3H), 3.86 (d, *J* = 5.8 Hz, 2H), 4.07 (q, *J* = 7.2 Hz, 2H), 4.42 (s, 2H), 5.82 (d, *J* = 7.2 Hz, 1H), 7.65 (d, *J* = 7.4 Hz, 1H), 7.90 (bs, 1H), 8.09 (bs, 1H), 8.61 (t, *J* = 5.8 Hz, 1H). ^{13}C-NMR (50 MHz, DMSO-d_6): δ 14.5, 41.2 (CH$_2$), 50.9 (CH$_2$), 61.0 (CH$_2$), 93.8, 148.8, 153.5, 164.4, 168.0, 170.0. HRMS (ESI$^+$): m/z calcd for C$_{10}$H$_{15}$N$_4$O$_4$ (M+H)$^+$: 255.1093; found: 255.1076.

[2-(4-Amino-2-oxo-2H-pyrimidin-1-yl) acetylamino] acetic acid (14): Compound **13** (0.09 g, 0.35 mmol) was transformed to compound **14** (0.05 g, 67%) following the general procedure in Section 3.5. R$_f$ = 0.3 [20% MeOH in DCM]. Yellowish white solid. M.P: > 200 °C. ^{1}H-NMR (200 MHz, DMSO-d_6): δ 3.61–3.95 (m, 2H), 4.54 (s, 2H), 6.22 (d, *J* = 7.2 Hz, 1H), 7.99 (d, *J* = 7.4 Hz, 1H), 8.77 (s, 1H), 8.94 (s, 1H), 10.17 (s, 1H). ^{13}C-NMR (50 MHz, DMSO-d_6): δ 40.8 (CH$_2$), 50.4 (CH$_2$), 93.1, 147.4, 150.6, 160.1, 166.4, 170.8. HRMS (ESI$^+$): m/z calcd for C$_8$H$_{11}$N$_4$O$_4$ (M+H)$^+$: 227.0780; found: 227.0784.

[4-[Bis(1,1-dimethylethoxy)carbonyl]amino-2-oxo-2H-pyrimidin-1-yl]-N,N-bis-[2(tertbutyldimethyl silyloxy) ethyl]acetamide (15): Compound **15** (0.55 g, 69%) was obtained from compound **11** (0.43 g, 1.16 mmol) following the general procedure in Section 3.2. R$_f$ = 0.3 [25% EtOAc in pet ether]. Colourless gum. ^{1}H-NMR (200 MHz, CDCl$_3$): δ −0.03 (s, 6H), 0.01 (s, 6H), 0.81 (s, 9H), 0.83 (s, 9H), 1.49 (s, 18H), 3.43 (t, *J* = 5.6 Hz, 2H), 3.64–3.74 (m, 6H), 4.70 (s, 2H), 7.00 (d, *J* = 7.4 Hz, 1H), 7.52 (d, *J* = 7.4 Hz, 1H). ^{13}C-NMR (50 MHz, CDCl$_3$): δ −5.4, 18.2, 18.3, 26.0, 27.7, 49.3 (CH$_2$), 49.5 (CH$_2$), 51.1 (CH$_2$), 61.0 (CH$_2$), 61.2 (CH$_2$), 84.8, 96.1, 149.1, 149.6, 155.1, 162.6, 167.1. HRMS (ESI$^+$): m/z calcd for C$_{32}$H$_{61}$N$_4$O$_8$Si$_2$ (M+H)$^+$: 685.4028; found: 685.4038.

2-(4-Amino-2-oxo-2H-pyrimidin-1-yl)-N,N-bis-(2-hydroxyethyl) acetamide (**16**): Compound **15** (0.48 g, 0.70 mmol) was transformed to compound **16** (0.13 g, 71%) following the general procedure in Section 3.4. R_f = 0.3 [15% MeOH in DCM]. White solid. M.P: 150 °C (decomposed). ^{1}H-NMR (200 MHz, DMSO-d_6): δ 2.99–3.04 (m, 1H), 3.34–3.68 (m, 8H), 4.65 (s, 2H), 5.78 (d, *J* = 7.2 Hz, 1H), 7.51 (d, *J* = 7.2 Hz, 1H), 7.75 (bs, 2H). ^{13}C-NMR (50 MHz, DMSO-d_6): δ 48.8 (CH$_2$), 49.3 (CH$_2$), 49.9 (CH$_2$), 58.7 (CH$_2$), 59.0 (CH$_2$), 93.2, 148.1, 154.4, 164.8, 167.3. HRMS (ESI$^+$): m/z calcd for C$_{10}$H$_{17}$N$_4$O$_4$ (M+H)$^+$: 257.1250; found: 257.1255.

{[4-[Bis(1,1-dimethylethoxy)carbonyl]amino-2-oxo-2H-pyrimidin-1-yl-acetyl]-[2-(tert-butyldimethylsilyloxy) ethyl] acetic acid ethyl ester (**17**): Compound **11** (0.46 g, 1.24 mmol) was converted to compound **17** (0.52 g, 69%) following the general procedure in Section 3.2. R_f = 0.4 [40% EtOAc in pet ether]. Colorless gum. The compound was obtained as a mixture of rotamers as indicated by the NMR spectra. ^{1}H-NMR (200 MHz, CDCl$_3$): δ −0.03–0.01 (m, 6H), 0.81 (s, 9H), 0.83 (s, 9H), 1.15–1.27 (m, 3H), 1.49 (s, 18H), 3.33–3.80 (m, 4H), 4.05–4.18 (m, 4H), 4.40 (s, 2H), 4.49 (s, 2H), 4.75 (s, 2H), 6.97–7.03 (m, 1H), 7.54–7.62 (m, 1H). ^{13}C-NMR (50 MHz, CDCl$_3$): δ −5.4, 14.2, 18.2, 25.9, 27.7, 49.0 (CH$_2$), 49.3 (CH$_2$), 51.1 (CH$_2$), 51.2 (CH$_2$), 61.2 (CH$_2$), 61.6 (CH$_2$), 61.8 (CH$_2$), 62.0 (CH$_2$), 84.8, 96.2, 149.0, 149.1, 149.5, 155.0, 162.7, 162.8, 167.4, 167.5, 168.8, 169.7. HRMS (ESI$^+$): m/z calcd for C$_{28}$H$_{49}$N$_4$O$_9$Si (M+H)$^+$: 613.3269; found: 613.3278.

{[2-(4-Amino-2-oxo-2H-pyrimidin-1-yl) acetyl]-(2-hydroxyethyl) amino] acetic acid (**18**): Compound **17** (0.40 g, 0.65 mmol) was hydrolyzed to corresponding acid following the general procedure in Section 3.5. The *crude* residue obtained was subjected to TFA treatment as mentioned in the general procedure in Section 3.4, and purified by column chromatography over silica gel to obtain **18** (0.10 g, 58%). R_f = 0.2 [30% MeOH in DCM]. White solid. M.P: 184–188 °C. The compound was obtained as a mixture of rotamers, as indicated by the NMR spectra. ^{1}H-NMR (200 MHz, DMSO-d_6): δ 3.35–3.59 (m, 4H), 4.01 (s, 2H), 4.24 (s, 2H), 4.54 (s, 2H) 4.75 (s, 2H), 5.84 (d, *J* = 7.0 Hz, 1H), 7.55 (d, *J* = 7.0 Hz, 1H), 7.97 (bs, 2H). ^{13}C-NMR (50 MHz, DMSO-d_6): δ 47.9 (CH$_2$), 49.1 (CH$_2$), 49.9 (CH$_2$), 58.9 (CH$_2$), 93.4, 148.6, 152.8, 163.6, 167.6, 170.7, 171.1. HRMS (ESI$^+$): m/z calcd for C$_{10}$H$_{15}$N$_4$O$_5$ (M+H)$^+$: 271.1042; found: 271.1039.

{[4-[Bis(1,1-dimethylethoxy)carbonyl]amino-2-oxo-2H-pyrimidin-2-yl-acetyl] ethoxycarbonylmethylamino} acetic acid ethyl ester (**19**): Compound **19** (0.29 g, 54%) was generated from compound **11** (0.37 g, 1.00 mmol) following the general procedure in Section 3.2. R_f = 0.4 [50% EtOAc in pet ether]. Colorless gum. ^{1}H-NMR (200 MHz, CDCl$_3$): δ 1.17–1.30 (m, 6H), 1.51 (s, 18H), 4.07–4.25 (m, 6H), 4.32 (s, 2H), 4.67 (s, 2H), 7.05 (d, *J* = 7.4 Hz, 1H), 7.62 (d, *J* = 7.4 Hz, 1H). ^{13}C-NMR (50 MHz, CDCl$_3$): δ 14.2, 27.8, 49.1 (CH$_2$), 49.4 (CH$_2$), 50.5 (CH$_2$), 61.6 (CH$_2$), 62.1 (CH$_2$), 85.0, 96.5, 149.0, 149.5, 155.1, 162.8, 167.9, 168.7, 168.9. HRMS (ESI$^+$): m/z calcd for C$_{24}$H$_{37}$N$_4$O$_{10}$ (M+H)$^+$: 541.2510; found: 541.2516.

{[2-(4-Amino-2-oxo-2H-pyrimidin-1-yl) acetyl] ethoxycarbonylmethylamino}acetic acid ethyl ester (**20**): Compound **19** (0.27 g, 0.50 mmol) was converted to compound **20** (0.09 g, 56%) following the general procedure in Section 3.4. R_f = 0.3 [3% MeOH in DCM]. White solid. M.P: 152–154 °C. ^{1}H-NMR (200 MHz, DMSO-d_6): δ 1.14–1.27 (m, 6H), 4.00–4.22 (m, 6H), 4.39 (s, 2H), 4.80 (s, 2H), 6.10 (d, *J* = 7.6 Hz, 1H), 7.81 (d, *J* = 7.6 Hz, 1H), 9.52 (bs, 1H), 9.67 (bs, 1H). ^{13}C-NMR (50 MHz, DMSO-d_6): δ 14.0, 48.5 (CH$_2$), 49.1 (CH$_2$), 60.6 (CH$_2$), 61.1 (CH$_2$), 93.5, 148.2, 150.2, 160.6, 167.3, 168.5, 168.8. HRMS (ESI$^+$): m/z calcd for C$_{14}$H$_{21}$N$_4$O$_6$ (M+H)$^+$: 341.1461; found: 341.1452.

{[2-(4-Amino-2-oxo-2H-pyrimidin-1-yl)-acetyl] carboxymethylamino}acetic acid (**21**): Compound **20** (0.07 g, 0.20 mmol) was transformed to compound **21** (0.04 g, 66%) following the general procedure in Section 3.5. R_f = 0.15 [40% MeOH in DCM]. Eluent: 30%–50% MeOH in DCM. White solid. M.P: 190–193 °C. ^{1}H-NMR (200 MHz, D$_2$O): δ 4.23 (s, 2H), 4.39 (s, 2H), 4.89 (s, 2H), 6.24 (d, *J* = 7.4 Hz, 1H), 7.79 (d, *J* = 7.8 Hz, 1H). ^{13}C-NMR (50 MHz, DMSO-d_6): δ 48.8 (CH$_2$), 49.3 (CH$_2$), 49.9 (CH$_2$), 93.4, 148.4, 152.2, 163.5, 167.8, 170.4, 170.7. HRMS (ESI$^+$): m/z calcd for C$_{10}$H$_{13}$N$_4$O$_6$ (M+H)$^+$: 285.0835; found: 285.0852.

3.6. Comparative Agarose Gel-Based Assay

Inhibition of RNase A was assayed qualitatively by the degradation of RNA in an agarose gel. In this method, 20 μL of RNase A (1 μM) was mixed with 20 μL (0.5 mM) of compounds **U-ol-ol (5)**, **C-ol-ol (16)**, **U-acid (3)**, **C-acid (14)**, **U-ol-acid (8)**, **C-ol-acid (18)**, **U-di-acid (10)**, and **C-di-acid (21)** separately to a final volume of 50 μL and the resulting solutions incubated for 3 h. Twenty-microliter aliquots of the incubated mixtures were then mixed with 20 μL of RNA solution (10.0 mg/mL RNA, freshly dissolved in RNase free water) and incubated for another 30 min. Then 10 μL of sample buffer (containing 10% glycerol and 0.025% bromophenol blue) were added to this mixture and 15 μL from each solution were extracted and loaded onto a 1.1% agarose gel. The gel was run using a 0.04-M Tris-Acetic acid-EDTA (TAE) buffer (pH 8.0). The residual RNA was visualized by ethidium bromide staining under UV light.

3.7. Inhibition Kinetics with RNase A

A quantitative account of RNase A inhibition by the individual inhibitors was obtained by a UV spectroscopic method described by Anderson and co-workers [22]. The assay was performed in a 0.1-M Mes-NaOH buffer, pH 6.0 containing 0.1 M NaCl using 2',3'-cCMP as the substrate. The inhibition constants were calculated from initial velocity data using a Lineweaver–Burk plot. The slopes from the Lineweaver–Burk double reciprocal plot were plotted against the corresponding inhibitor concentrations to get inhibition constants (K_i).

3.8. Circular Dichroism Measurements

Circular Dichroism (CD) was performed in order to monitor the changes in the secondary structure of the enzyme as a result of interaction with the inhibitors. In this method, 200 μL of RNase A (30 μM) were mixed separately with 200 μL of **U-di-acid (10)** (30 μM) and **C-di-acid (21)** (30 μM) and incubated for 3 h. From the resulting solutions, an aliquot of 300 μL was used for CD measurements, taking a 1 mm path length quartz cell. The spectra were recorded in the range of 190–240 nm with a scan rate of 50 nm/min. Three scans were accumulated for each spectrum. The secondary structure was determined using an online server, DICHROWEB [46].

3.9. Fluorescence Spectroscopy

Fluorescence quenching study was performed to garner an idea of the binding affinities of **U-di-acid (10)** and **C-di-acid (21)** towards RNase A. The emission spectra were recorded from 290 to 400 nm with excitation at 275 nm [47] using a 5 nm slit width. The interaction between the ligands and RNase A was investigated by titration of 3 mL solution of RNase A with successive addition of the respective ligands (0–15 μM) in a 20-mM phosphate buffer of pH 7.0. Binding constants (K_b) were calculated using double-logarithm plot [48].

3.10. FlexX Docking

The crystal structure of RNase A (PDB entry 1FS3) was downloaded from the Protein Data Bank [49]. The 3D structures of the inhibitors were generated in *Sybyl6.92* (Tripos Inc., St. Louis, MO, USA). Minimum energy conformations were obtained with the help of the MMFF94 force field using MMFF94 charges with a gradient of 0.005 kcal/mole by 1000 iterations with all other default parameters. The ligands were docked with the protein using *FlexX* software. The ranking of the generated solutions was performed using a scoring function that estimates the free binding energy (ΔG) of the protein-ligand complex considering various types of molecular interactions [50]. Docked conformations were visualized using *PyMol* [51].

3.11. Accessible Surface Area Calculations

Accessible surface area of uncomplexed RNase A and its docked complexes were calculated using the program *NACCESS*. The structures obtained from the *FlexX* analysis were used for the calculation. The change in ASA for a particular residue X was calculated using: $\Delta ASA^X = ASA^X_{RNase\ A} - ASA^X_{RNase\ A\ +\ inhibitor}$.

4. Conclusions

Strategically designed carboxylated acyclonucleosides have been established as moderate RNase A inhibitors. The experimental outcome points towards the possible contribution of a sugar ring in RNase A inhibition. Cytosine analogues have been proven to have better inhibitory properties than the corresponding uracil derivatives. **C-di-acid (21)** emerged as the most potent inhibitor of the series, having an inhibition constant (K_i) value of 235 µM. It was observed that an increment in the number of carboxylic acid groups resulted in better inhibitory properties. However, the absence of the rigid ribose ring in the molecules significantly affects the inhibition properties of the synthetic nucleosides. These findings should act as a guideline for future design of inhibitors for RNase A and other members of the ribonuclease superfamily.

Supplementary Materials: Supplementary materials can be accessed at: http://www.mdpi.com/1420-3049/20/04/5924/s1.

Acknowledgments: The authors thank the Department of Biotechnology, Ministry of Science and Technology, New Delhi for funding (project no. SB/S1/OC-30/2014). Kaustav Chakraborty thanks D. Tripathy and S. Ghosh for their help in some of the experiments and the Council for Scientific and Industrial Research, New Delhi for a fellowship.

Author Contributions: S.D.G and T.P conceived the idea and designed the study. K.C performed the synthetic and biological experiments, and analyzed the data. K.C, S.D.G, and T.P wrote the manuscript together. All authors read and approved the final version of the article.

Conflicts of Interest: The authors declare no conflict of interest.

References

1. D'Alessio, G. The superfamily of vertebrate-secreted ribonucleases. In *Ribonucleases*; Nicholson, A.W., Ed.; Springer: Heidelberg, Germany, 2011; pp. 1–34.
2. Loverix, S.; Steyaert, J. Ribonucleases: from prototypes to therapeutic targets. *Curr. Med. Chem.* **2003**, *10*, 779–785. [CrossRef] [PubMed]
3. Viola, M.; Libra, M.; Callari, D.; Sinatra, F.; Spada, D.; Noto, D.; Emmanuele, G.; Romano, F.; Averna, M.; Pezzino, F.M.; *et al.* Bovine seminal ribonuclease is cytotoxic for both malignant and normal telomerase-positive cells. *Int. J. Oncol.* **2005**, *27*, 1071–1077. [PubMed]
4. Zrinski, R.T.; Dodig, S. Eosinophil cationic protein-current concept and controversies. *Biochem. Med.* **2011**, *21*, 111–121. [CrossRef]
5. Li, S.; Ibaragi, S.; Hu, G. Angiogenin as a molecular target for the treatment of prostate cancer. *Curr. Cancer Ther. Rev.* **2011**, *7*, 83–90. [CrossRef] [PubMed]
6. Fang, E.F.; Ng, T.B. Ribonucleases of different origins with a wide spectrum of medicinal applications. *Biochim. Biophys. Acta* **2011**, *1815*, 65–74. [PubMed]
7. Leland, P.A.; Schultz, L.W.; Kim, B.; Raines, R.T. Ribonuclease A variants with potent cytotoxic activity. *Proc. Natl. Acad. Sci. USA* **1998**, *95*, 10407–10412. [CrossRef] [PubMed]
8. Leland, P.A.; Staniszewski, K.E.; Park, C.; Kelemen, B.R.; Raines, R.T. The ribonucleolytic activity of angiogenin. *Biochemistry* **2002**, *41*, 1343–1350. [CrossRef] [PubMed]
9. Raines, R.T. Ribonuclease A. *Chem. Rev.* **1998**, *98*, 1045–1065. [CrossRef] [PubMed]
10. Marshall, G.; Feng, J.A.; Kuster, D.J. Back to the future: ribonuclease A. *Biopolymers* **2007**, *90*, 259–277. [CrossRef]
11. Shapiro, R.; Weremowicz, S.; Riordan, J.F.; Vallee, B.L. Ribonucleolytic activity of angiogenin: Essential histidine, lysine, and arginine residues. *Proc. Natl. Acad. Sci. USA* **1987**, *84*, 8783–8787. [CrossRef] [PubMed]

12. Hamann, K.J.; Barker, R.L.; Loegering, D.A.; Pease, L.R.; Gleich, G.L. Sequence of human eosinophil-derived neurotoxin cDNA: Identity of deduced amino acid sequence with human nonsecretory ribonucleases. *Gene* **1989**, *83*, 161–167. [CrossRef] [PubMed]

13. Gagné, D.; Charest, L.; Morin, S.; Kovrigin, E.L.; Doucet, N. Conservation of flexible residue clusters among structural and functional enzyme homologues. *J. Biol. Chem.* **2012**, *287*, 44289–44300. [CrossRef] [PubMed]

14. Maiti, T.K.; De, S.; Dasgupta, S.; Pathak, T. 3'-*N*-Alkylamino-3'-deoxy-*ara*-uridines: A new class of potential inhibitors of ribonuclease A and angiogenin. *Bioorg. Med. Chem.* **2006**, *14*, 1221–1228. [CrossRef]

15. Leonidas, D.D.; Maiti, T.K.; Samanta, A.; Dasgupta, S.; Pathak, T.; Zographos, S.E.; Oikonomakos, N.G. The binding of 3'-*N*-piperidine-4-carboxyl-3'-deoxy-*ara*-uridine to ribonuclease A in the crystal. *Bioorg. Med. Chem.* **2006**, *14*, 6055–6066. [CrossRef] [PubMed]

16. Debnath, J.; Dasgupta, S.; Pathak, T. Inhibition of ribonuclease A by nucleoside-dibasic acid conjugates. *Bioorg. Med. Chem.* **2009**, *17*, 6491–6496. [CrossRef] [PubMed]

17. Debnath, J.; Dasgupta, S.; Pathak, T. Comparative inhibitory activity of 3'- and 5'-functionalized nucleosides on ribonuclease A. *Bioorg. Med. Chem.* **2010**, *18*, 8257–8263. [CrossRef] [PubMed]

18. Samanta, A.; Dasgupta, S.; Pathak, T. 5'-modified pyrimidine nucleosides as inhibitors of ribonuclease A. *Bioorg. Med. Chem.* **2009**, *17*, 6491–6496. [CrossRef] [PubMed]

19. Datta, D.; Samanta, A.; Dasgupta, S.; Pathak, T. 3'-Oxo-, amino-, thio-, and sulfone-acetic acid modified thymidines: Effect of increased acidity on ribonuclease A inhibition. *Bioorg. Med. Chem.* **2013**, *21*, 4634–4645. [CrossRef] [PubMed]

20. Datta, D.; Samanta, A.; Dasgupta, S.; Pathak, T. Synthesis of 5'-carboxymethylsulfonyl-5'-deoxyribonucleosides under mild hydrolytic conditions: A new class of acidic nucleosides as inhibitors of ribonuclease A. *RSC Adv.* **2014**, *4*, 2214–2218. [CrossRef]

21. Datta, D.; Dasgupta, S.; Pathak, T. Ribonuclease A inhibition by carboxymethylsulfonyl-modified xylo- and arabinopyrimidines. *ChemMedChem* **2014**, *9*, 2138–2149. [CrossRef] [PubMed]

22. Anderson, D.G.; Hammes, G.G.; Walz, F.G. Binding of phosphate ligands to ribonuclease A. *Biochemistry* **1968**, *7*, 1637–1645. [CrossRef] [PubMed]

23. Walz, F.G. Kinetic and equilibrium studies on the interaction of ribonuclease A and 2'-deoxyuridine 3'-phosphate. *Biochemistry* **1971**, *10*, 2156–2162. [CrossRef] [PubMed]

24. Russo, N.; Shapiro, R.; Vallee, B.L. 5'-Diphosphoadenosine-3'-phosphate is a potent inhibitor of bovine pancreatic ribonuclease A. *Biochem. Biophys. Res. Commun.* **1997**, *231*, 671–674. [CrossRef] [PubMed]

25. Leonidas, D.D.; Shapiro, R.; Irons, L.I.; Russo, N.; Acharya, K.R. Crystal structures of ribonuclease A complexes with 5'-diphosphoadenosine 3'-phosphate and 5'-diphosphoadenosine 2'-phosphate at 1.7 Å resolution. *Biochemistry* **1997**, *36*, 5578–5588. [CrossRef] [PubMed]

26. Leonidas, D.D.; Shapiro, R.; Irons, L.I.; Russo, N.; Acharya, K.R. Towards rational design of ribonuclease inhibitors: High resolution crystal structure of a ribonuclease A complex with a potent 3',5'-pyrophosphate-linked dinucleotide inhibitor. *Biochemistry* **1999**, *38*, 10287–10297. [CrossRef] [PubMed]

27. Russo, N.; Shapiro, R. Potent inhibition of mammalian ribonucleases by 3',5'-pyrophosphate-linked nucleotides. *J. Biol. Chem.* **1999**, *274*, 14902–14908. [CrossRef] [PubMed]

28. Leonidas, D.D.; Chavali, G.B.; Oikonomakos, N.G.; Chrysina, E.D.; Kosmopoulou, M.N.; Vlassi, M.; Frankling, C.; Acharya, K.R. High-resolution crystal structures of ribonuclease A complexed with adenylic and uridylic nucleotide inhibitors. Implications for structure-based design of ribonucleolytic inhibitors. *Protein Sci.* **2003**, *12*, 2559–2574. [CrossRef] [PubMed]

29. Kumar, K.; Jenkins, J.L.; Jardine, A.M.; Shapiro, R. Inhibition of mammalian ribonucleases by endogenous adenosine dinucleotides. *Biochem. Biophys. Res. Commun.* **2003**, *300*, 81–86. [CrossRef] [PubMed]

30. Jenkins, C.L.; Thiyagarajan, N.; Sweeney, R.Y.; Guy, M.P.; Kelemen, B.R.; Acharya, K.R.; Raines, R.T. Binding of non-natural 3'-nucleotides to ribonuclease A. *FEBS J.* **2005**, *272*, 744–755. [CrossRef] [PubMed]

31. Hatzopoulos, G.N.; Leonidas, D.D.; Kardakaris, R.; Kobe, J.; Oikonomakos, N.G. The binding of IMP to ribonuclease A. *FEBS J.* **2005**, *272*, 3988–4001. [CrossRef] [PubMed]

32. Yakovlev, G.I.; Mitkevich, V.A.; Makarov, A.A. Ribonuclease inhibitors. *Mol. Biol.* **2006**, *40*, 867–874. [CrossRef]

33. Nogués, M.V.; Vilanova, M.; Cuchillo, C.M. Bovine pancreatic ribonuclease A as a model of an enzyme with multiple substrate binding sites. *Biochim. Biophys. Acta* **1995**, *1253*, 16–24. [CrossRef] [PubMed]

34. Findlay, D.; Herries, D.G.; Mathias, A.P.; Rabin, B.R.; Ross, C.A. The active site and mechanism of action of bovine pancreatic ribonuclease. *Nature* **1961**, *190*, 781–784. [CrossRef] [PubMed]

35. Cuchillo, C.M.; Nogués, M.V.; Raines, R.T. Bovine pancreatic ribonuclease: Fifty years of the first enzymatic reaction mechanism. *Biochemistry* **2011**, *50*, 7835–7841. [CrossRef] [PubMed]

36. Silverman, R.B. *The Organic Chemistry of Drug Design and Drug Action*, 2nd ed.; Elsevier: San Diego, CA, USA, 2004; p. 126.

37. Liu, X.; Chen, R. Synthesis of novel phosphonotripeptides containing uracil or thymine group. *Phosphorus Sulfur Silicon Relat. Elem.* **2001**, *176*, 19–28. [CrossRef]

38. Rezazgui, O.; Boëns, B.; Teste, K.; Vergaud, J.; Trouillas, P.; Zerrouki, R. One-pot and catalyst-free amidation of ester: A matter of non-bonding interactions. *Tetrahedron Lett.* **2011**, *52*, 6796–6799. [CrossRef]

39. Porcheddu, A.; Giacomelli, G.; Piredda, I.; Carta, M.; Nieddu, G. A practical and efficient approach to PNA monomers compatible with Fmoc-mediated solid-phase synthesis protocols. *Eur. J. Org. Chem.* **2008**, *34*, 5786–5797. [CrossRef]

40. Gaudreau, S.; Novetta-dellen, A.; Neault, J.F.; Diamantoglou, S.; Tajmir-riahi, H.A. 3'-azido-3'-deoxythymidine binding to ribonuclease A: Model for drug-protein interaction. *Biopolymers* **2003**, *72*, 435–441. [CrossRef] [PubMed]

41. Ghosh, K.S.; Maiti, T.K.; Mandal, A.; Dasgupta, S. Copper complexes of (-)-epicatechin gallate and (-)-epigallactocatechin gallate act as inhibitors of ribonuclease A. *FEBS Lett.* **2006**, *580*, 4703–4708. [CrossRef] [PubMed]

42. Ghosh, K.S.; Debnath, J.; Dutta, P.; Sahoo, B.K.; Dasgupta, S. Exploring the potential of 3'-O-carboxy esters of thymidine as inhibitors of ribonuclease A and angiogenin. *Bioorg. Med. Chem.* **2008**, *16*, 2819–2828. [CrossRef] [PubMed]

43. Dutta, S.; Basak, A.; Dasgupta, S. Synthesis and ribonuclease A inhibition activity of resorcinol and phloroglucinol derivatives of catechin and epicatechin: Importance of hydroxyl groups. *Bioorg. Med. Chem.* **2010**, *18*, 6538–6546. [CrossRef] [PubMed]

44. Tripathy, D.R.; Roy, A.S.; Dasgupta, S. Complex formation of rutin and quercetin with copper alters the mode of inhibition of ribonuclease A. *FEBS Lett.* **2011**, *585*, 3270–3276. [CrossRef] [PubMed]

45. Sela, M.; Anfinsen, C.B. Some spectrophotometric and polarimetric experiments with ribonucleases. *Biochim. Biophys. Acta* **1957**, *24*, 229–235. [CrossRef] [PubMed]

46. Whitmore, L.; Wallace, B.A. DICHROWEB, an online server for protein secondary structure analyses from circular dichroism spectroscopic data. *Nucleic Acids Res.* **2004**, *32*, 668–673. [CrossRef]

47. Garcia-Borron, J.C.; Escribano, J.; Jimenez, M.; Iborra, J.L. Quantitative determination of tryptophanyl and tyrosyl residues of proteins by second-derivative fluorescence spectroscopy. *Anal. Biochem.* **1982**, *125*, 277–285. [CrossRef] [PubMed]

48. Jiang, M.; Xie, M.X.; Zheng, D.; Liu, Y.; Li, X.Y.; Cheng, X. Spectroscopic studies on the interaction of cinnamic acid and its hydroxyl derivatives with human serum albumin. *J. Mol. Struct.* **2004**, *692*, 71–80. [CrossRef]

49. Berman, H.M.; Westbrook, J.; Feng, Z.; Gilliland, G.; Bhat, T.N.; Weissig, H.; Shindyalov, I.N.; Bourne, P.E. The protein data bank. *Nucleic Acids Res.* **2000**, *28*, 235–242. [CrossRef] [PubMed]

50. Rarey, M.; Kramer, B.; Lengauer, T.; Klebe, G. A fast flexible docking method using an incremental construction algorithm. *J. Mol. Biol.* **1996**, *261*, 470–489. [CrossRef] [PubMed]

51. DeLano, W.L. *The PyMOL Molecular Graphics System*; DeLano Scientific: San Carlos, CA, USA, 2004. Available online: http://pymol.sourceforge.net/.

Sample Availability: *Sample Availability*: Samples are not available from authors.

molecules

Article

Non-Nucleosidic Analogues of Polyaminonucleosides and Their Influence on Thermodynamic Properties of Derived Oligonucleotides

Jolanta Brzezinska and Wojciech T. Markiewicz *

Institute of Bioorganic Chemistry, Polish Academy of Sciences, Noskowskiego 12/14, 61-704 Poznań, Poland; jabrzoza@ibch.poznan.pl
* Correspondence: markwt@ibch.poznan.pl; Tel.: +48-61-852-8503 (ext. 180); Fax: +48-61-852-0532

Academic Editors: Mahesh K. Lakshman and Fumi Nagatsugi
Received: 17 March 2015; Accepted: 9 July 2015; Published: 13 July 2015

Abstract: The rationale for the synthesis of cationic modified nucleosides is higher expected nuclease resistance and potentially better cellular uptake due to an overall reduced negative charge based on internal charge compensation. Due to the ideal distance between cationic groups, polyamines are perfect counterions for oligodeoxyribonucleotides. We have synthesized non-nucleosidic analogues built from units that carry different diol structures instead of sugar residues and functionalized with polyamines. The non-nucleosidic analogues were attached as internal or 5′-terminal modifications in oligodeoxyribonucleotide strands. The thermodynamic studies of these polyaminooligonucleotide analogues revealed stabilizing or destabilizing effects that depend on the linker or polyamine used.

Keywords: polyaminonucleoside analogue; non-nucleosidic analogue; spermine; putrescine; oligonucleotide; duplex DNA; thermodynamic stability

1. Introduction

The polyanionic character of nucleic acids and their synthetic fragments is a barrier for their introduction into cells. Many laboratories try to improve nucleic acid delivery to cells by synthesis of cationic bioconjugates [1–3], bioconjugate analogues with reduced polyanionic character [4–6], or a great variety of polymeric transfection media [7–12]. On the other hand, there are studies to elucidate a specific mode of small cationic molecular drugs interactions with biomolecules (e.g., nucleic acids, proteins) useful in developing their more active analogues [13]. Among this clinically important class of compounds are antibiotics carrying aminosugar residues [14].

Chemically modified oligonucleotides have been utilized as indispensable materials for DNA gene therapy [15,16], gene regulation [17,18], chip technology [19,20] and recent nanotechnology [21–23] because of their hybridization affinity for target DNA and/or RNA molecules [24]. The polyanionic character of antisense and siRNA oligonucleotides is a major cause of insufficient cellular uptake and side effects such as binding to serum proteins. The combination of nucleotides with aminoalkyl chains greatly enhances the variety of possible structures as well as their potential application [24]. Thus, oligonucleotides possessing cationic functionalities in addition to the anionic phosphate backbone have been shown to exhibit promising properties [25–28]. Polyamines, putrescine, spermine and spermidine, are involved in the regulation of gene function [29,30]. *In vitro*, they stabilize DNA and RNA duplexes [31,32], especially ones with imperfect base pairing [33]. The distance between amino groups of three and four carbon atoms is practically the same as the distance between phosphate anions in the backbone of DNA making polyamines the perfect compounds for creating "zwitterionic" oligonucleotides [27]. It is known that polyamines interact with DNA and RNA in different ways. The electrostatic binding is performed via water molecules, by hydrogen bonding with polar functional

Molecules **2015**, *20*, 12652–12669

groups or with hydrophobic surfaces of nucleobases. There have been numerous studies aimed at determining these interactions using NMR imaging, circular dichroism, Raman spectroscopy, IR spectroscopy, X-ray crystallography and differential scanning calorimetry. In spite of these extensive studies, precise mechanisms for the interaction between polyamine and DNA is still not fully understood *in vivo* [34].

The non-nucleosidic analogs of polyaminonucleosides have not been examined so far. Therefore, our aim was to elaborate versatile procedures for synthesis of non-nucleosidic polyamine derivatives and learn their properties within oligonucleotide chains. The choice of carbon chain skeletons of analogues (Figure 1) was based on the extent of their commercial availability.

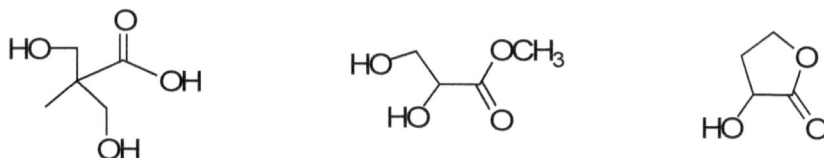

Figure 1. Substrates for the non-nucleosidic polyaminonucleoside analogues.

The obtained polyamine building blocks were incorporated at the 5′-end and internal positions within an oligodeoxyribonucleotide chain and the resulting conjugates were evaluated for their hybridization properties.

2. Results and Discussion

2.1. Chemistry

The synthesis of the non-nucleosidic polyamine derived phosphoramidite building blocks containing different linkers is shown in Scheme 1.

We decided to use 2,2-bis(hydroxymethyl)-propionic acid (bis-MPA, **1**) that was earlier applied in dendrimer synthesis [35,36].

Bis-MPA (**1**) was protected with benzaldehyde [37] as it is more lipophilic than a isopropylidene group preferred to ease the final workup procedure and increase yield when using an excess of polar polyamines [38,39]. The protection of hydroxyl groups in bis-MPA gives rise to formation of two stereoisomers in the ratio ca 2:1, however, further reactions were carried out without separation of isomers. Several approaches for activation of carboxyl function (**2**) were checked. Reactions with 1,4-butylamine as a model amine in the presence of carbonyldiimidazole or 2-chloro-4,6-dimethoxy-1,3,5-triazine gave the expected n-butylamide derivative (results not shown). The best results were obtained when activation was performed with thionyl chloride in the presence of pyridine and traces of DMF. Thus, reaction with molar excess of putrescine (1,4-diaminobutane) at ambient conditions led to N-(4-aminobutyl)-5-methyl-2-phenyl-1,3-dioxane-5-carboxyamide **2** in 60% yield (Scheme 1).

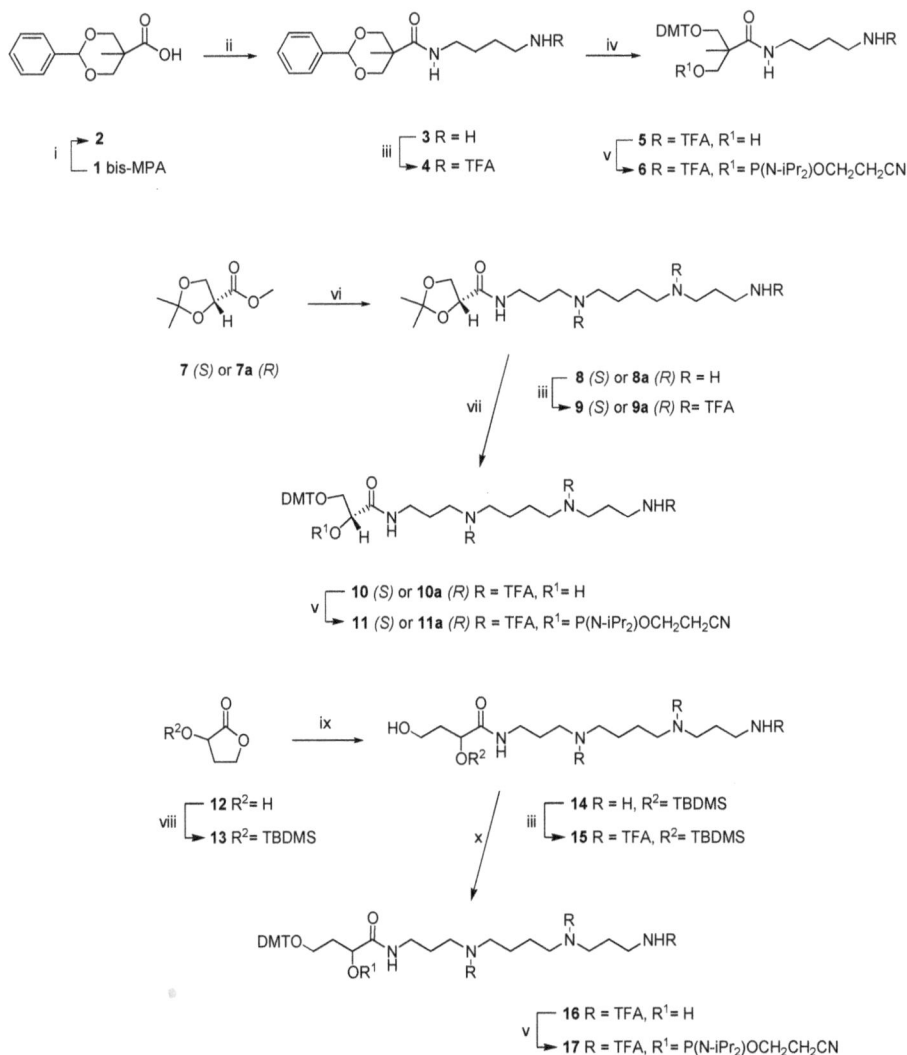

Scheme 1. Synthetic routes to non-nucleosidic polyaminonucleoside analogues: (i) benzaldehyde, *p*-TsOH, DMF; (ii) SOCl₂, putrescine, Et₃N, DCM; (iii) (F₃CCO)₂O, pyridine; (iv) (a) 6 M HCl, reflux, (b) DMTCl, pyridine; (v) (iPr₂N)₂POCH₂CH₂CN, 5-ethylthio-1*H*-tetrazole, DCM; (vi) spermine, MeOH; (vii) (a) *p*-TsOH, MeOH, (b) DMTCl, pyridine; (viii) TBDMSCl, DMAP, imidazole, CH₃CN; (ix) spermine, MW, 40 min; (x) (a) TEAHF, THF/dioxane, (b) DMTCl, pyridine.

We also synthesized the analogs based on the enantiomers (**7**) and (**7a**) which were coupled with excess of spermine in dry MeOH. Due to the lack of a chromophore in synthesized compounds, a reaction control was performed using TLC plates stained with a solution of fluorescein (free acid) in acetonitrile. Spermine derivatives (**8, 8a**) were isolated by several extractions with large amounts of organic solvent and used in the next step without additional purification. The configuration of the branching point in these linkers is retained throughout the synthetic procedure. After protection of the amine function with trifluoroacetyl groups, 1,3-dioxolane was hydrolyzed to prepare polyamine non-nucleosidic derivatives for selective 4,4'-dimethoxytritylation. In order to avoid transacetylation of trifluoroacetyl groups during the isopropylidene cleavage of **9** we used *p*-TsOH instead acetic acid. Otherwise, amino groups would be at least in part protected with Ac instead of TFA and their deprotection with ammonia would not be possible under the conditions of the final deprotection of oligonucleotides. Commercially available precursor α-hydroxy-γ-butyrolactone (**12**) was used as a starting material to obtain a polyaminonucleoside analogue (**17**). This compound provides a three carbon distance between the phosphate groups after introduction into the oligodeoxyribonucleotide strand. It also contains a chiral carbon center with a secondary hydroxyl group and the carbonyl group corresponding to the 3′-hydroxyl and 2′-carbon of nucleosidic residues, respectively. The lactone-spermine conjugate was subsequently obtained after protection of the secondary hydroxyl group of the lactone (**13**) with lipophilic and a bulky t-butyldimethylsilyl (TBDMS) group. Thus, this eased the isolation of the product (**14**) resulting from the coupling reaction with spermine (Scheme 1). The reaction of protected lactone **13** and unprotected spermine was performed by microwave synthesis to achieve the desired product **14** with 64% yield, even when both reagents were used in equimolar ratio. To eliminate side reactions during the condensation step of DNA synthesis, the protection of putrescine and spermine amino functions with trifluoroacetyl groups was ensured (as previously described [40]). For all non-nucleosidic analogue yields of trifluoroacetylation ranged from 58% (spermine derivatives) up to 70% (putrescine derivative) (Scheme 1). Conversion to the corresponding phosphoramidite derivative was preceded by removal of protecting groups of hydroxyl function with HCl (**4**) and *p*-TsOH (**9**, **9a**) acids or Et₃N*3HF (**15**). The protection 5′-OH-groups with a 4,4′-dimethoxytrityl group was used in a last step of the 3′-OH reaction with 2-cyanoethyl-*N,N,N′,N′*-tetraisopropylaminophosphane and 5-(ethylthio)-1*H*-tetrazole as the activator. However, due to similar physical properties, the separation of the products from impurities was difficult by silica gel column chromatography. This was resolved by precipitation from hexane. The spermine derivatives of (*R*) and (*S*) enantiomers of glyceric acid gained an additional chiral center as phosphoramidites (**11**, **11a**). In the case of the polyamine derivative of α-hydroxy-γ-butyrolactone, opening of the lactone ring **14** led to a mixture of enantiomers. Thus, after phosphitylation, compound **17** as a mixture of enantiomers was obtained. All amidites were lyophilized and were stable during long-term storage at −20 °C.

2.2. Oligonucleotides

The phosphoramidites (**6**, **11**, **11a**, **17**) were used for synthesis of polyamine analogs of oligodeoxyribonucleotides (Figure 2) using a 12-mer as a reference sequence (Table 1). The coupling efficiencies of these phosphoramidites were 53%–70% as determined by measuring detritylation. The modified oligodeoxyribonucleotides were obtained after a standard deprotection procedure, and their structures were confirmed by the MALDI-TOF (Table 1). The oligodeoxyribonucleotides **ON5–ON6** and **ON9** were used as pure enantiomers and **ON4** and **ON8** as mixtures of diastereoisomers.

Table 1. Sequences of oligodeoxyribonucleotides and MALDI-TOF MS data.

No.	Sequence 5′ → 3′ ᵃ	*m/z* [M − H]⁻	
		Calcd	Found
ON1	CTC AAG CAA GCT	3614.42	3613.08
ON2	CTC ACA TGC GCG	3606.40	3605.54
ON3	X0 CTC AAG CAA GCT	3994.42	3993.19
ON4	X1 CTC ACA TGC GCG	3973.12	3972.23
ON5	X2 CTC ACA TGC GCG	3959.54	3959.24
ON6	X3 CTC ACA TGC GCG	3959.54	3959.57
ON7	CTC AAG X0 CAA GCT	3994.42	3993.25
ON8	CTC ACA X1 TGC GCG	3973.12	3973.12
ON9	CTC ACA X2 TGC GCG	3959.54	3957.45

ᵃ R = H for X0–X3 in ON3–ON6 and R = 3′-end oligo for X0–X2 in ON7–ON9.

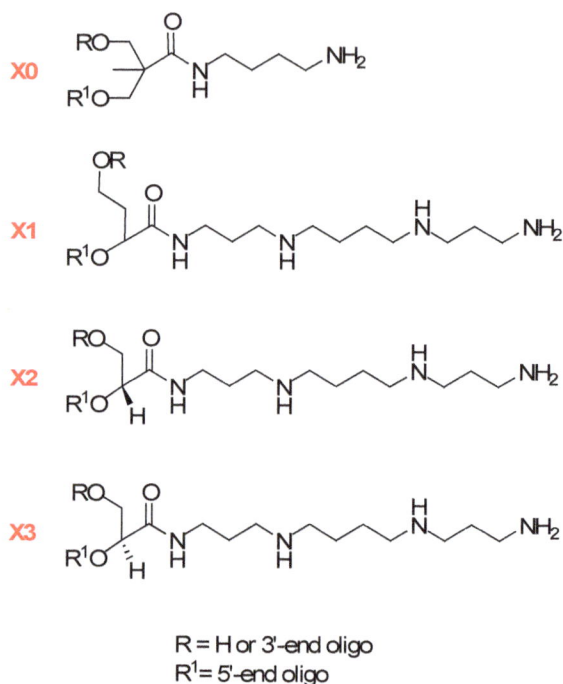

R = H or 3'-end oligo
R¹= 5'-end oligo

Figure 2. Non-nucleosidic polyamino-nucleoside analogues units in oligodeoxyribonucleotide strands.

2.3. Stability of Duplexes with Modified Oligodeoxyribonucleotides

The influence of conjugated polyamines on the stability of the DNA duplexes was studied using UV melting spectra. It was shown previously that polyamine conjugation to oligodeoxyribonucleotides results in stabilizing of DNA duplexes and triplexes [2–4,38,39,41–43]. Recently, we described the NMR structure of a DNA duplex carrying a single spermine modified deoxycytidine unit [34]. This modification moderately stabilizes the DNA duplex (Figure 3, dC^{Sp} *vs.* RF1) and does not perturb the DNA structure.

The 5'-end dangling modifications in oligodeoxyribonucleotide duplexes usually increase duplex stability and do not show large sequence dependence [44,45]. Since the terminal unpaired nucleotides are not involved in base pairing, stacking, electrostatic, perhaps to some extent, hydrophobic interactions are responsible for the thermodynamic effects of the 5'- and 3'-dangling ends. The nearest-neighborhood model assumes the duplex region for a dangling end to have the same calculated thermodynamic stability as a matching blunt-end duplex [32,33].

Thus, the stability was increased when **X1** and **X3** (Scheme 1) were incorporated at the 5' position (Table 1, entry **ON4** and **ON6**), but for incorporations of **X0** and **X2** (Scheme 1) a moderate destabilizing effect was observed (Table 1, entry **ON3** and **ON5**). Despite lack of a nucleobase and higher conformational flexibility at the 5'-end of the strand the observed stabilities are rather typical for 5'-end dangling units as described in the literature. Thermodynamic data (Table 2, Figure 1) show that **X3** results in $\Delta\Delta G$ (−0.3 kcal/mol) similar to unpaired nucleotide (−0.1 to −0.5 kcal/mol) at the 5'-end. The change of ΔG for **X1** is even higher ($\Delta\Delta G$ = −0.64 kcal/mol). These results suggest that the lack of a nucleobase does not affect the thermal stability of the duplexes. This might indicate that increased duplex stability is provided by the three positive charges in the spermine chain. In the case of the putrescine residue, **X0**, which introduces one positive charge only, a small decrease of duplex stability was observed. (Figure 1, **ON3**). One can conclude that the small decrease in duplex stability observed for **ON5** (Figure 1) containing spermine attached to the glycerol isomer (**X2**) is caused by the structure and configuration of this particular linker.

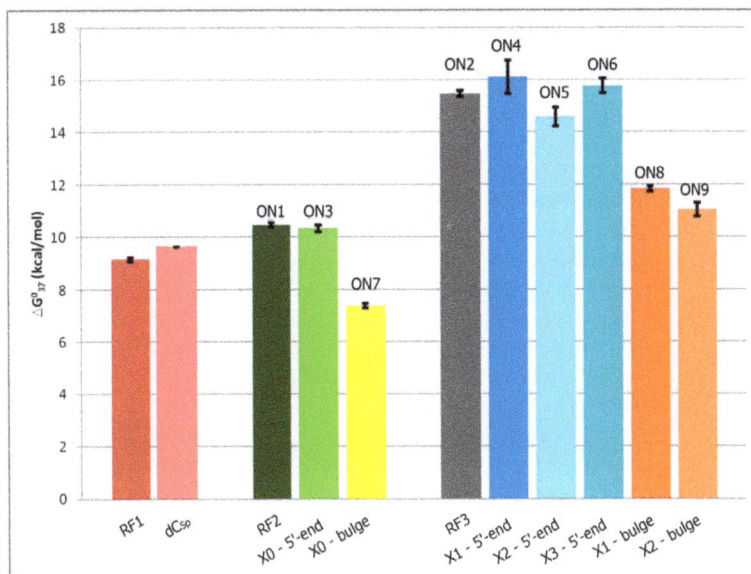

Figure 3. Changes in free energy (ΔG) of polyamine modified duplexes: RF-RF3- reference duplexes, dCSp—spermine modified (dC^{Sp} = 4-*N*-[4,9,13-triazatridecan-1-yl]-2′-deoxycytidine) duplex [34]; **ON3**, **ON7**—non-nucleosidic putrescine analogues (Table 2); **ON4–ON6** and **ON8–ON9**—non-nucleosidic spermine analogues (Table 2).

Table 2. DNA duplex stability of non-nucleosidic polyamine-modified oligodeoxyribonucleotides ^{a}.

Oligo	Average of Curve Fits				T_M^{-1} *vs.* log (C_T/4) Plots					
	−ΔH° [kcal/mol]	−ΔS° [eu]	−ΔG°$_{37}$ [kcal/mol]	T_M [°C]	−ΔH° [kcal/mol]	−ΔS° [eu]	−ΔG°$_{37}$ [kcal/mol]	T_M [°C]	ΔΔG°$_{37}$ [kcal/mol]	ΔT$_M$ ^{b} [°C]
A. Thermodynamic parameters of references duplexes										
ON1 ^{c}	67.8 ± 9.0	185.6 ± 27.8	10.27 ± 0.44	55.1	72.6 ± 2.0	200.6 ± 6.2	10.45 ± 0.08	54.7		
ON2 ^{d}	115.9 ± 4.8	322.1 ± 14.2	16.00 ± 041	64.6	108.5 ± 1.8	300.1 ± 5.4	15.46 ± 0.12	64.8		
B. Thermodynamic parameters of duplexes with the polyamine analog as a dangling end										
ON3 ^{c}	63.3 ± 6.8	171.5 ± 20.9	10.12 ± 0.37	55.7	69.09 ± 3.1	189.4 ± 9.7	10.33 ± 0.13	55.0	+0.12	−0.3
ON4 ^{d}	131.8 ± 10.4	368.8 ± 31.0	17.47 ± 0.80	65.1	114.0 ± 8.3	315.9 ± 24.9	16.10 ± 0.64	65.4	−0.64	+0.6
ON5 ^{d}	97.8 ± 9.4	268.4 ± 28.1	14.57 ± 0.68	64.8	98.9 ± 5.1	271.9 ± 15.4	14.59 ± 0.36	64.5	+0.87	−0.3
ON6 ^{d}	106.1 ± 4.7	292.4 ± 14.0	15.43 ± .037	65.4	110.9 ± 3.9	307.0 ± 11.7	15.76 ± 0.27	65.1	−0.3	+0.3
C. Thermodynamic parameters of duplexes with the polyamine analog as a bulge										
ON7 ^{c}	62.7 ± 16.7	178.2 ± 53.8	7.50 ± 0.20	41.9	59.0 ± 4.2	166.5 ± 13.7	7.38 ± 0.08	41.5	+3.07	−13.2
ON8 ^{d}	71.6 ± 6.0	194.4 ± 18.3	11.31 ± 0.35	59.2	82.0 ± 2..0	226.3 ± 6.2	11.83 ± 0.10	58.4	+3.63	−6.4
ON9 ^{d}	81.2 ± 11.5	225.0 ± 35.6	11.45 ± 0.48	57.0	70 ± 6.4	192.1 ± 20.0	11.03 ± 0.25	58.1	+4.43	−6.7

^{a} Solutions are 100 mM NaCl, 20 mM sodium cacodylate, 0.5 mM Na$_2$EDTA, pH 7; ^{b} Calculated for 10^{-4} M total strand concentration; ^{c} Complementary strand 5′-AGC TTG CTT GAG-3′; ^{d} Complementary strand 5′-CGC GCA TGT GAG-3′.

Next, we investigated the influence of polyamine derivatives **X0**, **X1** and **X2** on duplex stability, inserted as bulges in the middle of the sequence (Table 1, entry **ON7–ON9**). The changes of duplex stability caused by single nucleotide bulges differ and depend on the type of flanking bases [46,47]. Incorporation of **X0–X2** resulted in a lowering of melting temperature independently of polyamine residue. However, the melting data suggest that the duplex formation occurs with the proper W-C base pairing despite of a slight increase in free energy for **ON7–ON9** (Table 2, ΔΔG *ca.* 3–4.5 kcal/mol).

171

Moreover, the effect in **ON7** where the putrescine bulge (**X0**) is flanked by GC/CG pairs is practically the same as for **ON8** and **ON9** (spermine bulge, **X1** and **X2**) flanked by AT/TA pairs. This can be attributed to the stronger stabilizing effect of a spermine residue when compared to putrescine—three positive charges *vs.* one. Yet, even three positive charges of a spermine residue do not neutralize the bulge effect as such. The observed changes of ΔG for the studied duplexes (**ON7–ON9**) are in the range observed for duplexes with a single bulge (ΔΔG, 2–6 kcal/mol) [46,47].

Linkers X1 and X2 differ in length by one carbon and this additional carbon in X1 makes this linker somehow less rigid, allowing for more favorable placement of the polyamine residue. Thus, the energy gain of 1.5 kcal/mol in **ON4** when compared to **ON5**. Therefore, **ON4** duplex is more stable than the unmodified duplex. When the same modified linkers (X1 and X2) are inserted as bulges in a more spatially "demanding environment", the difference of free energies $\Delta\Delta G°_{37}$ of **ON8** (+3.63) and **ON9** (+4.43) is smaller (0.8 kcal/mol). The influence of linker structure is less profound when the modification is inserted in the middle of chain.

Linkers X2 and X3 carrying a spermine residue differ only in the configuration of carbon (*S* and *R* respectively). Comparison of $\Delta\Delta G°_{37}$ of ON6 (−0.3 kcal/mol) and ON5 (+0.87 kcal/mol) suggests that X3 allows more favorable placement of polyamine residue This seems to indicate that in this case the spermine residue attached to a glycerol linker with *R*-configuration (X3) is closer to the duplex charged surface.

We would like to conclude that the overall effect of DNA modification with polyamine analogues seems to offer a convenient way to modify nucleic acids properties. An attachment of polyamine residues via various open chain linkers allows to maintain the general scheme of base pairing. Moreover, the structure and configuration of the linkers seems to have a higher influence on stability when placed at the 5′-end of DNA duplexes.

3. Experimental Section

3.1. General Methods

All reagents were of analytical grade, obtained from commercial resources and used without further purification. For synthesis, solvents with quality pro analysis were used. Solvents were dried and distilled following standard methods and kept over molecular sieve. All reactions were carried at room temperature unless described otherwise. Column chromatography was performed with silica gel (Merck KGaA, Darmstadt, Germany, 200–630 mesh) and TLC was carried out on precoated plates (Merck silica gel 60, F_{254}). All NMR spectra were recorded at 298 K on Bruker AVANCE II (400 MHz, ^{1}H, ^{13}C) and Varian Unity (300 MHz, ^{31}P) spectrometers (Bruker BioSpin GmbH, Rheinstetten, Germany and Varian, Inc., Palo Alto, CA, USA). Chemical shifts (δ) are reported in parts per million (ppm). *J* values are given in Hz. Mass spectra were recorded on the MicroTofQ mass spectrometer with electrospray ionization (ESI) sources (ESI source voltage of 3.2 kV, nebulization with nitrogen at 0.4 bar, dry gas flow of 4.0 L/min at temperature 220 °C) and Bruker Autoflex MALDI-TOF (Bruker Daltonik GmbH, Bremen, Germany). Microwave reactions was performed in domestic microwave oven (800 W, Amica, Wronki, Poland). Some NMR (Figures S1–S16) and MS (Figures S17–S22) spectra are available in Supplementary.

3.2. Synthesis of Monomers

Benzylidene-2,2-bis(oxymethyl)propionic acid (**2**) [37]. Benzaldehyde (4.2 mL, 40.7 mmol) was added to a well-stirred solution of 2,2-bis(hydroxymethyl)-propionic acid (**1**) (5 g, 37 mmol) in DMF (30 mL), followed by catalytic *p*-TsOH (0.355 g, 1.85 mmol) with stirring at room temperature for 4 days. The reaction was quenched with NH_4OH/EtOH (1 mL, 1:1). The solvent was evaporated and the residue dissolved in DCM (100 mL) and washed with $NaHCO_3$ (2 × 100 mL). Organic extracts were dried over anhydrous $MgSO_4$ and filtered then evaporated under reduced pressure. The residue was purified by recrystallization from DCM to obtain product **2** as mixture of two isomers (77%). ^{1}H-NMR ($CDCl_3$);

isomer A: 7.28–7.46 (m, 5H, Ph), 5.45 (s, 1H, PhCHO$_2$), 4.65 (d, 2H, J = 11.7 Hz, OCH_2CCO, 3.67 (d, 2H, J = 11.7 Hz, OCH_2CCO), 1.02 (s, 3H, CH_3); isomer B: 7.28–7.51 (m, 5H, Ph), 5.41 (s, 1H, PhCHO$_2$), 4.14–4.06 (q, 4H, J = 11.23 Hz, J = 9.27 Hz, 2CH_2), 1.57 (s, 3H, CH_3). ^{13}C-NMR (CDCl$_3$): 178.89 (C=O), 137.58 (C, Ph), 129.08 (CH, Ph), 128.23 (CH, Ph), 128.23 (CH, Ph), 101.86 (CH), 73.43 (CH$_2$), 42.17 (CMe), 17.78 (CH$_3$). MS (ESI) *m*/*z*: calcd for C$_{12}$H$_{14}$O$_4$ 221.0819 [M − H]$^-$, found 221.182.

N-(4-Aminobutyl)-5-methyl-2-phenyl-1,3-dioxane-5-carboxamide (**3**). Thionyl chloride (1.26 mL, 17.8 mmol) was added dropwise to ice-cooled solution of **2** (2 g, 8.9 mmol) in 45 mL DCM containing 10% of pyridine and 3 drops of DMF. The mixture was stirred for 2 h at room temperature, and the excess thionyl chloride was removed by several co-evaporations with a mixture of DCM and toluene. The brown residue was dissolved in DCM (46 mL) and the temperature was lowered to −10 °C. Putrescine (4.47 mL, 44.5 mmol) and triethylamine (1.4 mL) in DCM (2 mL) was added after 15 min. The mixture was stirred for 1h at room temperature, and the reaction was quenched with saturated aqueous NaHCO$_3$ (3 mL). The mixture was extracted with DCM (2 × 100 mL). The combined organic extracts were dried over anhydrous MgSO$_4$ and concentrated under vacuum and purified by chromatography with 4% MeOH in DCM as the eluent to give **3** (0.910 g, 60% yield) as a yellow oil. ^{1}H-NMR (CDCl$_3$): 7.34–7.43 (m, 5H, Ph), 7.03 (t, 1H, NHCO), 5.48 (s, 1H, PhCHO), 4.33 (d, 2H, J = 12 Hz, OCH_2CCO), 3.78 (d, 2H, J = 12 Hz, OCH_2CCO), 3.33 (q, 2H, J = 6.3 Hz, J = 5.8 Hz, CONHCH_2), 3.21 (q, 2H, J = 6.3 Hz, J = 5.8 Hz, NH$_2$CH_2, putrescine), 1.5–1.56 (m, 4H, 2 × CH$_2$, putrescine), 1.04 (s, 3H, CH$_3$); ^{13}C-NMR (CDCl$_3$): δ (ppm) 178.89 (C=O), 137.45 (C, Ph), 128.7 (CH, Ph), 128.03 (CH, Ph), 101.22 (CH), 75.07 (CH$_2$), 47.6 (CMe), 47.6 (CH$_2$, putrescine), 26.53 (CH$_2$, putrescine), 18.03 (CH$_3$); MS (ESI) *m*/*z*: calcd for C$_{16}$H$_{24}$N$_2$O$_3$ 293.186 [M + H]$^+$, found 293.2037.

5-Methyl-2-phenyl-N-(4-(2,2,2-trifluoroacetamid)butyl)-1,3-dioxane-5-carboxamide (**4**) **3** (0.430 g, 1.47 mmol) was co-evaporated with pyridine (3 × 5 mL), dissolved in pyridine (15 mL), followed by the addition of N-methylimidazole (0.16 mL, 1.46 mmol). Then, trifluoroacetic anhydride (0.593 mL, 4.41 mmol) was added dropwise to the mixture cooled at 0 °C. The mixture was stirred for 30 min and then poured into saturated aqueous NaHCO$_3$ (30 mL), and extracted with DCM (3 × 50 mL). The combined organic extracts were dried over anhydrous Na$_2$SO$_4$ and concentrated under vacuum. The residue was purified by chromatography using DCM as the eluent to give **4** (0.500 g, 52% yield) as an oil. ^{1}H-NMR (CDCl$_3$): 7.33–7.45 (m, 5H, Ph), 7.03 (t, 1H, NHCO), 5.5 (s, 1H, PhCHO$_2$), 4.38 (d, 2H, J = 11.7 Hz, OCH_2CCO), 3.78 (d, 2H, J = 11.7 Hz, OCH_2CCO), 3.2–3.4 (m, 4H, 2NHCH_2), 1.49–1.54 (m, 4H, CH$_2$, putrescine), 1.06 (s, 3H, CH$_3$). ^{13}C-NMR (CDCl$_3$): 182.83 (C=O), 158.9 (C=O), 137.45 (C, Ph), 128.7 (CH, Ph), 128.03 (CH, Ph), 125.53 (C, CF$_3$), 101.22 (CH), 75.07 (CH$_2$), 47.4 (CMe), 45.3 (CH$_2$, putrescine), 26.53 (CH$_2$, putrescine), 18.03 (CH$_3$); ^{19}F NMR 1.78 (s, 3F, CF$_3$); MS (ESI) *m*/*z*: calcd for C$_{18}$H$_{23}$F$_3$N$_2$O$_4$ 389.1683 [M + H]$^+$, found 389.1627.

3-(4,4'-Dimethoxytrityl)-2-(hydroxymethyl)-2-methyl-N-(4-(2,2,2-trifluoroacetamido)butyl)propanamide (**5**). **4** (0.240 g, 0.61 mmol) was dissolved in a mixture conc. aq HCl/EtOH (1:5, *v*/*v*) and refluxed for 72 h. The excess of HCl was removed by several co-evaporations with a mixture of methanol and toluene and finally a brown oil with was dried by co-evaporation with anhydrous pyridine (2 × 5 mL) and dissolved in anhydrous pyridine (2.4 mL). To this solution 4,4'-dimethoxytrityl chloride (0.243 g, 0.72 mmol) was added and the reaction was quenched after 3 h by adding saturated aqueous NaHCO$_3$. The resulting solution was extracted with DCM and the combined organic extracts were washed with brine, dried over Na$_2$SO$_4$ and concentrated under vacuum. The residual yellow oil was purified by chromatography with 5% MeOH in DCM as the eluent to give 0.18 g (49%) **5** as a white foam. ^{1}H-NMR (CDCl$_3$): δ (ppm) 7.42–7.2 (m, 10H, Ph, NHCO), 7.04 (t, 1H, NHCO), 6.89 (m, 4H, Ph), 3.82 (s, 3H, OCH$_3$), 3.83–3.62 (m, 4H, OCH_2CCO), 3.43–3.23 (m, 4H, 2NHCH_2), 1.4–1.8 (m, 4H, 2CH$_2$, putrescine), 1.2 (s, 3H, CH$_3$). ^{13}C-NMR (CDCl$_3$): δ (ppm) 174.4 (C=O), 158.5 (2C, COCH$_3$), 158.4 (C=O, TFA), 144.6 (C, Ph), 135.5 (C, Ph), 130.0 (CH, Ph), 128.3 (CH, Ph), 128.0 (CH, Ph), 127.8 (CH, Ph), 126.8 (C, CF$_3$), 113.1 (CH, Ph), 86.2 (C), 66.9 (CH$_2$), 65.5 (CH$_2$), 55.1 (CH$_3$), 47.5 (C, CMe), 43.2 (CH$_2$, putrescine), 24.5 (CH$_2$, putrescine), 18.7 (CH$_3$); MS (ESI) *m*/*z*: calcd for C$_{32}$H$_{37}$F$_3$N$_2$O$_6$ 603.2676 [M + H]$^+$, found 603.261.

3-[(4,4'-Dimethoxytrityl)-2-(hydroxymethyl)-2-methyl-N-(4-(2,2,2-trifluoroacetamido)butyl)
propanamide]phosphoramidite (**6**). 2-Cyanoethyl-*N,N,N,N*-tetraisopropylphosphoramidite (0.9 mL, 0.3 mmol) was added to the solution of **5** (0.143 g, 0.234 mmol) and 5-(ethylthio)-1*H*-tetrazole (0.0273 g, 0.21 mmol) in dichloromethane (1.2 mL) and the mixture was stirred at room temperature. After 2 h, TLC revealed complete reaction. The mixture was diluted with dichloromethane, washed with saturated sodium bicarbonate solution and the organic extracts were dried over Na_2SO_4. The product was purified by silica gel chromatography with benzene–Et_3N (10%) and lyophilisation from benzene to give **6** (134 mg, 70% yield) as a white powder. ^{31}P-NMR (C_6H_6): δ (ppm) 148.30; 148.54. MS (ESI) m/z: calcd for, $C_{41}H_{54}F_3N_4O_7P$ [M + H]$^+$ 803.3755, found 803.371.

(S)-N-(4,9,13-Triazatridecan-1-yl)-2,2-dimethyl-1,3-dioxolane-4-carboxyamide (**8**). The methyl (*S*)-2,2-dimethyl-1,3-dioxalane-4-carboxylate (**7**) (0.45 mL, 3 mmol), spermine (2.4 g, 12 mmol) were dissolved in anhydrous methanol (1 mL) and the resulting solution was stirred at room temperature for 48–72 h at 25–35 °C. The solvent was evaporated under reduced pressure and the residue was purified by column chromatography over silica gel (MeOH/$MeNH_2$/H_2O) to give **8** (720 mg, 70% yield) as yellow oil. ^{1}H-NMR (CDCl$_3$): δ (ppm) 7.22 (m, 1H, NHCO), 4.45 (q, 1H, *J* = 5.37 Hz, *J* = 2.44 Hz, COCH), 4.25 (t, 1H, *J* =7.8, OCHCO), 4.05 (q, 1H, *J* = 5.37 Hz, *J* = 3.41 Hz, COCH), 3.38–3.24 (m, 2H, CONH*CH$_2$*), 2.53–2.65 (m, 6H, CH$_2$), 2.05 (2H, NH), 1.66 (m, 2H, spermine), 1.49 (m, 2H, spermine), 1.44 (s, 3H, CH$_3$), 1.44 (s, 3H, CH$_3$); ^{13}C-NMR (CDCl$_3$): δ (ppm) 171.3 (C=O), 110.6 (OCO), 74.8 (C), 67.6 (C), 49.1 (CH$_2$), 46.8 (CH$_2$), 37.1 (CH$_2$), 28.6 (CH$_2$), 27.1 (CH$_2$), 26.0 (CH$_2$), 24.8 (CH$_3$). MS (ESI) m/z: calcd for $C_{16}H_{34}N_4O_3$ [M + K]$^+$ 369.2262, found 370.2801.

(R)-N-(4,9,13-Triazatridecan-1-yl)-2,2-dimethyl-1,3-dioxolane-4-carboxyamide (**8a**). The methyl (*R*)-2,2-dimethyl-1,3-dioxalane-4-carboxylate (**7a**) (0.45 mL, 3 mmol) was converted into compound **8a** following the above procedure (566 mg, 55% yield). ^{1}H-NMR (CDCl$_3$): δ (ppm) 7.19 (m, 1H, NHCO), 4.45 (q, 1H, *J* = 5.37 Hz, *J* = 2.44 Hz, COCH), 4.26 (t, 1H, *J* = 7.8 Hz, OCHCO), 4.05 (q, 1H, *J* = 5.37 Hz, *J* = 3.41 Hz, COCH), 3.43–3.3 (m, 2H, CONH*CH$_2$*), 2.74–2.59 (m, 6H, CH$_2$), 1.76 (m, 2H, spermine), 1.63 (m, 2H, spermine), 1.47 (s, 3H, CH$_3$), 1.37 (s, 3H, CH$_3$); ^{13}C-NMR (CDCl$_3$): δ (ppm) 171.3 (C=O), 110.6 (OCO), 74.8 (C), 67.6 (C), 49.1 (CH$_2$), 46.8 (CH$_2$), 37.1 (CH$_2$), 28.6 (CH$_2$), 27.1 (CH$_2$), 26.0 (CH$_2$), 24.8 (CH$_3$). MS (ESI) m/z: calcd for $C_{16}H_{34}N_4O_3$ [M + H]$^+$ 331.2704, found 331.2701.

(S)-N-[Tris(2,2,2-trifluoroacet-1-yl)-4,9,13-triazatridecane]-2,2-dimethyl-1,3-dioxolane-4-carboxyamide (**9**). **8** (0.200 g, 0.6 mmol) was co-evaporated with anhydrous pyridine (3 × 5 mL) and redissolved in pyridine (6 mL). Trifluoroacetic anhydride (0.5 mL, 3.6 mmol) was added dropwise to a cooled and stirred solution at 0 °C. The solution was warmed to room temperature and stirred for 30 min, quenched with NaHCO$_3$ and extracted with CH$_2$Cl$_2$. The combined organic extracts were washed with saturated aqueous NaCl, dried over Na$_2$SO$_4$ and concentrated under vacuum. The residual oil was purified by chromatography with 2% MeOH in DCM as the eluent to give **9** (0.2 g, 54% yield) as a white foam. ^{1}H-NMR (CDCl$_3$): δ (ppm) 7.09 (m, 1H, NHCO), 6.72 (m, 1H, NHCO), 4.46 (q, 1H, *J* = 5.34 Hz, *J* = 3.78 Hz, COCH), 4.07 (q, 1H, *J* = 5.34 Hz, *J* = 3.78 Hz, COCH), 4.28 (t, 1H, *J* = 7.64 Hz, OCHCO), 3.49–3.25 (m, 8H, NH*CH$_2$CH$_2$*, spermine), 1.86–1.78 (m, 2H, spermine), 1.61 (m, 2H, spermine), 1.49 (s, 3H, CH$_3$), 1.39 (s, 3H, CH$_3$). ^{19}F-NMR: 6.98–7.22 (m, 9F, 3CF$_3$). MS (ESI) m/z: calcd for $C_{22}H_{30}F_9N_4O_6$ [M − H]$^-$ 617.2022, found 617.3021.

(R)-N-[Tris(2,2,2-trifluoroacet-1-yl)-4,9,13-triazatridecane]-2,2-dimethyl-1,3-dioxolane-4-carboxyamide (**9a**). **8a** (0.350 g, 1.05 mmol) was converted into **9a** (0.368 g, 58% yield) following procedure for **9**. ^{1}H-NMR (CDCl$_3$): δ (ppm) 7.17 (m, 1H, NHCO), 6.76 (m, 1H, NHCO), 4.47 (q, 1H, *J* = 7.2 Hz, *J* = 3.3 Hz, COCH), 4.07 (m, 1H, COCH), 4.29 (t, 1H, *J* = 8 Hz, OCHCO), 3.48–3.26 (m, 8H, NH*CH$_2$CH$_2$*, spermine), 1.88–1.82 (m, 2H, spermine), 1.62 (m, 2H, spermine), 1.49 (s, 3H, CH$_3$), 1.39 (s, 3H, CH$_3$). ^{19}F-NMR: 6.98–7.22 (m, 9F, 3CF$_3$). MS (ESI) m/z calcd for $C_{22}H_{31}F_9N_4O_3$ [M + K]$^+$ 657.1737, found 657.175.

(S)-N-[Tris(2,2,2-trifluoroacet-1-yl)-4,9,13-triazatridecane]-3-(4,4'-dimethoxytrityl)-3-hydroxypropanamide (**10**). To a solution of **9** (0.500 g, 0.75 mmol) in MeOH (4 mL), *p*-toluenosulfonic acid (0.03 g, 0.15 mmol)

was added at room temperature. The mixture was stirred for ca 30 h at room temperature, and solvent was evaporated. The mixture was diluted with ethyl acetate and washed with aq. saturated NaHCO$_3$. The combined organic extracts were dried over Na$_2$SO$_4$ and concentrated under vacuum. The resulting crude was co-evaporated with pyridine (4 × 5 mL) and dissolved in pyridine (2.5 mL). 4,4′-Dimethoxytrityl chloride (0.330 g, 0.975 mmol) was added in portions at room temperature. The mixture was stirred for 3 h at room temperature, and the reaction was quenched with methanol (1 mL) and saturated aq. NaHCO$_3$ (10 mL). The mixture was extracted with DCM. The combined organic extracts were dried over MgSO$_4$ and concentrated under vacuum. The resulting orange oil was purified by chromatography with 2%–3% MeOH in CH$_2$Cl$_2$ (with 0.2% of pyridine by vol.) as the eluent to give **10** (0.494 g, 69% yield) as a white solid. ^{1}H-NMR (CDCl$_3$): δ (ppm) 7.39–6.8 (m, 15H, DMT, NHCO), 4.19–4.13 (m, 1H, OCHCO), 3.75 (s, 6H, 2OCH$_3$), 3.51–3.22 (m, 14H), 1.99–175 (m, 4H, spermine), 1.68–1.55 (m, 4H, spermine). ^{13}C-NMR (CDCl$_3$): δ (ppm) 171.9 (C), 171.7 (C), 158.7 (C), 144.4 (C), 135 (C), 129.9 (CH), 127,9 (CH), 127 (CH), 125.8 (C), 121 (C), 113.2 (CH), 86.8 (C), 77.2 (CH), 67.7 (CH$_2$), 55.2 (CH$_3$), 46.3 (CH$_2$), 36.1 (CH$_2$), 35.1 (CH$_2$), 29.7 (CH$_2$), 27.1 (CH$_2$). MS (ESI) *m/z*: calcd for C$_{40}$H$_{45}$F$_9$N$_4$O$_8$ [M + Na]$^+$ 903.2991, found 903.2938; calcd for C$_{40}$H$_{45}$F$_9$N$_4$O$_8$ [M + K]$^+$ 919.2731, found 919.266.

(R)-N-[Tris(2,2,2-trifluoroacet-1-yl)-4,9,13-triazatridecane]-3-(4,4′-dimethoxytrityl)-3-hydroxypropanamide **(10a)**. **9a** (0.350 g, 1.05 mmol) was converted into **10a** (0.383 g, 54% yield) following procedure for **10**. ^{1}H-NMR (CDCl$_3$): δ (ppm) 7.39–6.8 (m, 15H, DMT, NHCO), 4.2–4.12 (m, 1H, OCHCO), 3.75 (s, 6H, 2OCH$_3$), 3.48–3.18 (m, 14H), 1.91–1.74 (m, 4H, spermine), 1.63–1.51 (m, 4H, spermine). MS (ESI) *m/z*: calcd for C$_{40}$H$_{45}$F$_9$N$_4$O$_8$ [M − H]$^-$ 879.3015, found 879.3028.

2-(S)-[(4,4′-Dimethoxytrityl)-3-(hydroxymethyl)-N-((2,2,2-trifluoroacet-1-yl)-4,9,13-triazatridecane) propanamide]phosphoramidite **(11)**. The compound **10** was converted into compound **11** following procedure for **6**. Except purification: crude phosphoramidite was trice precipitated from hexane and finally lyophilized from benzene to give **11** (0.210 g, 52% yield) as a white light solid. ^{31}P-NMR (C$_6$H$_6$): δ (ppm) 151.89; 151.68; 148.99; 148.68; 148.56; 146.81. MS (ESI) m/z: calcd for C$_{49}$H$_{62}$F$_9$N$_6$NaO$_9$P [M + Na]$^+$ 1103.4070, found 1103.4575; calcd for C$_{49}$H$_{62}$F$_9$KN$_6$O$_9$P [M + K]$^+$ found 1119.4509.

2-(R)-[(4,4′-Dimethoxytrityl)-3-(hydroxymethyl)-N-((2,2,2-trifluoroacet-1-yl)-4,9,13-triazatridecane) propanamide]phosphoramidite **(11a)**. **10a** (0.200 g, 0.226 mmol) was converted into compound **11a** (0.128 g, 52% yield) according to procedure for **11**. ^{1}H-NMR (CDCl$_3$): δ (ppm) 7.14–7.38 (m, 9H, DMT), 6.78 (m, 4H, DMT), 4.8 (m, 1H, COCH), 4.65 (m, 1H, OCHCO), 3.73 (s, 6H, 2OCH$_3$), 3.12–3.55 (m, 16H, CH$_2$OH, spermine, CH$_2$OP) 2.4–2.78 (m, 5H, 2CH, iPr, CH$_2$CN), 1.41–1.92 (m, 8H, spermine) 1.17–1.23 (m, 12H, iPr). ^{31}P-NMR (C$_6$H$_6$): δ (ppm) 151.76; 151.42; 149.24; 149.05; 148.96; 148.83. MS (ESI) m/z: calcd for C$_{49}$H$_{62}$F$_9$N$_6$O$_9$P [M − H]$^-$ 1079.4094, found 1079.63.

3-(tert-Butyldimethylsilyloxy)butyrolactone **(13)** [48]. tert-Butyldimethylchlorosilane (1.50 g, 9.63 mmol), imidazole (0.470 g, 3.21 mmol) and *N,N*-dimethylaminopyridine (0.392 g, 3.21 mmol) were added to a solution of α-hydroxy-γ-butyrolactone **(12)** (0.25 mL, 3.21 mmol) in anhydrous CH$_3$CN (16 mL). The reaction mixture was stirred at room temperature for 16 h. Then, solvent was evaporated under reduced pressure. The residue was partitioned between saturated aq. NaHCO$_3$ and Et$_2$O (3 × 50 mL). The organic extracts were washed with brine, dried over MgSO$_4$. The crude residue was purified by column chromatography (eluent hexane/Et$_2$O 4:1→2:1) to give compound **12** as colourless oil (0.485 g, 70% yield). ^{1}H-NMR (CDCl$_3$): δ (ppm) 4.35–4.4 (m, 2H, CH$_2$O), 4.15–4.2 (dt, 1H, OCH), 2.16–2.5 (m, 1H, OCHCH$_2$), 0.9 (s, 9H, 3CH$_3$, t-Bu), 0.16 (s, 3H, CH$_3$CSi), 0.14 (s, 3H, CH$_3$CSi); ^{13}C-NMR (CDCl$_3$): δ (ppm) 175.86 (C=O), 68.19 (CH), 64.72 (CH$_2$), 33.3 (CH$_2$), 25.61 (CH$_3$), 18.19 (SiC(CH$_3$)$_3$), 5.5 (CH$_3$); MS (ESI) *m/z*: calcd for C$_{10}$H$_{20}$O$_3$Si [M − H]$^-$ 215.1103, found 215.0295.

N-(4,9,13-Triazatridecan-1-yl)-2-(tert-butyldimethylsiloxy)-butyramide **(14)**. A solution of **13** (0.216 g, 1 mmol) and spermine (0.202 g, 1 mmol) in a small (10 mL) Erlenmeyer flask was kept in microwave oven for 45 min. The flask was cooled down and the mixture was diluted with DCM (20 mL), washed with NaHCO$_3$ (3 × 30 mL), brine (30 mL), and organic layers dried over MgSO$_4$. After evaporation, the residue was chromatographed using MeOH/H$_2$O/MeNH$_2$ as the eluent to give **14** (0.266 g, 64% yield) as oil. ^{1}H-NMR (CDCl$_3$): δ (ppm) 7.02 (t, 1H, NHCO), 4.25 (t, 1H, *J* = 6.01 Hz, COCHOSi), 3.71 (q, 2H, *J* = 5.89 *CH$_2$*OH), 3.38–3.32 (m, 2H, spermine), 2.65–2.58 (m, 4H), 1.99–1.94 (m, 2H, OCHCH$_2$), 1.76 (m, 2H, spermine), 1.63 (m, 2H, spermine), 0.91 (s, 9H, 3CH$_3$, t-Bu), 0.11 (s, 3H, CH$_3$CSi), 0.08 (s, 3H, CH$_3$CSi). MS (ESI) *m/z*: calcd for C$_{20}$H$_{47}$N$_4$O$_3$Si [M + H]$^+$ 419.3417, found 419.3539.

N-[Tris(2,2,2-trifluoroacet-1-yl)-4,9,13-triazatridecane]-2-(tert-butyldimethylsiloxy)-butyramide (**15**). **14** (0.266 g, 0.64 mmol) was co-evaporated with anhydrous pyridine (3 × 5 mL) and redissolved in pyridine (6 mL). Trifluoroacetic anhydride (0.517 mL, 3.83 mmol) was added dropwise at 0 °C. The solution was warmed to room temperature and stirred for 30 min., quenched with NaHCO$_3$ and extracted with DCM (3 × 30mL). The combined organic extracts were washed with brine, dried over MgSO$_4$ and purified by chromatography with 2%–3% MeOH in DCM as the eluent to give **15** (0.297 g, 66% yield) as an oil. ^{1}H-NMR (CDCl$_3$): δ (ppm) 7.09 (bs, 1H, NHCO), 6.83 (bs, 1H, NHCO), 4.27 (t, 1H, COCHOSi), 3.73 (q, 2H, *J* = 5.89 Hz, *CH$_2$*OH), 3.47–3.15 (m, 6H), 2.01–1.75 (m, 4H), 1.63 (m, 2H, spermine), 0.93 (s, 9H, 3CH$_3$, t-Bu), 0.12 (s, 3H, CH$_3$CSi), 0.10 (s, 3H, CH$_3$CSi). ^{19}F-NMR 6.98–7.22 (m, 9F, CF$_3$). MS (ESI) *m/z*: calcd for C$_{26}$H$_{43}$F$_9$N$_4$O$_6$Si [M − H]$^-$ 705.2730, found 705.2809.

N-[Tris(2,2,2-trifluoroacet-1-yl)-4,9,13-triazatridecane]-4-(4,4'-dimethoxytrityl)-butyramide (**16**). TEAHF [49,50] solution (1 M in THF/dioxane, 1.08 mmol, 1 mL) was added at room temperature to a solution of **15** (0.254 g, 0.36 mmol) in THF (1 mL) and kept overnight at room temperature. The reaction was quenched with NH$_4$Cl solution and extracted with ethyl acetate (3 × 15 mL). The combined extracts were dried over MgSO$_4$, concentrated under vacuum. The residue was co-evaporated with anhydrous pyridine (4 × 5 mL) and redissolved in pyridine (2 mL). 4,4'-Dimethoxytrityl chloride (1.2 eq) was added in portions at room temperature. The mixture was stirred for 3 h at room temperature, and the reaction was quenched with methanol (1 mL) and saturated aqueous NaHCO$_3$ (10 mL). The mixture was extracted with DCM. The combined organic extracts were dried over MgSO$_4$, concentrated under vacuum and the resulting orange oil was purified by chromatography with 2%–3% MeOH in DCM (with 0.2% of pyridine) as the eluent to give **16** (0.231 g, 72% yield) as a white solid. ^{1}H-NMR (CDCl$_3$): δ (ppm) 7.19–7.36 (m, 9H, DMT), 6.78 (d, 4H, *J* = 8.6 Hz, DMT), 4.21 (dt, 1H, COCHOH), 3.74 (s, 6H, 2OCH$_3$), 3.0–3.44 (m, 14H, CH$_2$OH, spermine), 2.18–2.19 (m, 2H, OCHCH$_2$), 1.56–1.88 (m, 8H, spermine). MS (ESI) *m/z*: calcd for C$_{40}$H$_{46}$F$_9$N$_4$O$_8$ [M + K]$^+$ 933.2887, found 933.279.

2-(4,4'-Dimethoxytrityl)-4-(hydroxymethyl)-N-((2,2,2-trifluoroacet-1-yl)-4,9,13-triazatridecane)-butyramide] phosphoramidite (**17**). **17** was synthesized according to procedure described for **11**. (0.155 g, 55% yield). ^{1}H-NMR (CDCl$_3$): δ (ppm) 6.76–7.4 (m, 13H, DMT), 4.22 (dt, 1H, COCHOH), 3.74 (s, 6H, 2OCH$_3$), 3.12–3.55 (m, 16H, *CH$_2$*OH, CONHCH$_2$ (spermine), CH$_2$OP), 2.56–2.62 (m, 2H, 2CH, i-Pr), 2.24–2.42 (m, 4H, OCHCH$_2$, CH$_2$CN), 1.57–1.84 (m, 8H, spermine), 1.1–1.2 (m, 12H, 4CH$_3$, i-Pr). ^{31}P-NMR (C$_6$H$_6$): δ (ppm) 151.73; 151.59; 148.25; 147.83. MS (ESI) *m/z*: calcd for C$_{50}$H$_{64}$F$_9$N$_6$O$_9$P [M + K]$^+$ 1133.3966, found 1133.303; calcd for C$_{50}$H$_{64}$F$_9$N$_6$NaO$_9$P [M + Na]$^+$ 1117.4226, found 1117.45.

3.3. Oligonucleotide Preparation

The DNA dodecamers (5'-CTC AAG CAA GCT-3', 5'-AGC TTG CTT GAG-3', 5'-CTC ACA TGC GCG-3', 5'-CGC GCA TGT GAG-3') were synthesized using DNA synthesizer Gene Assembler Plus from Pharmacia-LKB (Uppsala, Sweden) or K & A Laborgerate GbR DNA/RNA (Frankfurt am Main, Germany), using standard phosphoramidite chemistry. The non-nucleosidic analogues (**6, 11, 11a, 17**) were inserted at positions marked with an X as listed in the Table 1. For the modified phosphoramidites, two-fold excess of phosphoramidites (in comparison to the standard protocol) and a prolonged coupling step of 10 min were used. The oligomers were cleaved from the CPG-support with 32% aqueous ammonia (room temperature, 1 h). The deprotection under standard conditions using concentrated aqueous ammonia at 55 °C overnight allowed for removal of all protecting groups, including trifluoroacetyls [3,38]. The oligomers were purified by the TLC on Merck 60 F$_{254}$ TLC plates

Molecules **2015**, *20*, 12652–12669

with *n*-propanol/aqueous ammonia/water solution (55:35:10, by vol.) as an eluent. The product band (least mobile) was cut out, eluted with water and desalted with Waters Sep-pak C-18 cartridges. First, the solution containing the oligonucleotide was loaded onto the cartridge and the column was flushed with 10 mM ammonium acetate (10 mL). In the next step, the oligonucleotides were eluted by flushing the cartridge with 5 mL of 30% acetonitrile/water solution. The fraction with the product was evaporated to dryness and the purity of oligonucleotides was monitored using HPLC and confirmed (by MALDI-TOF spectrometry Autoflex, Bruker). Oligonucleotides were also purified in a subsequent reverse phase HPLC step (UFLC system with LC-20AD pump; C(18)2 100 Å column (15 cm × 4.6 mm); starting from 0.01 M triethylammonium acetate (pH 7.0) up to $CH_3CN:CH_3COOEt_3N$ (40%/60%). Identity was confirmed by mass spectroscopy on MALDI-TOF (Autoflex, Bruker Daltonik GmbH, Bremen, Germany). The isolated yields of modified oligonucleotides were at the range of 23%–57%: 26%–57% and 23%–26% for **ON3–ON6** and **ON7–ON9** respectively.

3.4. Thermodynamic Analysis

UV melting profiles of the DNA duplexes were obtained in a buffer containing 100 mM sodium chloride, 20 mM sodium cacodylate, 0.5 mM Na_2EDTA, pH 7.0. Duplexes were used in the 10^{-3}–10^{-6} M concentration range. Single strand concentrations were calculated from absorbance above 80 °C with single strand extinction coefficients approximated by the nearest-neighbor model [51]. The temperature range, in a heating-cooling cycle, was 0–90 °C with a temperature gradient of 1 °C/min. Thermal-induced transitions of each mixture were monitored at 260 nm with a Beckman DU 650 spectrophotometer with a temperature controller. The thermodynamic parameters were determined from fits of data acc. to a two-state model with the MeltWin 3.5 software [52].

4. Conclusions

Nucleic acids that carry polycationic modifications have many therapeutic and biotechnological advantages. Our studies corroborate that non-nucleosidic analogues of polyaminooligonucleosides maintain affinity and easily form duplexes with complementary strands. In some cases, this may increase the stability of modified complexes of nucleic acids. These new properties based on the rationale of charge masking do not change the scheme of the secondary structure of nucleic acids. Thus, they can ease transferring of nucleic acids past cellular barriers.

Supplementary Materials: Supplementary materials can be accessed at: http://www.mdpi.com/1420-3049/20/07/12652/s1.

Acknowledgments: This work was supported by the Polish Ministry of Science and Higher Education.

Author Contributions: W.T.M. designed research and supervised the project, J.B. performed all experiments. Both authors analyzed the data, wrote the manuscript and approved the final version of the manuscript.

Conflicts of Interest: The authors declare no conflict of interest.

References

1. Pack, D.W.; Hoffman, A.S.; Pun, S.; Stayton, P.S. Design and development of polymers for gene delivery. *Nat. Rev. Drug Discov.* **2005**, *4*, 581–593. [CrossRef] [PubMed]
2. Thomas, R.M.; Thomas, T.; Wada, M.; Sigal, L.H.; Shirahata, A.; Thomas, T.J. Facilitation of the cellular uptake of a triplex-forming oligonucleotide by novel polyamine analogues: Structure-activity relationships. *Biochemistry* **1999**, *38*, 13328–13337. [CrossRef] [PubMed]
3. Prakash, T.P.; Barawkar, D.A.; Vaijayanti, K.; Ganesh, K.N. Synthesis of site-specific oligonucleotide-polyamine conjugates. *Bioorg. Med. Chem. Lett.* **1994**, *4*, 1733–1738. [CrossRef]
4. Horn, T.; Chaturvedi, S.; Balasubramaniam, T.N.; Letsinger, R.L. Oligonucleotides with alternating anionic and cationic phosphoramidate linkages: Synthesis and hybridization of stereo-uniform isomers. *Tetrahedron Lett.* **1996**, *37*, 743–746. [CrossRef]

5. Letsinger, R.L.; Singman, C.N.; Histand, G.; Salunkhe, M. Cationic oligonucleotides. *J. Am. Chem. Soc.* **1988**, *110*, 4470–4471. [CrossRef]

6. Meade, B.R.; Gogoi, K.; Hamil, A.S.; Palm-Apergi, C.; Berg, A.V.D.; Hagopian, J.C.; Springer, A.D.; Eguchi, A.; Kacsinta, A.D.; Dowdy, C.F.; *et al.* Efficient delivery of RNAi prodrugs containing reversible charge-neutralizing phosphotriester backbone modifications. *Nat. Biotechnol.* **2014**, *32*, 1256–1261. [CrossRef] [PubMed]

7. Boussif, O.; Delair, T.; Brua, C.; Veron, L.; Pavirani, A.; Kolbe, H.V. Synthesis of polyallylamine derivatives and their use as gene transfer vectors *in vitro*. *Bioconjug. Chem.* **1999**, *10*, 877–883. [CrossRef] [PubMed]

8. Ahmed, M.; Bhuchar, N.; Ishihara, K.; Narain, R. Well-controlled cationic water-soluble phospholipid polymer-DNA nanocomplexes for gene delivery. *Bioconjug. Chem.* **2011**, *22*, 1228–1238. [CrossRef] [PubMed]

9. Zhang, S.; Zhao, B.; Jiang, H.; Wang, B.; Ma, B. Cationic lipids and polymers mediated vectors for delivery of siRNA. *J. Control. Release* **2007**, *123*, 1–10. [CrossRef] [PubMed]

10. Piest, M.; Engbersen, J.F.J. Effects of charge density and hydrophobicity of poly(amido amine)s for non-viral gene delivery. *J. Control. Release* **2010**, *148*, 83–90. [CrossRef] [PubMed]

11. Patil, S.P.; Yi, J.W.; Bang, E.K.; Jeon, E.M.; Kim, B.H. Synthesis and efficient siRNA delivery of polyamine-conjugated cationic nucleoside lipids. *Med. Chem. Commun.* **2011**, *2*, 505–508. [CrossRef]

12. Geihe, E.I.; Cooley, C.B.; Simon, J.R.; Kiesewetter, M.K.; Edward, J.A.; Hickerson, R.P.; Kaspar, R.L.; Hedrick, J.L.; Waymouth, R.M.; Wender, P.A. Designed guanidinium-rich amphipathic oligocarbonate molecular transporters complex, deliver and release siRNA in cells. *Proc. Natl. Acad. Sci. USA* **2012**, *109*, 13171–13176. [CrossRef] [PubMed]

13. Hamilton, P.L.; Arya, D.P. Natural product DNA major groove binders. *Nat. Prod. Rep.* **2012**, *29*, 134–143. [CrossRef] [PubMed]

14. Xi, H.; Davis, E.; Ranjan, N.; Xue, L.; Hyde-Volpe, D.; Arya, D.P. Thermodynamics of nucleic acid "shape readout" by an aminosugar. *Biochemistry* **2011**, *50*, 9088–9113. [CrossRef] [PubMed]

15. Bumcrot, D.; Manoharan, M.; Koteliansky, V.; Sah, D.W.Y. RNAi therapeutics: A potential new class of pharmaceutical drugs. *Nat. Chem. Biol.* **2006**, *2*, 711–719. [CrossRef] [PubMed]

16. Davis, S.; Lollo, B.; Freier, S.; Esau, C. Improved targeting of miRNA with antisense oligonucleotides. *Nucleic Acids Res.* **2006**, *34*, 2294–2304. [CrossRef] [PubMed]

17. Kim, D.H.; Behlke, M.A.; Rose, S.D.; Chang, M.S.; Choi, S.; Rossi, J.J. Synthetic dsRNA Dicer substrates enhance RNAi potency and efficacy. *Nat. Biotechnol.* **2005**, *23*, 222–226. [CrossRef] [PubMed]

18. Cekaite, L.; Furset, G.; Hovig, E.; Sioud, M. Gene Expression Analysis in Blood Cells in Response to Unmodified and 2′-Modified siRNAs Reveals TLR-dependent and Independent Effects. *J. Mol. Biol.* **2007**, *365*, 90–108. [CrossRef] [PubMed]

19. Raddatz, S.; Mueller-Ibeler, J.; Kluge, J.; Wäss, L.; Burdinski, G.; Havens, J.R.; Onofrey, T.J.; Wang, D.; Schweitzer, M. Hydrazide oligonucleotides: New chemical modification for chip array attachment and conjugation. *Nucleic Acids Res.* **2002**, *30*, 4793–4802. [CrossRef] [PubMed]

20. Yim, S.C.; Park, H.G.; Chang, H.N.; Cho, D.Y. Array-based mutation detection of BRCA1 using direct probe/target hybridization. *Anal. Biochem.* **2005**, *337*, 332–337. [CrossRef] [PubMed]

21. Wengel, J. Nucleic acid nanotechnology-towards Angstrom-scale engineering. *Org. Biomol. Chem.* **2004**, *2*, 277–280. [CrossRef] [PubMed]

22. Shu, W.; Liu, D.; Watari, M.; Riener, C.K.; Strunz, T.; Welland, M.E.; Balasubramanian, S.; McKendry, R.A. DNA molecular motor driven micromechanical cantilever arrays. *J. Am. Chem. Soc.* **2005**, *127*, 17054–17060. [CrossRef] [PubMed]

23. Mangraviti, A.; Tzeng, S.Y.; Kozielski, K.L.; Wang, Y.; Jin, Y.; Gullotti, D.; Pedone, M.; Buaron, N.; Liu, A.; Wilson, D.R.; *et al.* Polymeric Nanoparticles for Nonviral Gene Therapy Extend Brain Tumor Survival *in Vivo*. *ACS Nano* **2015**, *9*, 1236–1249. [CrossRef] [PubMed]

24. Saneyoshi, H.; Tamaki, K.; Ohkubo, A.; Seio, K.; Sekine, M. Synthesis and hybridization properties of 2′-*O*-(tetrazol-5-yl)ethyl-modified oligonucleotides. *Tetrahedron* **2008**, *64*, 4370–4376. [CrossRef]

25. Urban, E.; Noe, C.R. Structural modifications of antisense oligonucleotides. *Il Farmaco* **2003**, *58*, 243–258. [CrossRef]

26. Debart, F.; Abes, S.; Deglane, G.; Moulton, H. M.; Clair, P.; Gait, M.J.; Vasseur, J.J.; Lebleu, B. Chemical modifications to improve the cellular uptake of oligonucleotides. *Curr. Top. Med. Chem.* **2007**, *7*, 727–737. [CrossRef] [PubMed]

27. Winkler, J.; Saadat, K.; Díaz-Gavilán, M.; Urban, E.; Noe, C.R. Oligonucleotide-polyamine conjugates: Influence of length and position of 2'-attached polyamines on duplex stability and antisense effect. *Eur. J. Med. Chem.* **2009**, *44*, 670–677. [CrossRef] [PubMed]

28. Rahman, S.M.A.; Baba, T.; Kodama, T.; Islam, M.A.; Obika, S. Hybridizing ability and nuclease resistance profile of backbone modified cationic phosphorothioate oligonucleotides. *Bioorg. Med. Chem.* **2012**, *20*, 4098–5102. [CrossRef] [PubMed]

29. Bachrach, U. Polyamines and cancer: Minireview article. *Amino Acids* **2004**, *26*, 307–309. [CrossRef] [PubMed]

30. Childs, A.C.; Mehta, D.J.; Gerner, E.W. Polyamine-dependent gene expression. *Cell. Mol. Life Sci.* **2003**, *60*, 1394–1406. [CrossRef] [PubMed]

31. Venkiteswaran, S.; Vijayanathan, V.; Shirahata, A.; Thomas, T.; Thomas, T.J. Antisense recognition of the HER-2 mRNA: Effects of phosphorothioate substitution and polyamines on DNA·RNA, RNA·RNA, and DNA·DNA duplex stability. *Biochemistry* **2005**, *44*, 303–312. [CrossRef] [PubMed]

32. Antony, T.; Thomas, T.; Shirahata, A.; Thomas, T.J. Selectivity of polyamines on the stability of RNA-DNA hybrids containing phosphodiester and phosphorothioate oligodeoxyribonucleotides. *Biochemistry* **1999**, *38*, 10775–10784. [CrossRef] [PubMed]

33. Hou, M.H.; Lin, S.B.; Yuann, J.M.; Lin, W.C.; Wang, A.H.J.; Kan, L. Effects of polyamines on the thermal stability and formation kinetics of DNA duplexes with abnormal structure. *Nucleic Acids Res.* **2001**, *29*, 5121–5128. [CrossRef] [PubMed]

34. Brzezinska, J.; Gdaniec, Z.; Popenda, L.; Markiewicz, W.T. Polyaminooligonucleotide: NMR structure of duplex DNA containing a nucleoside with spermine residue, *N*-[4,9,13-triazatridecan-1-yl]-2'-deoxycytidine. *Biochim. Biophys. Acta Gen. Subj.* **2014**, *1840*, 1163–1170. [CrossRef] [PubMed]

35. Malkoch, M.; Schleicher, K.; Drockenmuller, E.; Hawker, C.J.; Russell, T.P.; Wu, P.; Fokin, V.V. Structurally Diverse Dendritic Libraries: A Highly Efficient Functionalization Approach Using Click Chemistry. *Macromolecules* **2005**, *38*, 3663–3678. [CrossRef]

36. Asanuma, H.; Liang, X.; Komiyama, M. meta-Aminoazobenzene as a thermo-insensitive photo-regulator of DNA-duplex formation. *Tetrahedron Lett.* **2000**, *41*, 1055–1058. [CrossRef]

37. Ihre, H.; Padilla De Jesús, O.L.; Fréchet, J.M.J. Fast and convenient divergent synthesis of aliphatic ester dendrimers by anhydride coupling. *J. Am. Chem. Soc.* **2001**, *123*, 5908–5917. [CrossRef] [PubMed]

38. Markiewicz, W.T.; Godzina, P.; Markiewicz, M.; Astriab, A. Synthesis of A Polyaminooligonucleotide Combinatorial Library. *Nucleos. Nucleot.* **1998**, *17*, 1871–1880. [CrossRef]

39. Godzina, P.; Adrych-Rozek, K.; Markiewicz, W.T. Synthetic Oligonucleotide Combinatorial Libraries. 3. Synthesis of Polyaminonucleosides. *Nucleos. Nucleot.* **1999**, *18*, 2397–2414. [CrossRef]

40. Wuts, P.G.M.; Greene, T.W. *Greene's Protective Groups in Organic Synthesis*, 4th ed.; John Wiley & Sons, Inc.: Hoboken, NJ, USA, 2007.

41. Sund, C.; Puri, N.; Chattopadhyaya, J. The Chemistry of C-Branched Spermine Tethered Oligo-DNAs and Their Properties in Forming Duplexes and Triplexes. *Nucleos. Nucleot.* **1997**, *16*, 755–760. [CrossRef]

42. Sund, C.; Puri, N.; Chattopadhyaya, J. Synthesis of C-branched spermine tethered oligo-DNA and the thermal stability of the duplexes and triplexes. *Tetrahedron* **1996**, *52*, 12275–12290. [CrossRef]

43. Moriguchi, T.; Sakai, H.; Suzuki, H.; Shinozuka, K. Spermine moiety attached to the C-5 position of deoxyuridine enhances the duplex stability of the phosphorothioate DNA/complementary DNA and shows the susceptibility of the substrate to RNase H. *Chem. Pharm. Bull. (Tokyo)* **2008**, *56*, 1259–1263. [CrossRef] [PubMed]

44. Breslauer, K.J.; Frank, R.; Blöcker, H.; Marky, L.A. Predicting DNA duplex stability from the base sequence. *Proc. Natl. Acad. Sci. USA* **1986**, *83*, 3746–3750. [CrossRef] [PubMed]

45. Bommarito, S.; Peyret, N.; SantaLucia, J. Thermodynamic parameters for DNA sequences with dangling ends. *Nucleic Acids Res.* **2000**, *28*, 1929–1934. [CrossRef] [PubMed]

46. Longfellow, C.E.; Kierzek, R.; Turner, D.H. Thermodynamic and spectroscopic study of bulge loops in oligoribonucleotides. *Biochemistry* **1990**, *29*, 278–285. [CrossRef] [PubMed]

47. Minetti, C.A.; Remeta, D.P.; Dickstein, R.; Breslauer, K.J. Energetic signatures of single base bulges: Thermodynamic consequences and biological implications. *Nucleic Acids Res.* **2009**, *38*, 97–116. [CrossRef] [PubMed]

48. Dauben, W.G.; Hendricks, R.T.; Pandy, B.; Wu, S.C.; Zhang, Xiaoming; Luzzio, M.J. Stereoselective intramolecular cyclopropanations: Enantioselective syntheses of 1α,25-dihydroxyvitamin D3 A-ring precursors. *Tetrahedron Lett.* **1995**, *36*, 2385–2388. [CrossRef]

49. Markiewicz, W.T.; Biała, E.; Adamiak, R.W.; Grześkowiak, K.; Kierzek, R.; Kraszewski, A.; Stawiński, J.; Wiewiórowski, M. Further studies on oligoribonucleotide synthesis. *Nucleic Acids Symp. Ser.* **1980**, 115–127.

50. Markiewicz, W.T.; Biała, E.; Kierzek, R. Application of the Tetraisopropyldisiloxane-1,3-diyl Group in the Chemical Synthesis of Oligoribonucleotides. *Bull. Pol. Acad. Sci. Chem.* **1984**, *32*, 433–451.

51. Borer, P.N.; Dengler, B.; Tinoco, I.; Uhlenbeck, O.C. Stability of ribonucleic acid double-stranded helices. *J. Mol. Biol.* **1974**, *86*, 843–853. [CrossRef]

52. McDowell, J.A.; Turner, D.H. Investigation of the structural basis for thermodynamic stabilities of tandem GU mismatches: Solution structure of (rGAGGUCUC)2 by two-dimensional NMR and simulated annealing. *Biochemistry* **1996**, *35*, 14077–14089. [CrossRef] [PubMed]

Sample Availability: *Sample Availability:* Samples are not available from authors.

molecules

MDPI

Article

Cladribine Analogues via O^6-(Benzotriazolyl) Derivatives of Guanine Nucleosides

Sakilam Satishkumar [1,†], Prasanna K. Vuram [1,†], Siva Subrahmanyam Relangi [2,†], Venkateshwarlu Gurram [2], Hong Zhou [3], Robert J. Kreitman [3], Michelle M. Martínez Montemayor [4], Lijia Yang [1], Muralidharan Kaliyaperumal [2], Somesh Sharma [2], Narender Pottabathini [2] and Mahesh K. Lakshman [1,*]

[1]   Department of Chemistry, The City College and The City University of New York, 160 Convent Avenue, New York, NY 10031, USA; ssakilam@ccny.cuny.edu (S.S.); vkumar@ccny.cuny.edu (P.K.V.); lyang@sci.ccny.cuny.edu (L.Y.)
[2]   Discovery and Analytical Services, GVK Biosciences Pvt. Ltd., 28A IDA Nacharam, Hyderabad 500076, India; siva.relangi@gvkbio.com (S.S.R.); venkateshwarlu.gurram@gvkbio.com (V.G.); muralidharan.k@gvkbio.com (M.K.); somesh.sharma@gvkbio.com (S.S.); narender.pottabathini@gvkbio.com (N.P.)
[3]   Clinical Immunotherapy Section, Laboratory of Molecular Biology, National Cancer Institute, National Institutes of Health, 9000 Rockville Pike, Bethesda, MD 20892, USA; hong.zhou@nih.gov (H.Z.); kreitmar@mail.nih.gov (R.J.K.)
[4]   Department of Biochemistry, Universidad Central del Caribe-School of Medicine, P. O. Box 60327, Bayamón, PR 00960, USA; mmmtz92@gmail.com
*   Correspondence: mlakshman@ccny.cuny.edu; Tel.: +1-212-650-7835; Fax: +1-212-650-6107
†   These authors contributed equally to this work.

Academic Editor: Fumi Nagatsugi
Received: 17 August 2015; Accepted: 22 September 2015; Published: 9 October 2015

Abstract: Cladribine, 2-chloro-2′-deoxyadenosine, is a highly efficacious, clinically used nucleoside for the treatment of hairy cell leukemia. It is also being evaluated against other lymphoid malignancies and has been a molecule of interest for well over half a century. In continuation of our interest in the amide bond-activation in purine nucleosides via the use of (benzotriazol-1yl-oxy)tris(dimethylamino)phosphonium hexafluorophosphate, we have evaluated the use of O^6-(benzotriazol-1-yl)-2′-deoxyguanosine as a potential precursor to cladribine and its analogues. These compounds, after appropriate deprotection, were assessed for their biological activities, and the data are presented herein. Against hairy cell leukemia (HCL), T-cell lymphoma (TCL) and chronic lymphocytic leukemia (CLL), cladribine was the most active against all. The bromo analogue of cladribine showed comparable activity to the ribose analogue of cladribine against HCL, but was more active against TCL and CLL. The bromo ribose analogue of cladribine showed activity, but was the least active among the C6-NH2-containing compounds. Substitution with alkyl groups at the exocyclic amino group appears detrimental to activity, and only the C6 piperidinyl cladribine analogue demonstrated any activity. Against adenocarcinoma MDA-MB-231 cells, cladribine and its ribose analogue were most active.

Keywords: cladribine; nucleoside; guanosine; benzotriazole; (benzotriazol-1yl-oxy)-tris (dimethylamino)phosphonium hexafluorophosphate; BOP

---

1. Introduction

Cladribine, 2-chloro-2′-deoxyadenosine, has been a molecule of interest for well over five decades. The history of this compound dates back to 1960, when it was used in the synthesis of 2′-deoxyguanosine and 2′-deoxyinosine [1]. A decade later, cladribine was shown to be a poor substrate for adenosine deaminase that underwent phosphorylation by deoxycytidine kinase, finally resulting in the triphosphate, and inhibiting DNA synthesis rather than RNA synthesis [2,3]. Later still, it was shown to be a substrate for deoxyguanosine kinase, which is responsible for phosphorylation of purine nucleosides in mitochondria [3]. Several mechanisms have been proposed by which cladribine can cause mitochondrial damage and apoptotic cell death [4–8].

In contemporary medicine, cladribine is used in the treatment of lymphoid malignancies, most notably for its efficacy against hairy cell leukemia [9]. Cladribine is also being evaluated against several other indolent lymphoid malignancies, also in combination with other drug candidates [10,11].

The synthesis of cladribine has primarily relied on three major methods: (a) glycosylation reactions of a nucleobase with a sugar [12–18], and its variations; (b) deoxygenation of the C2′ hydroxyl group of a suitable nucleoside derivative [12,15,19,20]; (c) enzymatic glycosyl transfer reactions [21–24]; and (d) conversion of readily available nucleoside precursors (some utilizing nucleosides for glycosyl transfer reactions) [21,22,24–26]. Each of these methods has been used with varying levels of convenience and success. Among the many approaches, one convenient method is the selective displacement of a leaving group (chloride or aryl sulfonate) from the C6 position of a suitable purine nucleoside precursor. Despite the availability of this selective S_NAr displacement, we find that no other N6-substituted cladribine analogues have been synthesized by such a method. Because of our interest in broadening the utilities of O^6-(benzotriazol-1-yl)purine nucleoside derivatives, we elected to evaluate the synthesis of N6-substituted cladribine analogues via amide-bond activation of guanine nucleosides with (benzotriazol-1yl-oxy)tris(dimethylamino)phosphonium hexafluorophosphate (BOP).

2. Results and Discussion

For the modification of the nucleobases of inosine, and 2′-deoxyinosine (3 examples), BOP had been used for *in situ* activation of the amide moieties in these substrates, followed by S_NAr displacement with amines [27,28]. It was proposed that reaction of the amide group with BOP proceeded via a phosphonium intermediate, which could be directly captured by a reactive amine [27]. On the other hand, with less reactive amines, the O-(benzotriazol-1-yl) intermediate can be formed by competitive capture of the phosphonium intermediate by the benzotriazol-1-yloxy anion [27].

In 2007, we first reported the isolation of O^6-(benzotriazol-1-yl)inosine and -2′-deoxyinosine by amide-bond activation with BOP, in the absence of a nucleophile [29], an observation that was later reconfirmed by others [30,31]. These electrophilic nucleosides, which are stable to storage, are exceptionally good partners in S_NAr reactions with oxygen, nitrogen, and sulfur nucleophiles [29]. Subsequently, we demonstrated that O^6-(benzotriazol-1-yl)inosine and -2′-deoxyinosine can also be prepared via the use of PPh_3/I_2 and 1-hydroxybenzotriazole [32]. The amide-bond activation protocol was then modified to tether the O^6-(benzotriazol-1-yl) nucleosides onto a polymer support for high-throughput type of applications [33]. Interestingly, the amide bond activation when applied to the urea functionality of O^6-benzyl-protected 2′-deoxyxanthosine did not yield the O^2-(benzotriazol-1-yl) derivative but rather terminated in an isolable and synthetically useful phosphonium salt [34]. We also showed that guanosine and 2′-deoxyguanosine undergo facile reactions with BOP, and that O^6-(benzotriazol-1-yl) guanine nucleosides are effective substrates for S_NAr reactions as well [35]. In the combined course of these investigations we had ascertained plausible operative mechanisms of these amide-activation reactions, results that were later applied to a one-pot etherification protocol for purine nucleosides and pyrimidines [36].

Considering only the nucleoside modification literature, our BOP-mediated amide-activation methodology has found wide application [37–43]. Of specific interest to our current work was a report wherein some of our results on guanosine derivatives were reevaluated [44]. In addition to 2′,3′,5′-tri-O-(*t*-butyldimethylsilyl)guanosine, which we had originally investigated, five other compounds were also studied: unprotected 2′-deoxyguanosine, its 2′,3′,5′-triacetyl and the 2′,3′-isopropylidene derivatives, and two 2′,3′-isopropylidene 5′-carboxamides [44]. These products were subsequently tested in diazotization-halogenation reactions, *en route* to 2-chloro and 2-iodo adenosine analogues [44]. 2,6-Dihalopurine ribonucleosides are more readily accessible for this purpose, as compared to 2′-deoxy derivatives. Thus, we were intrigued by the fact that diazotization-halogenation reactions of O^6-(benzotriazol-1yl)-2′-deoxyguanosine have not been reported. This prompted use to investigate the chlorination and bromination of this compound, and in this context we decided to evaluate the synthesis of cladribine and other cladribine analogues via such an approach.

In 2001, diazotization reactions leading to cladribine and other halo derivatives have been performed on unprotected 2,6-diaminopurine 2′-deoxyribonucleoside [45]. However, no substituents other than NH$_2$ were introduced into the C6 position, and the synthesis of the precursor is not particularly convenient. For the present study, we anticipated the need for saccharide protection, and both acetyl and *t*-butyldimethysilyl (TBS) protecting groups were considered. Between these, TBS was selected because acetyl groups can be susceptible to cleavage with amines, which could be a complicating problem in the method development stage. Thus, nucleosides **1a** and **1b** were silylated to give the corresponding products **2a** and **2b**, which were converted to the O^6-(benzotriazol-1-yl) guanosine derivatives **3a** and **3b**, respectively (Scheme 1).

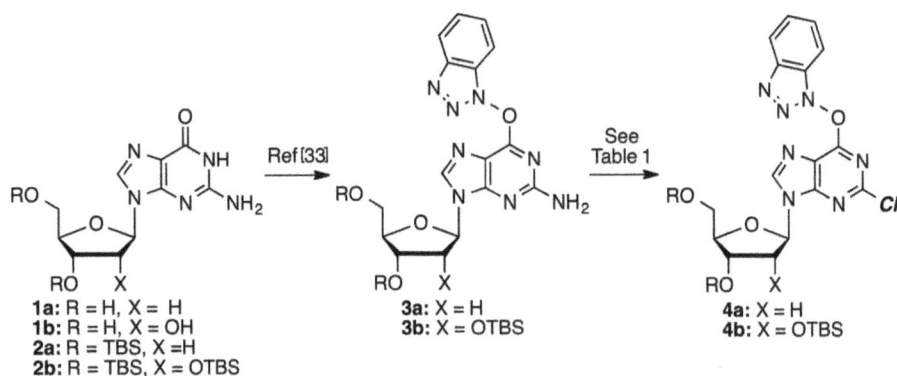

1a: R = H, X = H
1b: R = H, X = OH
2a: R = TBS, X = H
2b: R = TBS, X = OTBS

3a: X = H
3b: X = OTBS

4a: X = H
4b: X = OTBS

Scheme 1. Preparation of O^6-(benzotriazol-1-yl) guanosine derivatives and the chlorination reaction.

We opted for diazotization-chlorination conditions using *t*-BuONO/TMSCl, a reagent combination initially introduced for nucleoside modification in 2003 [46]. We [47] and others [44] have previously used non-aqueous conditions for halogenation at the C2 position of purine nucleoside derivatives. Under these conditions, reactions of substrate **3a** proceeded modestly. However, the obtained product was contaminated with ~30% of the C2 protio O^6-(benzotriazol-1-yl)-3′,5′-di-*O*-(*t*-butyldimethylsilyl)-2′-deoxyinosine (Table 1, entries 1 and 2). With the combination of *t*-BuONO/TMSCl/(BnNEt$_3$)$^+$Cl$^-$, no C2 protio product was observed, but only a low product yield was obtained (entry 3).

These results compelled us to consider other conditions. SbCl$_3$ and SbBr$_3$ have previously been used for the halogenation of nucleosides [48,49] Thus, the next series of experiments involved SbCl$_3$, (BnNEt$_3$)$^+$Cl$^-$, and combinations of these reagents (entries 4–11). Whereas most experiments yielded only 35%–39% of product **4a**, a reasonable yield improvement was observed in entry 9. It was noted that efficient filtration of the reaction mixture after workup is critical to obtaining a good product recovery due to the pasty nature of the mixture (see the Experimental Section). On larger scales, better product recoveries were observed (entries 10 and 11). By comparison, diazotization/chlorination of the ribose derivative **3b** proceeded well with both *t*-BuONO/SbCl$_3$ (entry 12) and *t*-BuONO/TMSCl (entry 13), with the latter providing a better yield of compound **4b**. No C2 protio product was apparent in the reactions of precursor **3b**.

Table 1. Evaluation of conditions for the diazotization/chlorination of O^6-(benzotriazol-1-yl)-3′,5′-di-O-(t-butyldimethylsilyl)-2′-deoxyguanosine.

Entry	Conditions	Scale	Yield
1	t-BuONO (2.6 equiv), TMSCl (2.6 equiv), CH$_2$Cl$_2$, −10 °C to 0 °C, 2 h	73 μmol of **3a**	43% [a]
2	t-BuONO (10 equiv), TMSCl (5 equiv), CH$_2$Cl$_2$, 0 °C, 1 h	73 μmol of **3a**	58% [a]
3	t-BuONO (10 equiv), TMSCl (0.4 equiv), (BnNEt$_3$)$^+$Cl$^-$ (10 equiv), CH$_2$Cl$_2$, −10 °C to 0 °C, 2 h	49 μmol of **3a**	26%
4	t-BuONO (3.5 equiv), SbCl$_3$ (1.4 equiv), ClCH$_2$CH$_2$Cl, −10 °C to −15 °C, 1.5 h	49 μmol of **3a**	35%
5	t-BuONO (10 equiv), (BnNEt$_3$)$^+$Cl$^-$ (10 equiv), CH$_2$Cl$_2$, −78 °C and then rt, 1 h	49 μmol of **3a**	39%
6	t-BuONO (20 equiv), SbCl$_3$ (0.4 equiv), (BnNEt$_3$)$^+$Cl$^-$ (20 equiv), CH$_2$Cl$_2$, −10 °C to 0 °C, 5 h	49 μmol of **3a**	39%
7	t-BuONO (10 equiv), (BnNEt$_3$)$^+$Cl$^-$ (10 equiv), (Me$_3$Si)$_2$NH (10 equiv), CH$_2$Cl$_2$, −10 °C to 0 °C, 4 h	49 μmol of **3a**	35%
8	t-BuONO (3.5 equiv), SbCl$_3$ (1.4 equiv), CH$_2$Cl$_2$, −10 °C, 2 h	49 μmol of **3a**	39% [b]
9	t-BuONO (3.5 equiv), SbCl$_3$ (1.4 equiv), CH$_2$Cl$_2$, −10 °C to −15 °C, 2 h	73 μmol of **3a**	51%
10	t-BuONO (3.5 equiv), SbCl$_3$ (1.4 equiv), CH$_2$Cl$_2$, −10 °C to −15 °C, 2 h	0.14 mmol of **3a**	60%
11	t-BuONO (3.5 equiv), SbCl$_3$ (1.4 equiv), CH$_2$Cl$_2$, −10 °C to −15 °C, 2 h	1.6 mmol of **3a**	65%
12	t-BuONO (3.5 equiv), SbCl$_3$ (1.4 equiv), CH$_2$Cl$_2$, −10 °C to −15 °C, 3.5 h	0.27 mmol of **3b**	61%
13	t-BuONO (10 equiv), TMSCl (5 equiv), CH$_2$Cl$_2$, 0 °C, 1 h	2.69 mmol of **3b**	68%

[a] Yield was calculated on the basis of the molecular weight of compound **4a**. However, by ^{1}H-NMR (500 MHz, CDCl$_3$), the chromatographically homogenous product band was observed to contain a 2:1 ratio of compound **4a** and the C2 protio O^6-(benzotriazol-1-yl)-3′,5′-di-O-(t-butyldimethylsilyl)-2′-deoxyinosine. In addition to these, a minor uncharacterized nucleoside byproduct was also formed; [b] Compound **3a** dissolved in CH$_2$Cl$_2$ was added to the mixture of reagents in CH$_2$Cl$_2$.

The next stage in the chemistry involved S_NAr reactions at the C6 positions of substrates **4a** and **4b**. Cladribine and its ribose analogue were prepared by reactions with aqueous ammonia (see the Experimental Section for details). In order to prepare other analogues, reactions were conducted with 1.5 equiv each of methylamine, dimethylamine, pyrrolidine, piperidine, morpholine, and *N,N,N′*-trimethylethylenediamine (products, reaction times, and yields are shown in Figure 1).

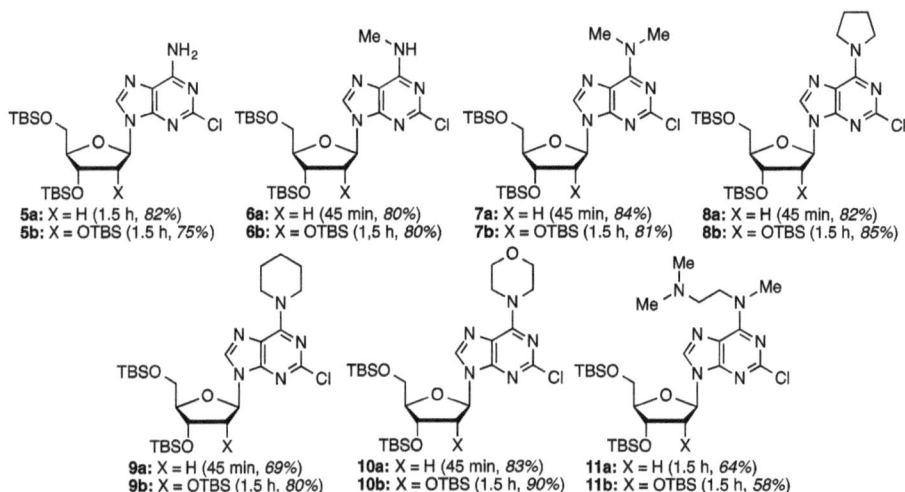

Figure 1. Structures of the products, reaction times, and yields obtained in the S_NAr reactions.

Most reactions proceeded smoothly and in good to high yields. Methylamine (2 M in THF) was used for the synthesis of **6a** and **6b**, whereas a 40% aqueous solution of dimethylamine was used for the synthesis of **7a** and **7b**. As we have previously shown, water is not generally detrimental to reactions of these benzotriazolyl purine nucleosides [35]. In reactions of **4a** and **4b** with *N,N,N′*-tri-methylethylenediamine, yields were lower. In each ~10%–15% of compounds **7a** and **7b** were isolated as byproducts. The source of dimethylamine is currently unknown but its origin can be linked to *N,N,N′*-trimethylethylenediamine.

Finally, for biological testing, desilylation was performed. Because we did not anticipate decomposition of starting materials or products, we only tested the use of KF (2 equiv/silyl group) in MeOH at 80 °C (Scheme 2). The results of the desilylation reactions are shown in Table 2.

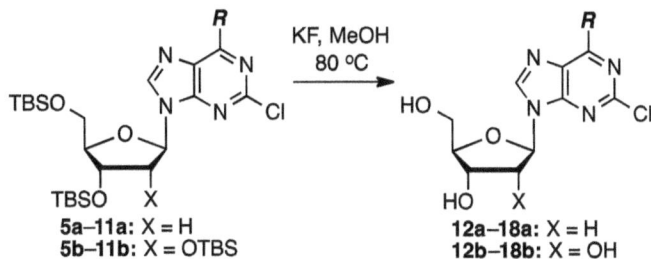

Scheme 2. Desilylation of the protected nucleosides.

Table 2. Structures of the products, reaction times, and yields of the desilylated compounds.

Entry	R	Reaction Times/KF Equiv	Yields
1	NH₂	24 h with 4 equiv KF	12a: 72%
2		24 h with 6 equiv KF	12b: 74%
3	Me–NH	24 h with 4 equiv KF	13a: 85%
4		16 h with 6 equiv KF	13b: 75%
5	Me–N–Me	24 h with 4 equiv KF	14a: 82%
6		16 h with 6 equiv KF	14b: 75%
7	(pyrrolidine)	20 h with 4 equiv KF	15a: 81%
8		16 h with 6 equiv KF	15b: 85%
9	(piperidine)	24 h with 4 equiv KF	16a: 71%
10		16 h with 6 equiv KF	16b: 84%
11	(morpholine)	26 h with 4 equiv KF	17a: 74%
12		16 h with 6 equiv KF	17b: 78%
13	Me–N–N–Me (with Me)	24 h with 4 equiv KF	18a: 84%
14		24 h with 6 equiv KF	18b: 65%

Because there is currently no known method for the diazotization/bromination of compound **3a**, we investigated a route similar to that for chlorination. Results from these experiments are listed in Table 3. What is notable with the diazotization/bromination, in contrast to the chlorination, was that use of 3.5 equiv of *t*-BuONO led to incomplete conversion over 2 h at −10 to −15 °C. Addition of another 3.5 equiv of *t*-BuONO then led to complete conversion over an additional 1 h.

A similar conversion of **3b** led to the ribose analogue **19b** in a comparable 64% yield (Table 3, entry 3). Compound **19a** was then converted to the bromo analogue of cladribine as shown in Scheme 3. S$_N$Ar displacement with aqueous NH$_3$ proceeded in 83% yield producing compound **20a**, which was finally desilylated with KF in anhydrous MeOH at 80 °C (28 h) to yield 2-bromo-2-deoxyadenosine (**21a**) in 66% yield. Corresponding conversions of the ribose analog **19b**, via intermediate **20b**, gave compound **21b**. Yields for these conversions were comparable to the deoxyribose series.

Table 3. Diazotization-bromination of silyl-protected O^6-(benzotriazol-1-yl) guanine nucleosides.

Entry	Conditions	Scale	Yield
1	*t*-BuONO (7.0 equiv), SbBr₃ (1.4 equiv), CH₂Br₂, −10 °C to −15 °C, 3 h	73 μmol of **3a**	69%
2	*t*-BuONO (7.0 equiv), SbBr₃ (1.4 equiv), CH₂Br₂, −10 °C to −15 °C, 3 h	0.49 mmol of **3a**	63%
3	*t*-BuONO (7.0 equiv), SbBr₃ (1.4 equiv), CH₂Br₂, −10 °C to −15 °C, 3 h	0.49 mmol of **3b**	64%

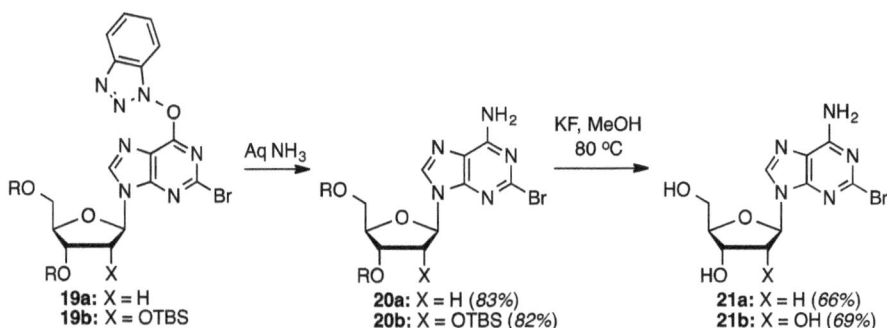

Scheme 3. Synthesis of 2-bromo-2′-deoxyadenosine.

Notably, previously unknown compound **19a** and **19b** are relatively easily prepared, *new bifunctional reactive nucleosides* that can undergo S_NAr reactions at the C6 and metal-mediated reactions at the C2 position. To the extent that C-Cl bonds can be activated by metal catalysts compounds **4a** and **4b** also offer this type of orthogonal reactivity. Results from the orthogonal reactivities of these new halo nucleosides will be reported in the future.

Results of Tests against HCL, TCL, CLL, and MDA-MB-231 Breast Cancer Cells

The newly synthesized compounds as well as cladribine and its bromo analogue were tested against HCL, TCL, and CLL. Data from these assays are shown in Table 4.

From these data, across the entire series, cladribine (**12a**) was best. The bromo analogue of cladribine (**21a**) showed lower activities. A comparison of the ribose analogue of cladribine (**12b**) and the bromo analogue of cladribine (**21a**) is interesting. Both compounds **12b** and **21a** show comparable activities against HCL, but the latter shows higher activities against TCL and CLL. The bromo ribose derivative **21b** showed activity but was inferior to compounds **12a**, **12b**, and **21a**. In this series, the only other compound to demonstrate any activity was the piperidinyl derivative **16a**.

Table 4. IC$_{50}$ values (µM) of the cladribine and its analogues on HCL, TCL, and CLL [a].

Compound	HCL		TCL		CLL	
	ATP	³H-Leu	ATP	³H-Leu	ATP	³H-Leu
12a	0.065	0.090	0.020	0.039	0.004	0.011
12b	0.81	0.85	4.81	9.40	1.33	0.88
13a	>33	>33	>33	>33	>33	>33
13b	>32	>32	>32	>32	ND [b]	>32
14a	>32	>32	>32	>32	>32	>32
14b	>30	>30	>30	>30	>30	>30
15a	>29	>29	>29	>29	>29	>29
15b	>28	>28	>28	>28	>28	>28
16a	11.5	16.3	9.6	9.0	4.3	5.4
16b	>27	>27	>27	>27	ND [b]	ND [b]
17a	>28	>28	>28	>28	>28	>28
17b	>27	>27	>27	>27	>27	>27
18a	>27	>27	>27	>27	>27	>27
18b	>26	>26	>26	>26	>26	>26
21a	0.76	0.96	0.13	0.16	0.13	0.14
21b	1.8	7.2	8.6	10.2	2.0	4.0

[a] One patient per disease type; [b] ND = not determined.

The compounds were also tested against adenocarcinoma MDA-MB-231 breast cancer cells. These cells were treated with three-fold serial dilutions of the compounds ranging from 0–1.8 mM for 24 h. Viable cells were fixed with cold methanol and nuclei were stained with 0.4% propidium iodide (PI). Cell viability was calculated as the percentage of surviving cells after treatment as measured by differences in fluorescence units between treated and untreated wells (Figure 2).

IC$_{50}$ values were obtained from dose response curve fittings using the non-linear regression function of GraphPad Prism® (La Jolla, CA, USA). Dashed horizontal line represents 50% cell viability. Columns represent means ± SEM of at least three independent experiments. Significant differences are described with * $p \leq 0.05$, ** $p \leq 0.01$, *** $p \leq 0.001$, **** $p \leq 0.0001$ compared to control.

From the IC$_{50}$ values shown in Table 5, the two compounds that emerged as most promising were cladribine **12a** and its ribose analogue **12b**, followed by the bromo derivatives **21a** and **21b**, which were about 10 times less active.

Table 5. IC$_{50}$ values (mM) of the compounds synthesized on MDA-MB-231 breast cancer cells.

2′-Deoxyribose Series								Ribose Series							
12a	13a	14a	15a	16a	17a	18a	21a	12b	13b	14b	15b	16b	17b	18b	21b
0.05	2.15	6.28	2.54	4.18	2.32	1.88	0.64	0.06	2.39	1.70	1.54	2.45	2.57	2.91	0.55

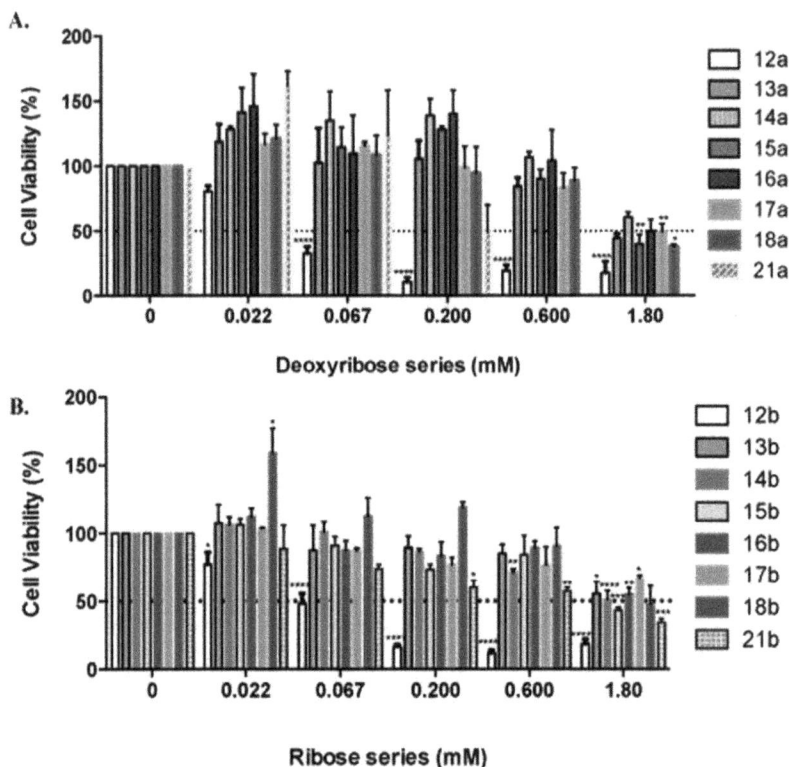

Figure 2. Effects of cladribine analogues on breast cancer cell viability. (**A**): Deoxyribose series and (**B**): Ribose series. Significant differences are described with * $p \leq 0.05$, ** $p \leq 0.01$, *** $p \leq 0.001$, **** $p \leq 0.0001$ compared to control.

3. Experimental Section

3.1. General Considerations

Thin-layer chromatography was performed on 200 μm aluminum-foil-backed silica gel plates for the 2'-deoxynucleosides and on Merck 60F$_{254}$ (Merck, Billerica, MA, USA) for the ribose analogues. Column chromatographic purifications were performed on 100–200 mesh silica gel. CH$_2$Cl$_2$ for the chlorination reactions was distilled over CaH$_2$. Precursors **3a** and **3b** were prepared as described previously [35]. The yield of **3a** was 71% on a 2.52 mmol scale and the yield of **3b** was 71% on a 2.1 mmol scale. TMSCl was redistilled prior to use and all other commercially available compounds were used without further purification. ^{1}H-NMR spectra were recorded at 500 MHz or at 400 MHz in the solvents indicated under the individual compound headings and are referenced to residual protonated solvent resonances. ^{13}C-NMR spectra were recorded at 125 MHz or at 100 MHz in the solvents indicated under the individual compound headings and are referenced to the solvent resonances (Supplementary Materials). Chemical shifts (δ) are reported in parts per million (ppm), and coupling constants (*J*) are in hertz (Hz). Standard abbreviations are used to designate resonance multiplicities (s = singlet, d = doublet, t = triplet, dd = doublet of doublet, ddd = doublet of doublet of doublet, quint = quintet, m = multiplet, br = broad, app = apparent). The saccharide carbons of the nucleoside are numbered 1' through 5' starting at the anomeric carbon atom and proceeding via the carbon chain to the primary carbinol carbon atom. The purinyl proton is designated as H-8 and the saccharide protons are designated on the basis of the carbon atom they are attached to.

O^6-(*Benzotriazol-1-yl*)-2-*chloro-9-[2-deoxy-3,5-di-O-(t-butyldimethylsilyl)-β-D-ribofuranosyl]purine* (**4a**): A mixture of compound **3a** (1.0 g, 1.63 mmol) and SbCl$_3$ (520.6 mg, 2.28 mmol) in dry CH$_2$Cl$_2$ (16.3 mL) was cooled to −15 °C using dry ice and acetone, in a nitrogen atmosphere. *t*-BuONO (0.678 mL, 5.70 mmol) was added dropwise and the mixture was stirred at −10 to −15 °C for 3.5 h, at which time TLC indicated the reaction to be complete. The reaction mixture was poured into ice-cold, saturated aqueous NaHCO$_3$ (25 mL) with stirring. The mixture was filtered using vacuum (*note*: use of vacuum for this filtration is critical for maximizing product recovery) and the residue was washed with CH$_2$Cl$_2$ (25 mL). The organic layer was separated and the aqueous layer was back extracted with CH$_2$Cl$_2$ (2 × 15 mL). The combined organic layer was washed with water (15 mL) and brine (15 mL), dried over anhydrous Na$_2$SO$_4$, and evaporated under reduced pressure. Purification of the crude material on a silica gel column sequentially eluted with hexanes, 5% EtOAc in hexanes, followed by 20% EtOAc in hexanes gave 0.67 g (65% yield) of compound **4a** as a white foam. R_f (SiO$_2$ and 30% EtOAc in hexanes) = 0.55. ^{1}H-NMR (500 MHz, CDCl$_3$): δ 8.47 (s, 1H, H-8), 8.09 (d, *J* = 8.3 Hz, 1H, Ar-H), 7.52–7.41 (m, 3H, Ar-H), 6.45 (t, *J* = 5.8 Hz, 1H, H-1′), 4.62 (m, 1H, H-3′), 4.01 (m, 1H, H-4′), 3.89 (dd, *J* = 3.4, 11.2 Hz, 1H, H-5′), 3.76 (dd, *J* = 2.4, 11.2 Hz, 1H, H-5′), 2.61 (app quint, J_{app} ~6.2 Hz, 1H, H-2′), 2.49–2.48 (app quint, J_{app} ~5.8 Hz, 1H, H-2′), 0.88 (s, 18H, *t*-Bu), 0.08 and 0.07 (2s, 12H, SiCH$_3$). ^{13}C-NMR (125 MHz, CDCl$_3$): δ 159.2, 154.9, 152.6, 144.4, 143.5, 129.0, 128.8, 125.0, 120.7, 119.1, 108.6, 88.4, 85.3, 71.6, 62.6, 41.6, 26.1, 25.9, 18.5, 18.1, −4.5, −4.7, −5.2, −5.3. HRMS (TOF) calcd for C$_{28}$H$_{43}$ClN$_7$O$_4$Si$_2$ [M + H]$^+$ 632.2598, found 632.2583.

2-*Chloro-3′,5′-di-O-(t-butyldimethylsilyl)-2′-deoxyadenosine* (**5a**): To a solution of compound **4a** (126.0 mg, 0.2 mmol) in 1,2-DME (2 mL), 28%–30% aqueous ammonia (32 μL) was added, and the mixture was stirred at room temperature for 1.5 h. The mixture was diluted with EtOAc (5 mL) and washed with 5% aqueous NaCl (5 mL). The organic layer was separated and the aqueous layer was back extracted with EtOAc (5 mL). The combined organic layer was dried over anhydrous Na$_2$SO$_4$ and evaporated under reduced pressure. The crude material was chromatographed on a silica gel column by sequential elution with 20% EtOAc in hexanes followed by 5% MeOH in CH$_2$Cl$_2$ to afford 85.0 mg (82% yield) of compound **5a** as a white solid. R_f (SiO$_2$ and 10% MeOH in CH$_2$Cl$_2$) = 0.40. ^{1}H-NMR (500 MHz, CDCl$_3$): δ 8.12 (s, 1H, H-8), 6.54 (br s, 2H, NH$_2$), 6.38 (t, *J* = 6.1 Hz, 1H, H-1′), 4.61 (m, 1H, H-3′), 3.99 (app q, J_{app} ~3.4 Hz, 1H, H-4′) 3.88 (dd, *J* = 3.9, 11.2 Hz, 1H, H-5′), 3.76 (dd, *J* = 3.0, 11.2 Hz, 1H, H-5′), 2.61 (app quint, J_{app} ~6.4 Hz, 1H, H-2′), 2.44–2.39 (ddd, *J* = 3.1, 6.3, 13.2 Hz, 1H, H-2′), 0.90 (s, 18H, *t*-Bu), 0.09 and 0.08 (2s, 12H, SiCH$_3$). ^{13}C-NMR (125 MHz, CDCl$_3$): δ 156.2, 154.0, 150.4, 139.5, 118.7, 87.9, 84.5, 71.7, 62.6, 41.3, 25.9, 25.7, 18.4, 18.0, −4.7, −4.8, −5.4, −5.5. HRMS (TOF) calcd for C$_{22}$H$_{40}$ClN$_5$O$_3$Si$_2$Na [M + Na]$^+$ 536.2250, found 536.2252.

3.2. General Procedure for the Synthesis of Cladribine Analogues

To a solution of compound **4a** (126.0 mg, 0.2 mmol) in 1,2-DME (2 mL) the appropriate amine (1.5 equiv) was added, and the mixture was stirred at room temperature. The mixture was diluted with EtOAc (5 mL for compounds **6a**, **9a**, **11a**, and 15 mL for compounds **7a**, **8a**, **10a**) and washed 5% aqueous NaCl (5 mL for compounds **6a**, **9a**, **11a**, and 15 mL for compounds **7a**, **8a**, **10a**). The organic layer was separated and the aqueous layer was back extracted with EtOAc (5 mL for compounds **6a**, **9a**, **11a**, and 15 mL for compounds **7a**, **8a**, **10a**). The combined organic layer was dried over anhydrous Na$_2$SO$_4$ and evaporated under reduced pressure. The crude material was chromatographed on a silica gel column; see individual compound headings for details.

2-*Chloro-N^6-methyl-3′,5′-di-O-(t-butyldimethylsilyl)-2′-deoxyadenosine* (**6a**): Compound **6a** (85.0 mg, 80% yield) was obtained as a white, foamy solid after chromatography on a silica gel column by sequential elution with 5% EtOAc in hexanes and 15% EtOAc in hexanes. R_f (SiO$_2$ and 30% EtOAc in hexanes) = 0.26. ^{1}H-NMR (500 MHz, CDCl$_3$): δ 8.02 (s, 1H, H-8), 6.37 (t, *J* = 6.4 Hz, 1H, H-1′), 6.17 (br s, 1H, NH),

4.60 (m, 1H, H-3'), 3.97 (app q, J_{app} ~3.4 Hz, 1H, H-4'), 3.87 (dd, J = 4.1, 11.2 Hz, 1H, H-5'), 3.75 (dd, J = 3.0, 11.2 Hz, 1H, H-5'), 2.61 (app quint, J_{app} ~6.4 Hz, 1H, H-2'), 2.42–2.37 (ddd, J = 2.8, 6.1, 12.7 Hz, 1H, H-2'), 0.90 (s, 18H, *t*-Bu), 0.09 and 0.08 (2s, 12H, SiCH₃). ¹³C-NMR (125 MHz, CDCl₃): δ 155.9, 154.5, 149.2; 138.7, 119.1, 87.9, 84.4, 71.8, 62.7, 41.2, 27.5, 25.9, 25.7, 18.4, 18.0, −4.7, −4.8, −5.4, −5.5. HRMS (TOF) calcd for $C_{23}H_{42}ClN_5O_3Si_2Na$ [M + Na]⁺ 550.2407, found 550.2419.

2-Chloro-N^6,N^6-dimethyl-3',5'-di-O-(t-butyldimethylsilyl)-2'-deoxyadenosine (**7a**): Compound **7a** (91.1 mg, 84% yield) was obtained as a colorless, thick gum after chromatography on a silica gel column by sequential elution with hexanes, 5% EtOAc in hexanes, and 20% EtOAc in hexanes. R_f (SiO₂ and 30% EtOAc in hexanes) = 0.70. ¹H-NMR (500 MHz, CDCl₃): δ 7.93 (s, 1H, H-8), 6.36 (t, J = 6.3 Hz, 1H, H-1'), 4.58 (m, 1H, H-3'), 3.95 (m, 1H, H-4'), 3.83 (dd, J = 4.4, 11.2 Hz, 1H, H-5'), 3.73 (dd, J = 2.9, 11.2 Hz, 1H, H-5'), 3.50 (br s, 6H, N(CH₃)₂), 2.58 (app quint, J_{app} ~6.5 Hz, 1H, H-2'), 2.39–2.35 (m, 1H, H-2'), 0.88 (s, 18H, *t*-Bu), 0.08 and 0.06 (2s, 12H, SiCH₃). ¹³C-NMR (125 MHz, CDCl₃): δ 155.2, 153.8, 151.4, 137.2, 119.6, 87.9, 84.4, 72.1, 62.9, 41.1, 26.1, 26.0, 25.9, 25.8, 18.6, 18.2, −4.5, −4.6, −5.2, −5.3. HRMS (TOF) calcd for $C_{24}H_{45}ClN_5O_3Si_2$ [M + H]⁺ 542.2744, found 542.2749.

2-Chloro-6-(pyrrolidin-1-yl)-9-[2-deoxy-3,5-di-O-(t-butyldimethylsilyl)-β-D-ribofuranosyl]purine (**8a**): Compound **8a** (93.3 mg, 82% yield) was obtained as a colorless, thick gum after chromatography on a silica gel column by sequential elution with hexanes, 5% EtOAc in hexanes, and 20% EtOAc in hexanes. R_f (SiO₂ and 30% EtOAc in hexanes) = 0.65. ¹H-NMR (500 MHz, CDCl₃): δ 7.92 (s, 1H, H-8), 6.37 (t, J = 6.6 Hz, 1H, H-1'), 4.59 (m, 1H, H-3'), 4.13 (br s, 2H, pyrrolidinyl NCH), 3.96 (m, 1H, H-4'), 3.83 (dd, J = 4.9, 11.2 Hz, 1H, H-5'), 3.74 (dd, J = 3.9, 11.2 Hz, 1H, H-5'), 3.79–3.73 (br m, 2H, pyrrolidinyl NCH), 2.61 (app quint, J_{app} ~6.5 Hz, 1H, H-2'), 2.39–2.35 (m, 1H, H-2'), 2.10–2.0 (br m, 2H, pyrrolidinyl CH), 1.96–1.85 (br m, 2H, pyrrolidinyl CH), 0.90 (s, 18H, *t*-Bu), 0.09 and 0.07 (2s, 12H, SiCH₃). ¹³C-NMR (125 MHz, CDCl₃): δ 154.3, 153.5, 151.0, 137.8, 119.8, 88.1, 84.4, 72.3, 63.1, 49.1, 41.08, 26.4, 26.2, 25.9, 24.3, 18.6, 18.2, −4.5, −4.6, −5.2, −5.3. HRMS (TOF) calcd. for $C_{26}H_{47}ClN_5O_3Si_2$ [M + H]⁺ 568.2900, found 568.2906.

2-Chloro-6-(piperidin-1-yl)-9-[2-deoxy-3,5-di-O-(t-butyldimethylsilyl)-β-D-ribofuranosyl]purine (**9a**): Compound **9a** (99.0 mg, 85% yield) was obtained as a white, foamy solid after chromatography on a silica gel column by sequential elution with 5% EtOAc in hexanes and 15% EtOAc in hexanes. R_f (SiO₂ and 30% EtOAc in hexanes) = 0.65. ¹H-NMR (500 MHz, CDCl₃): δ 7.92 (s, 1H, H-8), 6.37 (t, J = 6.6 Hz, 1H, H-1'), 4.59 (m, 1H, H-3'), 4.45–3.73 (br m, 4H, piperidinyl N(CH₂)₂), 3.96 (app q, J_{app} ~3.6 Hz, 1H, H-4'), 3.83 (dd, J = 4.9, 11.2 Hz, 1H, H-5'), 3.74 (dd, J = 3.4, 11.2 Hz, 1H, H-5'), 2.59 (app quint, J_{app} ~6.3 Hz, 1H, H-2'), 2.40–2.35 (ddd, J = 3.3, 6.2, 13.3 Hz, 1H, H-2'), 1.68 (br m, 6H, piperidinyl CH), 0.90 (s, 18H, *t*-Bu), 0.09 and 0.07 (2s, 12H, SiCH₃). ¹³C-NMR (125 MHz, CDCl₃): δ 153.8, 151.4, 136.7, 119.0, 109.9, 87.8, 84.2, 72.0, 62.8, 46.3, 40.9, 26.1, 25.9, 25.7, 24.6, 18.4, 18.0, −4.7, −4.8, −5.4, −5.5. HRMS (TOF) calcd for $C_{27}H_{49}ClN_5O_3Si_2$ [M + H] ⁺ 582.3057 found 582.3074.

2-Chloro-6-(morpholin-4-yl)-9-[2-deoxy-3,5-di-O-(t-butyldimethylsilyl)-β-D-ribofuranosyl]purine (**10a**): Compound **10a** (97.4 mg, 83% yield) was obtained as a colorless, thick gum after chromatography on a silica gel column by sequential elution with hexanes, 5% EtOAc in hexanes, and 20% EtOAc in hexanes. R_f (SiO₂ and 30% EtOAc in hexanes) = 0.68. ¹H-NMR (500 MHz, CDCl₃): δ 7.98 (s, 1H, H-8), 6.37 (t, J = 6.3 Hz, 1H, H-1'), 4.58 (m, 1H, H-3'), 4.55–3.98 (br m, 4H, morpholinyl N(CH₂)₂), 3.96 (m, 1H, H-4'), 3.84 (dd, J = 4.4, 11.2 Hz, 1H, H-5'), 3.78 (t, J = 4.6 Hz, 4H, morpholinyl CH₂OCH₂), 3.74 (dd, J = 2.9, 11.2 Hz, 1H, H-5'), 2.56 (app quint, J_{app} ~6.5 Hz, 1H, H-2'), 2.40–2.36 (m, 1H, H-2'), 0.89 (s, 18H, *t*-Bu), 0.08 and 0.07 (2s, 12H, SiCH₃). ¹³C-NMR (125 MHz, CDCl₃): δ 154.0, 153.9, 151.8, 137.6, 119.3, 88.0, 84.4, 72.0, 67.1, 62.9, 43.5 (br), 41.3, 26.1, 26.0, 25.9, 18.6, 18.2, −4.5, −4.6, −5.2, −5.3. HRMS (TOF) calcd for $C_{26}H_{47}ClN_5O_4Si_2$ [M + H]⁺ 584.2850, found 584.2855.

2-Chloro-6-[(2-dimethylamino)ethyl)(methyl)amino)-9-[2-deoxy-3,5-di-O-(t-butyldimethylsilyl)-β-D-ribofuranosyl]purine (**11a**): Compound **11a** (77.0 mg, 64% yield) was obtained as a pale-yellow, thick gum after chromatography on a silica gel column by sequential elution with 10% EtOAc in hexanes,

20% EtOAc in hexanes, and 10% MeOH in CH_2Cl_2. R_f (SiO_2 and 10% MeOH in CH_2Cl_2) = 0.10. ^1H-NMR (500 MHz, $CDCl_3$): δ 7.94 (s, 1H, H-8), 6.36 (t, *J* = 6.3 Hz, 1H, H-1′), 4.62 (m, 1H, H-3′), 4.28–3.84 (br m, 2H, NCH_2), 3.99 (app q, J_{app} ~3.7 Hz, 1H, H-4′), 3.85 (dd, *J* = 4.4, 11.2 Hz, 1H, H-5′), 3.78 (dd, *J* = 3.4, 11.2 Hz, 1H, H-5′), 3.47 (br s, 3H, NCH_3), 2.65–2.59 (m, 2H, NCH_2 and 1H, H-2′), 2.41–2.37 (ddd, *J* = 4.2, 6.2, 13.1 Hz, 1H, H-2′), 2.34 (s, 6H, $N(CH_3)_2$), 0.92 and 0.91 (s, 18H, *t*-Bu), 0.11, 0.083, and 0.079 (3s, 12H, $SiCH_3$). ^{13}C-NMR (125 MHz, $CDCl_3$): δ 154.9, 153.8, 151.4, 137.2, 119.4, 87.9, 84.4, 72.0, 62.9, 57.0, 48.6, 45.7, 41.0, 36.9, 26.0, 25.8, 18.4, 8.0, −4.6, −4.8, −5.4, −5.5. HRMS (TOF) calcd for $C_{27}H_{52}ClN_6O_3Si_2$ [M + H]$^+$ 599.3322 found 599.3313.

O^6-(Benzotriazol-1-yl)-2-chloro-9-[2,3,5-tri-O-(t-butyldimethylsilyl)-β-D-ribofuranosyl]purine (**4b**): To a solution of *t*-BuONO (2.78 g, 26.91 mmol) in CH_2Cl_2 (197 mL) was added TMSCl (1.46 g, 13.45 mmol) at 0 °C. To this mixture a solution of **3b** (2 g, 2.69 mmol) in CH_2Cl_2 (197 mL) was added dropwise, and then the reaction mixture was stirred at 0 °C for 1 h at which time the reaction was completed as indicated by TLC. The reaction mixture was diluted with CH_2Cl_2 (200 mL), washed with saturated $NaHCO_3$ (100 mL), H_2O (100 mL), and brine (100 mL). The organic layer was dried over anhydrous Na_2SO_4 and evaporated under reduced pressure. Purification of the crude material on a silica gel column sequentially eluted with hexanes, 5% EtOAc in hexanes, and 20% EtOAc in hexanes gave 1.4 g (68% yield) of compound **4b** as a white foam. R_f (SiO_2 and 30% EtOAc in hexanes) = 0.55. ^1H-NMR (400 MHz, $CDCl_3$): δ 8.56 (s, 1H, H-8), 8.14 (d, *J* = 8.0 Hz, 1H, Ar-H), 7.55–7.45 (m, 3H, Ar-H), 6.05 (d, *J* = 4.0 Hz, 1H, H-1′), 4.57 (t, *J* = 4.2 Hz, 1H, H-2′), 4.34 (t, *J* = 4.4 Hz, 1H, H-3′), 4.18–4.15 (m, 1H, H-4′), 4.08 (dd, *J* = 3.6, 11.6 Hz, 1H, H-5′), 3.83 (dd, *J* = 2.4, 11.6 Hz, 1H, H-5′), 0.96, 0.93, and 0.88 (3s, 27H, *t*-Bu), 0.15, 0.11, 0.02, and −0.11 (4s, 18H, $SiCH_3$). ^{13}C-NMR (100 MHz, $CDCl_3$): δ 159.0, 154.9, 152.6, 144.4, 143.4, 128.9, 128.7, 124.9, 120.6, 119.0, 108.5, 89.4, 85.3, 71.2, 62.0, 26.1, 25.8, 25.6, 18.5, 18.0, 17.9, −4.3, −4.6, −4.7, −4.9, −5.3, −5.4. HRMS (TOF) calcd for $C_{34}H_{57}ClN_7O_5Si_3$ [M + H]$^+$ 762.3412, found 762.3427.

2-Chloro-2′,3′,5′-tri-O-(t-butyldimethylsilyl)adenosine (**5b**): To a solution of compound **4b** (500 mg, 0.655 mmol) in 1,2-DME (8 mL), 28%–30% aqueous ammonia (64 µL) was added and the mixture was stirred at room temperature for 1.5 h. The reaction mixture was diluted with EtOAc (25 mL) and washed 5% aqueous NaCl (25 mL). The organic layer was separated and the aqueous layer was back extracted with EtOAc (25 mL). The combined organic layer was dried over anhydrous Na_2SO_4 and evaporated under reduced pressure. The crude material was chromatographed on a silica gel column by sequential elution with 20% EtOAc in hexanes and 5% MeOH in EtOAc to afford 310 mg (75% yield) of compound **5b** as an off-white solid. R_f (SiO_2 and 10% MeOH in EtOAc) = 0.21. ^1H-NMR (400 MHz, $CDCl_3$): δ 8.12 (s, 1H, H-8), 6.17 (br s, 2H, NH_2), 5.93 (d, *J* = 4.8 Hz, 1H, H-1′), 4.69 (t, *J* = 4.6 Hz, 1H, H-2′), 4.32 (t, *J* = 4.6 Hz, H-3′), 4.14–4.06 (m, 1H, H-4′), 4.06 (dd, *J* = 4.8, 11.6 Hz, 1H, H-5′), 3.80 (dd, *J* = 2.8, 11.2 Hz, 1H, H-5′), 0.94, 0.91, and 0.84 (3s, 27H, *t*-Bu), 0.14, 0.09, −0.01, and −0.17 (4s, 18H, $SiCH_3$). ^{13}C-NMR (100 MHz, $CDCl_3$): δ 156.1, 154.0, 150.7, 140.2, 118.9, 88.9, 85.4, 75.5, 71.7, 62.3, 26.0, 25.8, 25.7, 18.5, 18.0, 17.9, −4.3, −4.5, −4.7, −5.0, −5.3, −5.4. HRMS (TOF) calcd. for $C_{28}H_{55}ClN_5O_4Si_3$ [M + H]$^+$ 644.3245 found 644.3217.

3.3. General Procedure for the Synthesis of Ribose Analogues of Cladribine

To a solution of compound **4b** (500 mg, 0.655 mmol) in 1,2-DME (8 mL) was added the appropriate amine (1.5 equiv) and the mixture was stirred at room temperature. The mixture was diluted with EtOAc (25 mL) and washed with 5% aqueous NaCl (15 mL). The organic layer was separated and the aqueous layer was back extracted with EtOAc (15 mL). The combined organic layers were dried over anhydrous Na_2SO_4 and evaporated under reduced pressure. The crude material was chromatographed on a silica gel column. See individual compound headings for details.

2-Chloro-N^6-methyl-2′,3′,5′-tri-O-(t-butyldimethylsilyl)adenosine (**6b**): Compound **6b** (340 mg, 80% yield) was obtained as a pale-yellow, thick gum after chromatography on a silica gel column by sequential elution with 10% EtOAc in hexanes, 20% EtOAc in hexanes, and 50% EtOAc in hexanes. R_f (SiO_2 and

EtOAc) = 0.20. ^{1}H-NMR (400 MHz, CDCl$_3$): δ 8.02 (s, 1H, H-8), 6.04 (br s, 1H, NH), 5.91 (d, J = 5.2 Hz, 1H, H-1′), 4.72 (t, J = 4.6 Hz, 1H, H-2′), 4.32 (t, J = 4.0 Hz, 1H, H-3′), 4.12–4.10 (m, 1H, H-4′), 4.06 (dd, J = 5.2, 11.6 Hz, 1H, H-5′), 3.79 (dd, J = 2.8, 11.2 Hz, 1H, H-5′), 3.17 (s, 3H, CH$_3$), 0.94, 0.92, and 0.81 (3s, 27H, t-Bu), 0.13, 0.06, −0.02, and −0.19 (4s, 18H, SiCH$_3$). ^{13}C-NMR (100 MHz, CDCl$_3$): δ 155.9, 154.5, 139.3, 119.3, 88.8, 85.4, 75.3, 71.5, 65.2, 62.4, 26.0, 25.8, 25.7, 18.4, 18.0, 17.8, 14.0, 11.0, −4.4, −4.7, −5.1, −5.4, −5.7. HRMS (TOF) calcd for C$_{29}$H$_{57}$ClN$_5$O$_4$Si$_3$ [M + H]$^+$ 658.3401, found 658.3416.

2-Chloro-N^6,N^6-dimethyl- 2′,3′,5′-tri-O-(t-butyldimethylsilyl)adenosine (**7b**): Compound **7b** (360 mg, 81% yield) was obtained as a pale-yellow, thick gum after chromatography on a silica gel column by sequential elution with 10% EtOAc in hexanes and 20% EtOAc in hexanes. R$_f$ (SiO$_2$ and 20% EtOAc in hexanes) = 0.40. ^{1}H-NMR (400 MHz, CDCl$_3$): δ 7.93 (s, 1H, H-8), 5.90 (d, J = 5.2 Hz, 1H, H-1′), 4.77 (t, J = 4.8 Hz, 1H, H-2′), 4.31 (t, J = 3.8 Hz, 1H, H-3′), 4.04–4.09 (m, 1H, H-4′), 4.06 (dd, J = 5.6, 11.2 Hz, 1H, H-5′), 3.78 (dd, J = 2.8, 11.2 Hz, 1H, H-5′), 3.58 (br s, 6H, CH$_3$), 0.94, 0.92, and 0.82 (3s, 27H, t-Bu), 0.13, 0.10, −0.02, and −0.18 (4s, 18H, SiCH$_3$). ^{13}C-NMR (100 MHz, CDCl$_3$): δ 155.9, 153.7, 151.4, 137.9, 119.7, 88.8, 85.4, 74.8, 72.0, 62.5, 38.4, 26.0, 25.8, 25.7, 18.4, 18.0, 17.9, −4.3, −4.7, −5.0, −5.4. HRMS (TOF) calcd for C$_{30}$H$_{59}$ClN$_5$O$_4$Si$_3$ [M + H]$^+$ 672.3558, found 672.3561.

2-Chloro-6-(pyrrolidin-1-yl)-9-[2,3,5-tri-O-(t-butyldimethylsilyl)-β-D-ribofuranosyl]purine (**8b**): Compound **8b** (390 mg, 85% yield) was obtained as a pale-yellow, thick gum after chromatography on a silica gel column by sequential elution with 10% EtOAc in hexanes and 20% EtOAc in hexanes. R$_f$ (SiO$_2$ and 20% EtOAc in hexanes) = 0.30. ^{1}H-NMR (400 MHz, CDCl$_3$): δ 7.90 (s, 1H, H-8), 5.90 (d, J = 5.2 Hz, 1H, H-1′), 4.81 (t, J = 5.0 Hz, 1H, H-2′), 4.32 (t, J = 4.0 Hz, 1H, H-3′), 4.14–4.08 (m, 1H, H-4′ and br m, 2H, pyrrolidinyl NCH), 4.06 (dd, J = 5.6, 11.2 Hz, 1H, H-5′), 3.77–3.73 (m, 3H, H-5′ and pyrrolidinyl NCH), 2.04 (br m, 2H, pyrrolidinyl CH), 1.98 (br m, 2H, pyrrolidinyl CH), 0.93, 0.89, and 0.81 (3s, 27H, t-Bu), 0.12, 0.11, −0.03, and −0.20 (4s, 18H, SiCH$_3$). ^{13}C-NMR (100 MHz, CDCl$_3$): δ 154.0, 153.4, 151.0, 138.0, 119.9, 88.7, 85.5, 74.6, 72.2, 62.6, 48.9, 47.6, 26.0, 25.8, 25.7, 18.4, 18.1, 17.9, −4.4, −4.6, −4.7, −5.0, −5.3. HRMS (TOF) calcd for C$_{32}$H$_{61}$ClN$_5$O$_4$Si$_3$ [M + H]$^+$ 698.3714 found 698.3731.

2-Chloro-6-(piperidin-1-yl)-9-[2,3,5-tri-O-(t-butyldimethylsilyl)-β-D-ribofuranosyl]purine (**9b**): Compound **9b** (375 mg, 80% yield) was obtained as a pale-yellow, thick gum after chromatography on a silica gel column by sequential elution with 10% EtOAc in hexanes and 20% EtOAc in hexanes. R$_f$ (SiO$_2$ and 20% EtOAc in hexanes) = 0.50. ^{1}H-NMR (400 MHz, CDCl$_3$): δ 7.91 (s, 1H, H-8), 5.89 (d, J = 5.6 Hz, 1H, H-1′), 4.79 (t, J = 4.8 Hz, 1H, H-2′), 4.32 (t, J = 4.0 Hz, 1H, H-3′), 4.22 (br s, 4H, piperidinyl NCH$_2$), 4.11–4.09 (m, 1H, H-4′), 4.06 (dd, J = 5.2, 10.8 Hz, 1H, H-5′), 3.77 (dd, J = 2.8, 10.8 Hz, 1H, H-5′), 1.69 (br m, 6H, piperidinyl CH$_2$), 0.94, 0.90, and 0.84 (3s, 27H, t-Bu), 0.12, 0.10, −0.02, and −0.17 (4s, 18H, SiCH$_3$). ^{13}C-NMR (100 MHz, CDCl$_3$): δ 153.9, 153.8, 151.6, 137.6, 119.4, 88.8, 85.3, 74.7, 71.9, 62.5, 29.6, 25.8, 25.7, 24.6, 22.6, 18.4, 18.0, 17.9, 14.0, −4.3, −4.7, −5.0, −5.3. HRMS (TOF) calcd for C$_{33}$H$_{63}$ClN$_5$O$_4$Si$_3$ [M + H]$^+$ 712.3871, found 712.3875.

2-Chloro-6-(morpholin-4-yl)-9-[2,3,5-tri-O-(t-butyldimethylsilyl)-β-D-ribofuranosyl]purine (**10b**): Compound **10b** (430 mg, 90% yield) was obtained as a pale-yellow, thick gum after chromatography on a silica gel column by sequential elution with 10% EtOAc in hexanes and 20% EtOAc in hexanes. R$_f$ (SiO$_2$ and 20% EtOAc in hexanes) = 0.47. ^{1}H-NMR (400 MHz, CDCl$_3$): δ 7.98 (s, 1H, H-8), 5.92 (d, J = 4.8 Hz, 1H, H-1′), 4.73 (t, J = 4.8 Hz, 1H, H-2′), 4.31–4.20 (t, J = 4.2 Hz, 1H, H-3′and br m, 4H, morpholinyl N(CH$_2$)$_2$), 4.12–4.09 (m, 1H, H-4′), 4.06 (dd, J = 5.2, 11.2 Hz, 1H, H-5′), 3.83 (t, J = 5.0 Hz, 4H, morpholinyl CH$_2$OCH$_2$), 3.77 (dd, J = 3.2, 11.2 Hz, 1H, H-5′), 0.94, 0.92, and 0.83 (3s, 27H, t-Bu), 0.13, 0.10, −0.01, and −0.16 (4s, 18H, SiCH$_3$). ^{13}C-NMR (100 MHz, CDCl$_3$): δ 153.9, 153.7, 151.8, 138.1, 119.4, 88.8, 85.3, 77.3, 77.0, 76.6, 75.0, 71.8, 66.9, 62.3, 45.60, 29.6, 29.3, 26.0, 25.8, 25.7, 22.6, 18.4, 18.0, 17.8, 14.0, −4.3, −4.7, −5.0, −5.4. HRMS (TOF) calcd for C$_{32}$H$_{61}$ClN$_5$O$_5$Si$_3$ [M + H]$^+$ 714.3664 found 714.3668.

2-Chloro-6-[(2-dimethylamino)ethyl)(methyl)amino)]-9-[2,3,5-tri-O-(t-butyldimethylsilyl)-β-D-ribofuranosyl]purine (**11b**): Compound **11b** (280 mg, 58%) was obtained as a pale-yellow, thick gum after chromatography on a silica gel column by sequential elution with 10% MeOH in CH$_2$Cl$_2$, 20% MeOH in CH_2Cl_2, and 30% MeOH in CH_2Cl_2. R_f (SiO_2 and 10% MeOH in CH_2Cl_2) = 0.10. ^{1}H-NMR (500 MHz, CDCl$_3$): δ 7.97 (s, 1H, H-8), 5.90 (d, J = 4.8 Hz, 1H, H-1'), 4.72 (t, J = 4.8 Hz, 1H, H-2'), 4.32 (t, J = 3.8 Hz, 1H, H-3'), 4.15–4.10 (m, 1H, H-4' and br s, 2H, NCH$_2$), 4.07 (dd, J = 5.2, 11.2 Hz, 1H, H-5'), 3.78 (dd, J = 2.8, 11.2 Hz, 1H, H-5' and br m, 2H, NCH$_2$), 2.59 (br s, 2H, NCH$_2$), 2.32 (s, 6H, N(CH$_3$)$_2$), 0.94, 0.92, and 0.82 (3s, 27H, *t*-Bu), 0.13, 0.10, −0.01, and −0.15 (4s, 18H, SiCH$_3$). ^{13}C-NMR (100 MHz, CDCl$_3$): δ 154.8, 153.7, 151.4, 137.9, 119.5, 88.8, 85.2, 75.0, 71.9, 62.4, 45.7, 29.6, 26.0, 25.8, 25.7, 18.4, 18.0, 17.9, −4.3, −4.7, −5.0, −5.4. HRMS (TOF) calcd for $C_{33}H_{66}ClN_6O_4Si_3$ [M + H]$^+$ 729.4136, found 729.4157.

3.4. General Procedure for the Desilylation of Cladribine Analogues

To a 0.1 M solution of the silylated compound in anhydrous MeOH, KF (2 equiv/silyl group) was added. The mixture was heated at 80 °C for 20–26 h, cooled, and silica gel was added. The mixture was evaporated to dryness and the compound-impregnated silica gel was loaded onto a wet-packed silica gel column. The products were obtained by elution with appropriate solvents (see the individual compound headings for details).

2-Chloro-2'-deoxyadenosine (**12a**): Prepared from compound **5a** (60.0 mg, 0.117 mmol) and KF (27.0 mg, 0.467 mmol) in MeOH (1.17 mL). Chromatography on a silica gel column sequentially eluted with 5% MeOH in EtOAc and 10% MeOH in EtOAc gave 24.0 mg (72% yield) of compound **12a** as an off-white solid. R_f (SiO_2 and 10% MeOH in EtOAc) = 0.13. ^{1}H-NMR (500 MHz, CD$_3$OD): δ 8.28 (s, 1H, H-8), 6.36 (t, J = 6.8 Hz, 1H, H-1'), 4.57 (m, 1H, H-3'), 4.04 (m, 1H, H-4'), 3.84 (dd, J = 2.9, 12.2 Hz, 1H, H-5'), 3.74 (dd, J = 3.4, 12.2 Hz, 1H, H-5'), 2.76 (app quint, J_{app} ~6.7 Hz, 1H, H-2'), 2.43–2.39 (ddd, J = 2.9, 5.9, 13.2 Hz, 1H, H-2'). ^{13}C-NMR (125 MHz, CD$_3$OD): δ 158.3, 155.3, 151.4, 141.9, 119.8, 89.9, 87.0, 73.0, 63.6, 41.6. HRMS (TOF) calcd for $C_{10}H_{12}ClN_5O_3Na$ [M + Na]$^+$ 308.0521, found 308.0523.

2-Chloro-N^6-methyl-2'-deoxyadenosine (**13a**) [50]: Prepared from compound **6a** (80.0 mg, 0.154 mmol) and KF (36.0 mg, 0.618 mmol) in MeOH (1.54 mL). Chromatography on a silica gel column sequentially eluted with 2.5% MeOH in EtOAc and 5% MeOH in EtOAc gave 39.0 mg (85% yield) of compound **13a** as a white, foamy solid. R_f (SiO_2 and 10% MeOH in EtOAc) = 0.21. ^{1}H-NMR (500 MHz, CD$_3$OD): δ 8.18 (s, 1H, H-8), 6.34 (t, J = 6.8 Hz, 1H, H-1'), 4.56 (m, 1H, H-3'), 4.04 (m, 1H, H-4'), 3.84 (dd, J = 2.4, 12.2 Hz, 1H, H-5'), 3.74 (dd, J = 3.4, 12.2 Hz, 1H, H-5'), 3.05 (br s, 3H, NCH$_3$), 2.76 (app quint, J_{app} ~ 6.8 Hz, 1H, H-2'), 2.42–2.38 (ddd, J = 2.9, 5.8, 13.7 Hz, 1H, H-2'). ^{13}C-NMR (125 MHz, CD$_3$OD): δ 157.3, 155.5, 150.1, 141.2, 120.4, 89.9, 87.0, 73.0, 63.7, 41.6, 27.8. HRMS (TOF) calcd for $C_{11}H_{14}ClN_5O_3Na$ [M + Na]$^+$ 322.0677, found 322.0682.

2-Chloro-N^6,N^6-dimethyl-2'-deoxyadenosine (Cladribine, **14a**): Prepared from compound **7a** (80.0 mg, 0.147 mmol) and KF (34.3 mg, 0.590 mmol) in MeOH (1.4 mL). Chromatography on a silica gel column sequentially eluted with 50% EtOAc in hexanes, EtOAc, and 10% MeOH in EtOAc gave 38.0 mg (82% yield) of compound **14a** as a white, foamy solid. R_f (SiO_2 and 5% MeOH in EtOAc) = 0.29. ^{1}H-NMR (500 MHz, CD$_3$OD): δ 8.16 (s, 1H, H-8), 6.35 (t, J = 6.8 Hz, 1H, H-1'), 4.56 (m, 1H, H-3'), 4.04 (m, 1H, H-4'), 3.84 (dd, J = 2.9, 12.2 Hz, 1H, H-5'), 3.73 (dd, J = 3.4, 12.2 Hz, 1H, H-5'), 3.70–3.10 (br m, 6H, N(CH$_3$)$_2$), 2.74 (app quint, J_{app} ~6.8 Hz, 1H, H-2'), 2.41–2.36 (ddd, J = 2.9, 5.8, 13.2 Hz, 1H, H-2'). ^{13}C-NMR (125 MHz, CD$_3$OD): δ 156.4, 154.6, 152.2, 140.0, 120.72, 89.8, 86.9, 73.0, 63.7, 41.5, 39.0 (br). HRMS (TOF) calcd for $C_{12}H_{16}ClN_5O_3Na$ [M + Na]$^+$ 336.0834, found 336.0823.

2-Chloro-6-(pyrrolidin-1-yl)-9-(2-deoxy-β-D-ribofuranosyl)purine (**15a**): Prepared from compound **8a** (60.0 mg, 0.105 mmol) and KF (24.5 mg, 0.422 mmol) in MeOH (1.0 mL). Chromatography on a silica gel column sequentially eluted with 50% EtOAc in hexanes, EtOAc, and 10% MeOH in EtOAc gave 29.1 mg (81% yield) of compound **15a** as a white solid. R_f (SiO_2 and 5% MeOH in EtOAc) = 0.19. ^{1}H-NMR (500 MHz, CD$_3$OD): δ 8.19 (s, 1H, H-8), 6.36 (t, J = 7.1 Hz, 1H, H-1'), 4.59 (m, 1H, H-3'), 4.14–4.07 (br m, 2H, pyrrolidinyl NCH), 4.04 (m, 1H, H-4'), 3.84 (dd, J = 3.4, 12.7 Hz, 1H, H-5'), 3.74 (dd, J = 3.4, 12.2 Hz, 1H, H-5'), 3.71–3.62 (br m, 2H, pyrrolidinyl NCH), 2.75 (app quint, J_{app} ~6.7 Hz, 1H, H-2'), 2.40–2.36 (ddd, J = 2.9, 5.8, 13.2 Hz, 1H, H-2'), 2.15–2.06 (br m, 2H, pyrrolidinyl CH), 2.04–1.92

(br m, 2H, pyrrolidinyl CH). ^{13}C-NMR (125 MHz, CD$_3$OD): δ 155.0, 154.7, 151.7, 140.6, 120.8, 89.9, 87.0, 73.0, 63.7, 50.2, 49.6, 41.6, 27.3, 25.2. HRMS (TOF) calcd for C$_{14}$H$_{19}$ClN$_5$O$_3$ [M + H]$^+$ 340.1171, found 340.1149.

2-Chloro-6-(piperidin-1-yl)-9-(2-deoxy-β-D-ribofuranosyl)purine (**16a**): Prepared from compound **9a** (70.0 mg, 0.120 mmol) and KF (28.0 mg, 0.481 mmol) in MeOH (1.20 mL). Chromatography on a silica gel column sequentially eluted with EtOAc and 2% MeOH in EtOAc gave 30.0 mg (71% yield) of compound **16a** as a white, foamy solid. R_f (SiO$_2$ and 10% MeOH in EtOAc) = 0.46. ^{1}H-NMR (500 MHz, CD$_3$OD): δ 8.15 (s, 1H, H-8), 6.34 (t, *J* = 6.6 Hz, 1H, H-1′), 4.55 (m, 1H, H-3′), 4.18 (br s, 4H, piperidinyl N(CH$_2$)$_2$), 4.03 (m, 1H, H-4′), 3.83 (dd, *J* = 2.4, 12.2 Hz, 1H, H-5′), 3.73 (dd, *J* = 3.4, 12.2 Hz, 1H, H-5′), 2.74 (app quint, J_{app} ~6.8 Hz, 1H, H-2′), 2.40–2.35 (ddd, *J* = 2.7, 5.9, 13.5 Hz, 1H, H-2′), 1.74 (br m, 2H, piperidinyl CH$_2$), 1.65 (br m, 4H, piperidinyl CH$_2$). ^{13}C-NMR (125 MHz, CD$_3$OD): δ 155.3, 154.9, 152.7, 139.6, 120.5, 89.8, 86.8, 73.0, 63.7, 47.8, 41.6, 27.3, 25.7. HRMS (TOF) calcd for C$_{15}$H$_{20}$ClN$_5$O$_3$Na [M + Na]$^+$ 376.1147, found 376.1148.

2-Chloro-6-(morpholin-4-yl)-9-(2-deoxy-β-D-ribofuranosyl)purine (**17a**): Prepared from compound **10a** (80.0 mg, 0.137 mmol) and KF (31.8 mg, 0.548 mmol) in MeOH (1.4 mL). Chromatography on a silica gel column sequentially eluted with 50% EtOAc in hexanes, EtOAc, and 10% MeOH in EtOAc gave 36.1 mg (74% yield) of compound **17a** as a white, foamy solid. R_f (SiO$_2$ and 5% MeOH in EtOAc) = 0.21. ^{1}H-NMR (500 MHz, CD$_3$OD): δ 8.22 (s, 1H, H-8), 6.37 (t, *J* = 7.1 Hz, 1H, H-1′), 4.56 (m, 1H, H-3′), 4.40–4.12 (br m, 4H, morpholinyl N(CH$_2$)$_2$), 4.04 (m, 1H, H-4′), 3.83 (dd, *J* = 2.9, 12.2 Hz, 1H, H-5′), 3.79 (t, *J* = 4.9 Hz, 4H, morpholinyl CH$_2$OCH$_2$), 3.74 (dd, *J* = 3.4, 12.2 Hz, 1H, H-5′), 2.74 (app quint, J_{app} ~6.7 Hz, 1H, H-2′), 2.41–2.37 (ddd, *J* = 2.9, 5.8, 13.2 Hz, 1H, H-2′). ^{13}C-NMR (125 MHz, CD$_3$OD): δ 155.2, 154.7, 152.7, 140.3, 120.6, 89.8, 86.8, 72.9, 68.0, 63.6, 47.0 (br), 41.5. HRMS (TOF) calcd for C$_{14}$H$_{19}$ClN$_5$O$_4$ [M + H]$^+$ 356.1120, found 356.1107.

2-Chloro-6-[(2-dimethylamino)ethyl)(methyl)amino]-9-(2-deoxy-β-D-ribofuranosyl)purine (**18a**): Prepared from compound **11a** (70.0 mg, 0.117 mmol) and KF (27.0 mg, 0.467 mmol) in MeOH (1.17 mL). Chromatography on a silica gel column sequentially eluted with 10% MeOH in EtOAc and 20% MeOH in EtOAc gave 36.0 mg (84% yield) of compound **18a** as a white, foamy solid. R_f (SiO$_2$ and 10% MeOH in EtOAc) = 0.10. ^{1}H-NMR (500 MHz, CD$_3$OD): δ 8.14 (s, 1H, H-8), 6.35 (t, *J* = 6.9 Hz, 1H, H-1′), 4.56 (m, 1H, H-3′), 4.12 (br s, 2H, NCH$_2$), 4.03 (m, 1H, H-4′), 3.83 (dd, *J* = 3.4, 12.2 Hz, 1H, H-5′), 3.74 (dd, *J* = 3.9, 12.2 Hz, 1H, H-5′), 3.48 (br s, 3H, NCH$_3$), 2.76–2.71 (br m, 3H, NCH$_2$ and 1H, H-2′), 2.42–2.41 (m, 1H, H-2′), 2.39 (s, 6H, N(CH$_3$)$_2$). ^{13}C-NMR (125 MHz, CD$_3$OD): δ 156.4, 154.7, 152.5, 140.1, 120.8, 89.8, 86.8, 73.0, 63.7, 57.5, 45.8, 41.6, 37.6, (one broadened resonance could not be identified). HRMS (TOF) calcd for C$_{15}$H$_{24}$ClN$_6$O$_3$ [M + H]$^+$ 371.1598, found 371.1584.

3.5. General Procedure for the Desilylation of Ribose Cladribine Analogues

To a 0.1 M solution of the silylated compound in anhydrous MeOH, KF (2 equiv/silyl group) was added. The mixture was heated at 80 °C for 24 h, cooled, and silica gel was added. The mixture was evaporated to dryness and the compound-impregnated silica gel was loaded onto a wet-packed silica gel column. The products were obtained by elution with appropriate solvents (see the individual compound headings for details).

2-Chloroadenosine (**12b**): Prepared from compound **5b** (100 mg, 0.155 mmol) and KF (54.0 mg, 0.93 mmol) in MeOH (1.55 mL). Chromatography on a silica gel column sequentially eluted with 5% MeOH in EtOAc and 10% MeOH in EtOAc gave 35.0 mg (74% yield) of compound **12b** as an off-white solid. R_f (SiO$_2$ and 10% MeOH in EtOAc) = 0.13. ^{1}H-NMR (400 MHz, CD$_3$OD): δ 8.27 (s, 1H, H-8), 5.92 (d, *J* = 6.0 Hz, 1H, H-1′), 4.72 (t, *J* = 5.6, 1H, H-2′), 4.32 (dd, *J* = 2.8, 5.2 Hz, 1H, H-3′), 4.14 (m, 1H, H-4′), 3.90 (dd, *J* = 2.8, 12.4 Hz, 1H, H-5′), 3.76 (dd, *J* = 2.8, 12.4 Hz, 1H, H-5′). ^{13}C-NMR (100 MHz, DMSO-d_6): δ 174.5, 156.7, 152.9, 150.3, 140.0, 118.1, 87.4, 85.7, 73.8, 70.3, 61.3, 25.3. HRMS (TOF) calcd for C$_{10}$H$_{13}$ClN$_5$O$_4$ [M + H]$^+$ 302.0651, found 302.0627.

2-Chloro-N⁶-methyladenosine (**13b**): Prepared from compound **6b** (100 mg, 0.152 mmol) and KF (52.9 mg, 0.912 mmol) in MeOH (1.52 mL). Chromatography on a silica gel column sequentially eluted with 2.5% MeOH in EtOAc and 5% MeOH in EtOAc gave 36.0 mg (75% yield) of compound **13b** as a white, foamy solid. R_f (SiO$_2$ and 10% MeOH in EtOAc) = 0.21. ^{1}H-NMR (400 MHz, CD$_3$OD): δ 8.20 (s, 1H, H-8), 5.90 (d, J = 6.0 Hz, 1H, H-1′), 4.68 (t, J = 5.6 Hz, 1H, H-2′), 4.31 (dd, J = 2,8, 4.8 Hz, 1H, H-3′), 4.14 (m, 1H, H-4′), 3.96 (dd, J = 2.4, 12.4 Hz, 1H, H-5′), 3.76 (dd, J = 2.8, 12.4 Hz, 1H, H-5′), 3.07 (br s, 3H, NCH$_3$). ^{13}C-NMR (100 MHz, DMSO-d_6): δ 155.5, 153.2, 149.2, 139.7, 118.6, 87.3, 85.6, 73.6, 70.6, 70.3, 61.6, 61.3, 53.9, 27.1. HRMS (TOF) calcd for C$_{11}$H$_{15}$ClN$_5$O$_4$ [M + H]$^+$ 316.0807, found 316.0808.

2-Chloro-N⁶,N⁶-dimethyladenosine (**14b**): Prepared from compound **7b** (100 mg, 0.148 mmol) and KF (51.8 mg, 0.892 mmol) in MeOH (1.48 mL). Chromatography on a silica gel column sequentially eluted with 50% EtOAc in hexanes, EtOAc, and 10% MeOH in EtOAc gave 36.8 mg (75% yield) of compound **14b** as an off-white, foamy solid. R_f (SiO$_2$ and 5% MeOH in EtOAc) = 0.32. ^{1}H-NMR (400 MHz, CD$_3$OD): δ 8.17 (s, 1H, H-8), 5.91 (d, J = 6.4 Hz, 1H, H-1′), 4.67 (t, J = 5.4 Hz, 1H, H-2′), 4.31 (dd, J = 2.8, 4.8 Hz, 1H, H-3′), 4.14 (m, 1H, H-4′), 3.90 (dd, J = 2.8, 12.4 Hz, 1H, H-5′), 3.76 (dd, J = 2.4, 12.4 Hz, 1H, H-5′), 3.60–3.49 (br m, H, N(CH$_3$)$_2$). ^{13}C-NMR (100 MHz, DMSO-d_6): δ 154.4, 152.5, 151.1, 138.6, 118.5, 87.2, 85.6, 73.6, 70.2, 61.2, 37.5 (br). HRMS (TOF) calcd for C$_{12}$H$_{17}$ClN$_5$O$_4$ [M + H]$^+$ 330.0964, found 330.0964.

2-Chloro-6-(pyrrolidin-1-yl)-9-(β-D-ribofuranosyl)purine (**15b**): Prepared from compound **8b** (100 mg, 0.143 mmol) and KF (49.8 mg, 0.858 mmol) in MeOH (1.43 mL). Chromatography on a silica gel column sequentially eluted with 50% EtOAc in hexanes, EtOAc, and 10% MeOH in EtOAc gave 43.3 mg (85% yield) of compound **15b** as a white solid. R_f (SiO$_2$ and 5% MeOH in EtOAc) = 0.38. ^{1}H-NMR (400 MHz, CD$_3$OD): δ 8.18 (s, 1H, H-8), 5.91 (d, J = 6.4 Hz, 1H, H-1′), 4.67 (t, J = 5.4 Hz, 1H, H-2′), 4.32 (dd, J = 2.8, 4.8 Hz, 1H, H-3′), 4.14 (m, 1H, H-4′), 4.13–4.10 (br m, 2H, pyrrolidinyl NCH), 3.90 (dd, J = 2.8, 12.8 Hz, 1H, H-5′), 3.74 (dd, J = 2.4, 12.4 Hz, 1H, H-5′), 3.72–3.65 (br m, 2H, pyrrolidinyl NCH), 2.13–2.07 (br m, 2H, pyrrolidinyl CH), 2.05–1.95 (br m, 2H, pyrrolidinyl CH). ^{13}C-NMR (100 MHz, DMSO-d_6): δ 152.7, 152.7, 150.7, 139.1, 118.7, 87.2, 85.6, 73.7, 70.2, 61.2, 48.6, 47.3, 25.6, 23.6. HRMS (TOF) calcd for C$_{14}$H$_{19}$ClN$_5$O$_4$ [M + H]$^+$ 356.1120, found 356.1127.

2-Chloro-6-(piperidin-1-yl)-9-(β-D-ribofuranosyl)purine (**16b**): Prepared from compound **9b** (100 mg, 0.140 mmol) and KF (48.9 mg, 0.842 mmol) in MeOH (1.40 mL). Chromatography on a silica gel column sequentially eluted with EtOAc and 2% MeOH in EtOAc gave 43.6 mg (84% yield) of compound **16b** as a white, foamy solid. R_f (SiO$_2$ and 10% MeOH in EtOAc) = 0.41. ^{1}H-NMR (400 MHz, CD$_3$OD): δ 8.13 (s, 1H, H-8), 5.87 (d, J = 6.0 Hz, 1H, H-1′), 4.64 (t, J = 5.4 Hz, 1H, H-2′), 4.27 (dd, J = 2.8, 5.2 Hz, 1H, H-3′), 4.19 (br s, 4H, piperidinyl N(CH$_2$)$_2$), 4.11 (m, 1H, H-4′), 3.89 (dd, J = 2.8, 12.2 Hz, 1H, H-5′), 3.73 (dd, J = 2.4, 12.2 Hz, 1H, H-5′), 1.74 (br m, 2H, piperidinyl CH$_2$), 1.64 (br m, 4H, piperidinyl CH$_2$). ^{13}C-NMR (100 MHz, DMSO-d_6): δ 153.1, 152.6, 151.4, 138.5, 118.2, 87.2, 85.6, 73.6, 70.4, 70.2, 70.1, 61.2, 44.8 (br), 25.6, 25.2, 23.9, 14.0. HRMS (TOF) calcd for C$_{15}$H$_{21}$ClN$_5$O$_4$ [M + H]$^+$ 370.1277, found 370.1302.

2-Chloro-6-(morpholin-4-yl)-9-(β-D-ribofuranosyl)purine (**17b**): Prepared from compound **10b** (100 mg, 0.139 mmol) and KF (48.7 mg, 0.839 mmol) in MeOH (1.40 mL). Chromatography on a silica gel column sequentially eluted with 50% EtOAc in hexanes, EtOAc, and 10% MeOH in EtOAc gave 40.6 mg (78% yield) of compound **17b** as a white, foamy solid. R_f (SiO$_2$ and 5% MeOH in EtOAc) = 0.21. ^{1}H-NMR (400 MHz, CD$_3$OD): δ 8.21 (s, 1H, H-8), 5.92 (d, J = 6.0 Hz, 1H, H-1′), 4.66 (t, J = 5.6 Hz, 1H, H-2′), 4.32 (dd, J = 2.8, 4.8 Hz, 1H, H-3′), 4.30–4.24 (br m, 4H, morpholinyl N(CH$_2$)$_2$), 4.14 (m, 1H, H-4′), 3.90 (dd, J = 2.8, 12.4 Hz, 1H, H-5′), 3.80 (t, J = 4.8 Hz, 4H, morpholinyl CH$_2$OCH$_2$), 3.76 (dd, J = 2.8, 12.4 Hz, 1H, H-5′). ^{13}C-NMR (100 MHz, DMSO-d_6): δ 153.3, 152.5, 151.6, 139.0, 118.3, 87.3, 85.6, 85.4, 73.7, 70.4, 70.2, 66.0, 61.4, 61.1, 48.5, 45.0 (br). HRMS (TOF) calcd for C$_{14}$H$_{19}$ClN$_5$O$_5$ [M + H]$^+$ 372.1069, found 372.1056.

2-Chloro-6-[(2-dimethylamino)ethyl)(methyl)amino)]-9-(β-D-ribofuranosyl)purine (**18b**): Prepared from compound **11b** (100 mg, 0.137 mmol) and KF (47.7 mg, 0.822 mmol) in MeOH (1.37 mL). Chromatography on a silica gel column sequentially eluted with 10% MeOH in EtOAc and 20% MeOH in EtOAc gave 34.5 mg (65% yield) of compound **18b** as a white, foamy solid. R_f (SiO$_2$ and 15% MeOH in EtOAc) = 0.10. ^{1}H-NMR (400 MHz, CD$_3$OD): δ 8.19 (s, 1H, H-8), 5.92 (d, J = 6 Hz, 1H, H-1′), 4.67 (t, 1H, J = 5.6 Hz, 1H, H-2′), 4.34 (dd, J = 3.2, 5.2 Hz 1H, H-3′), 4.14 (d, J = 2.4 Hz, 3H, H-4′), 3.90 (dd, J = 2.4, 12.4 Hz, 1H, H-5′), 3.76 (dd, J = 2.8, 12.4 Hz, 1H, H-5′), 3.49 (br s, 3H, NCH$_3$), 2.83 (t, 2H, NCH$_2$, J = 6.8 Hz), 2.47 (s, 6H, N(CH$_3$)$_2$). ^{13}C-NMR (100 MHz, CD$_3$OD): δ 156.1, 154.7, 152.3, 140.5, 120.7, 90.8, 87.7, 75.3, 72.3, 63.2, 57.0, 49.6, 49.4, 49.2, 49.0, 48.7, 48.5, 48.3, 45.5, 37.4. HRMS (TOF) calcd for C$_{15}$H$_{24}$ClN$_6$O$_4$ [M + H]$^+$ 387.1542, found 387.1513.

O^6-(Benzotriazol-1-yl)-2-bromo-9-[2-deoxy-3,5-di-O-(t-butyldimethylsilyl)-β-D-ribofuranosyl]purine (**19a**): A mixture of compound **3a** (300.0 mg, 0.489 mmol) and SbBr$_3$ (247.7 mg, 0.685 mmol) in dry CH$_2$Br$_2$ (4.9 mL) was cooled to −15 °C using dry ice and acetone, in a nitrogen atmosphere. *t*-BuONO (203.8 μL, 1.713 mmol) was added dropwise and the mixture was stirred at −10 °C to −15 °C for 2 h. Because TLC indicated the presence of starting material, another aliquot of *t*-BuONO (203.8 μL, 1.713 mmol) was added and the reaction proceeded for 1 h at −15 °C, at which time TLC indicated the reaction to be complete. The reaction mixture was poured into ice-cold, saturated aqueous NaHCO$_3$ (5 mL) with stirring. The mixture was filtered using vacuum (*note: use of vacuum for this filtration is critical for maximizing product recovery*) and the residue was washed with CH$_2$Cl$_2$ (5 mL). The organic layer was separated and the aqueous layer was back extracted with CH$_2$Cl$_2$ (2 × 5 mL). The combined organic layer was washed with water (5 mL) and brine (5 mL), dried over anhydrous Na$_2$SO$_4$, and evaporated under reduced pressure. Purification of the crude material on a silica gel column sequentially eluted with hexanes, 5% EtOAc in hexanes, and 30% EtOAc in hexanes gave 208.5 mg (63% yield) of compound **19a** as a white foam. R_f (SiO$_2$ and 30% EtOAc in hexanes) = 0.60. ^{1}H-NMR (500 MHz, CDCl$_3$): δ 8.45 (s, 1H, H-8), 8.13 (d, J = 8.3 Hz, 1H, Ar-H), 7.56–7.45 (m, 3H, Ar-H), 6.47 (t, J = 5.9 Hz, 1H, H-1′), 4.63 (m, 1H, H-3′), 4.03 (m, 1H, H-4′), 3.90 (dd, J = 3.4, 11.2 Hz, 1H, H-5′), 3.78 (dd, J = 1.5, 11.2 Hz, 1H, H-5′), 2.61 (app quint, J_{app} ~6.2 Hz, 1H, H-2′), 2.49 (app quint, J_{app} ~5.8 Hz, 1H, H-2′), 0.91 (s, 18H, *t*-Bu), 0.11 and 0.09 (2s, 12H, SiCH$_3$). ^{13}C-NMR (125 MHz, CDCl$_3$): δ 158.7, 154.9, 144.3, 143.6, 142.6, 129.1, 128.9, 125.1, 120.8, 119.5, 108.7, 88.5, 85.4, 71.7, 62.7, 41.8, 26.2, 25.9, 18.6, 18.2, −4.4, −4.6, −5.2, −5.3. HRMS (TOF) calcd for C$_{28}$H$_{43}$BrN$_7$O$_4$Si$_2$ [M + H]$^+$ 676.2093, found 676.2078.

2-Bromo-3′,5′-di-O-(t-butyldimethylsilyl)-2′-deoxyadenosine (**20a**): To a solution of compound **19a** (135.3 mg, 0.20 mmol) in 1,2-DME (2 mL), 28%–30% aqueous ammonia (48.6 μL) was added, and the mixture was stirred at room temperature for 45 min. The mixture was diluted with EtOAc (15 mL) and washed with 5% aqueous NaCl (10 mL). The organic layer was separated and the aqueous layer was back extracted with EtOAc (15 mL). The combined organic layer was dried over anhydrous Na$_2$SO$_4$ and evaporated under reduced pressure. The crude material was chromatographed on a silica gel column by sequential elution with hexanes, 20% EtOAc in hexanes and 40% EtOAc in hexanes to afford 93.1 mg (83% yield) of compound **20a** as a white foam. R_f (SiO$_2$ and EtOAc) = 0.50. ^{1}H-NMR (500 MHz, CDCl$_3$): δ 8.08 (s, 1H, H-8), 6.63 (br s, 2H, Ar-NH$_2$), 6.37 (t, J = 6.1 Hz, 1H, H-1′), 4.61 (m, 1H, H-3′), 3.98 (m, 1H, H-4′), 3.88 (dd, J = 3.9, 11.2 Hz, 1H, H-5′), 3.75 (dd, J = 2.4, 11.2 Hz, 1H, H-5′), 2.61 (app quint, J_{app} ~6.3 Hz, 1H, H-2′), 2.43–2.38 (m, 1H, H-2′), 0.90 (s, 18H, *t*-Bu), 0.09 and 0.08 (2s, 12H, SiCH$_3$). ^{13}C-NMR (125 MHz, CDCl$_3$): δ 156.3, 150.4, 144.9, 139.6, 119.3, 88.2, 84.7, 71.9, 62.9, 41.4, 26.1, 25.9, 18.6, 18.2, −4.4, −4.6, −5.2, −5.3. HRMS (TOF) calcd for C$_{22}$H$_{41}$BrN$_5$O$_3$Si$_2$ [M + H]$^+$ 558.1926, found 558.1902.

2-Bromo-2′-deoxyadenosine (**21a**): As described in the general desilylation procedures, compound **21a** was prepared from compound **20a** (80.0 mg, 0.143 mmol) and KF (33.2 mg, 0.572 mmol) in MeOH (1.4 mL), over 28 h. Chromatography on a silica gel column sequentially eluted with 50% EtOAc in hexanes, EtOAc, and 10% MeOH in EtOAc gave 31.1 mg (66% yield) of compound **21a** as a white/off-white solid. R_f (SiO$_2$ and 10% MeOH in EtOAc) = 0.40. ^{1}H-NMR (500 MHz, CD$_3$OD): δ 8.25

(s, 1H, H-8), 6.36 (t, J = 6.8 Hz, 1H, H-1′), 4.57 (m, 1H, H-3′), 4.04 (m, 1H, H-4′), 3.84 (dd, J = 2.9, 12.2 Hz, 1H, H-5′), 3.74 (dd, J = 3.4, 12.2 Hz, 1H, H-5′), 2.76 (app quint, J_{app} ~6.7 Hz, 1H, H-2′), 2.43–2.39 (ddd, J = 2.9, 5.9, 13.2 Hz, 1H, H-2′). ^{13}C-NMR (125 MHz, CD$_3$OD): δ 158.1, 151.4, 145.9, 141.7, 120.3, 89.9, 86.9, 72.9, 63.7, 41.6. HRMS (TOF) calcd for C$_{10}$H$_{12}$BrN$_5$O$_3$Na [M + Na]$^+$ 352.0016, found 352.0021.

O^6-(Benzotriazol-1-yl)-2-bromo-9-[2,3,5-tri-O-(t-butyldimethylsilyl)-β-D-ribofuranosyl]purine (**19b**): A mixture of compound **3b** (500.0 mg, 0.67 mmol) and SbBr$_3$ (340.8 mg, 0.94 mmol) in dry CH$_2$Br$_2$ (8.3 mL) was cooled to −15 °C using dry ice and acetone, in a nitrogen atmosphere. *t*-BuONO (280 μL, 2.352 mmol) was added dropwise and the mixture was stirred at −10 °C to −15 °C for 2 h. Because TLC indicated the presence of starting material, another aliquot of *t*-BuONO (280 μL, 2.352 mmol) was added and the reaction was allowed to progress for 1 h at −15 °C, at which time TLC indicated the reaction to be complete. The reaction mixture was poured into ice-cold, saturated aqueous NaHCO$_3$ (10 mL) with stirring. The mixture was filtered using vacuum (*note: use of vacuum for this filtration is critical for maximizing product recovery*) and the residue was washed with CH$_2$Cl$_2$ (15 mL). The organic layer was separated and the aqueous layer was back extracted with CH$_2$Cl$_2$ (2 × 15 mL). The combined organic layer was washed with water (10 mL) and brine (10 mL), dried over anhydrous Na$_2$SO$_4$, and evaporated under reduced pressure. Purification of the crude material on a silica gel column sequentially eluted with hexanes, 5% EtOAc in hexanes, and 30% EtOAc in hexanes gave 350 mg (64% yield) of compound **19b** as a white foam. R_f (SiO$_2$ and 30% EtOAc in hexanes) = 0.80. ^{1}H-NMR (400 MHz, CDCl$_3$): δ 8.54 (s, 1H, H-8), 8.15 (d, J = 8.4 Hz, 1H, Ar-H), 7.58–7.45 (m, 3H, Ar-H), 6.05 (d, J = 4.0 Hz, 1H, H-1′), 4.57 (t, J = 4.2 Hz, 1H, H-2′), 4.33 (t, J = 4.6 Hz, 1H, H-3′), 4.17 (m, 1H, H-4′), 4.08 (dd, J = 3.6, 11.6 Hz, 1H, H-5′), 3.83 (dd, J = 2.4, 11.6 Hz, 1H, H-5′), 0.96, 0.93, and 0.85 (3s, 27H, *t*-Bu), 0.16, 0.11, −0.03, and −0.09 (4s, 18H, SiCH$_3$). ^{13}C-NMR (100 MHz, CDCl$_3$): δ 158.5, 154.8, 144.2, 143.4, 142.5, 128.8, 128.7, 124.8, 120.6, 119.4, 108.5, 89.5, 85.2, 76.1, 71.1, 61.9, 29.6, 26.1, 25.8, 25.6, 18.5, 18.0, 17.9, −4.2, −4.6, −4.7, −4.9, −5.3, −5.4. HRMS (TOF) calcd for C$_{34}$H$_{57}$BrN$_7$O$_5$Si$_3$ [M + H]$^+$ 806.2907, found 806.2912.

2-Bromo-2′,3′,5′-tri-O-(t-butyldimethylsilyl)-adenosine (**20b**): To a solution of compound **19b** (200 mg, 0.247 mmol) in 1,2-DME (4 mL), 28%–30% aqueous ammonia (60 μL) was added, and the mixture was stirred at room temperature for 45 min. The mixture was diluted with EtOAc (20 mL) and washed with 5% aqueous NaCl (20 mL). The organic layer was separated and the aqueous layer was back extracted with EtOAc (25 mL). The combined organic layer was dried over anhydrous Na$_2$SO$_4$ and evaporated under reduced pressure. The crude material was chromatographed on a silica gel column by sequential elution with hexanes, 20% EtOAc in hexanes, and 40% EtOAc in hexanes to afford 140 mg (82% yield) of compound **20b** as a white foam. R_f (SiO$_2$ and EtOAc) = 0.55. ^{1}H-NMR (400 MHz, CDCl$_3$): δ 8.10 (s, 1H, H-8), 5.92 (d, J = 4.8 Hz, 1H, H-1′), 5.77 (br s, 2H, Ar-NH$_2$), 4.69 (t, J = 4.6 Hz, 1H, H-2′), 4.32 (t, J = 4.2 Hz, 1H, H-3′), 4.13 (m, 1H, H-4′), 4.06 (dd, J = 4.8, 11.2 Hz, 1H, H-5′), 3.80 (dd, J = 2.8, 11.2 Hz, 1H, H-5′), 0.95, 0.93, and 0.83 (3s, 27H, *t*-Bu), 0.14, 0.11, −0.00, and −0.15 (4s, 18H, SiCH$_3$). ^{13}C-NMR (100 MHz, CDCl$_3$): δ 155.8, 150.5, 144.6, 140.0, 119.4, 110.0, 89.0, 85.3, 75.4, 71.7, 62.3, 26.0, 25.8, 25.7, 18.4, 18.0, 17.9, −4.3, −4.7, −5.0, −5.3, −5.4. HRMS (TOF) calcd for C$_{28}$H$_{55}$BrN$_5$O$_4$Si$_3$ [M + H]$^+$ 688.2740, found 688.2728.

2-Bromoadenosine (**21b**): As described in the general desilylation procedures, compound **21b** was prepared from compound **20b** (100 mg, 0.145 mmol) and KF (50.5 mg, 0.870 mmol) in MeOH (1.9 mL), over 24 h. Chromatography on a silica gel column sequentially eluted with 50% EtOAc in hexanes, EtOAc, and 10% MeOH in EtOAc which gave 34.9 mg (69% yield) of compound **21b** as a pale yellow solid. R_f (SiO$_2$ and 10% MeOH in EtOAc) = 0.39. ^{1}H-NMR (400 MHz, CD$_3$OD): δ 8.25 (s, 1H, H-8), 5.92 (d, J = 6.4 Hz, 1H, H-1′), 4.67 (d, J = 5.6 Hz, 1H, H-2′), 4.32 (dd, J = 2.8, 4.8 Hz, 1H, H-3′), 4.15 (m, 1H, H-4′), 3.90 (dd, J = 2.8, 12.4 Hz, 1H, H-5′), 3.76 (dd, J = 2.8, 12.4 Hz, 1H, H-5′). ^{13}C-NMR (100 MHz, CD$_3$OD): δ 156.5, 150.2, 144.1, 139.8, 118.4, 87.2, 85.7, 73.6, 70.3, 61.3. HRMS (TOF) calcd for C$_{10}$H$_{13}$BrN$_5$O$_4$ [M + H]$^+$ 346.0145, found 346.0154.

3.6. Protocols for Tests against HCL, TCL, and CCL

Blood was obtained in sodium heparin tubes from patients as part of protocols with consent forms approved by the investigators review board (IRB) of the National Cancer Institute. The blood was diluted 1:1 with PBS without calcium or magnesium, layered over 15 mL Ficoll in 50 mL tubes, and centrifuged to obtain mononuclear cells. Patients with high lymphocytosis had leukemic cells >80%–90% pure after Ficoll. The cells were viably frozen in 7.5% DMSO in leucine-poor media (LPM, 88% leucine-free RPMI, 2% RPMI, and 10% FBS) in cryovials and stored under liquid nitrogen. LPM also contained penicillin, streptomycin, glutamine, gentamycin, and doxycycline. To assay, thawed cells were washed, suspended in LPM, added to 96-well round-bottom plates (15 μL/well), and treated with 15 μL of purine analogues diluted in LPM. The aliquots were incubated 3 days, then treated with 10 μL of either ATP (CellTiter-Glo, Promega, Madison, WI, USA) or {^{3}H}-leucine (Perkin-Elmer, Waltham, MA, USA) diluted in leucine-free RPMI. After 30 min of ATP, the plate is read for bioluminescence. After 6 h of {^{3}H}-leucine, the cells were liberated by freeze-thaw, harvested on to glass-fiber filters, counted either by a beta scintillation counter. The number of cells cultured in 30 μL aliquots for ATP assay was 20,000, 20,000, and 100,000 for HCL, TCL, and CLL, respectively. The cell number for {^{3}H}-leucine assay was 60,000, 60,000 and 200,000 for HCL, TCL, and CLL, respectively. HCL and TCL cells were pulsed with 1 μCi of {^{3}H}-leucine, while CLL cells were pulsed with 1.5–2 μCi/well. The IC$_{50}$ was the calculated concentration needed for 50% inhibition, defined as the ATP uptake or {^{3}H}-leucine incorporation corresponding to halfway between that of control (cells with LPM alone) and that of cycloheximide 10 μg/mL. Reported IC$_{50}$ values were the means of 3 triplicate experiments.

3.7. Protocols for Tests against Breast Cancer Cells

MDA-MB-231 breast cancer (BC) cells were obtained from the American Type Culture Collection (ATCC, Manassas, VA, USA), and were cultured at 37 °C in 5% CO$_2$ using culture medium recommended by the supplier. Each tested compound was dissolved in sterile DMSO, then diluted (4.8–57.7 mM) in sterile 1X PBS for a final DMSO concentration of 0.1%. Then, 2×10^5 cells/well, were seeded and cultured for 24 h as described [51,52]. Afterwards, the media was changed to 5% FBS for 1 h and cells were treated in duplicate with dilutions of each treatment (0–1.8 mM) for 24 h. Cells were fixed (cold methanol), and nuclei stained (0.4% propidium iodide, (PI)) (Sigma-Aldrich, St Louis, MO, USA), and measured using a GloMax® Microplate Reader (Promega, Madison, WI, USA). Cell viability was calculated as percent of surviving cells after treatment relative to vehicle wells. The IC$_{50}$ was obtained from dose response curve fittings using the non-linear regression function of GraphPad Prism® version 6.0b for Mac (GraphPad Software, San Diego, CA, USA) [53].

4. Conclusions

We have demonstrated, for the first time, the synthesis of O^6-(benzotriazol-1-yl)-2-chloro-9-[2-deoxy-3,5-di-*O*-(*t*-butyldimethylsilyl)-β-D-ribofuranosyl]purine (**4a**) by diazotization-chlorination of O^6-(benzotriazol-1-yl)-3′,5′-di-*O*-(*t*-butyldimethylsilyl)-2′-deoxyguanosine (**3a**) with *t*-BuONO and SbCl$_3$ in CH$_2$Cl$_2$. This procedure afforded better yields than the chlorination using *t*-BuONO and Me$_3$SiCl. This compound and its ribose analogue, O^6-(benzotriazol-1-yl)-2-chloro-9-[2,3,5-tri-*O*-(*t*-butyldimethylsilyl)-β-D-ribofuranosyl]purine, both undergo smooth reactions with ammonia, and primary, and secondary amines to produce cladribine (**12a**), its N-modified analogues (**13a–18a**), and the corresponding ribose derivatives (**12b–18b**), after a simple desilylation with KF in MeOH. These compounds were tested against HCL, TCL, and CLL, but none of the new compounds was more active than cladribine itself. The bromo as well as ribose analogues of cladribine displayed activity but the bromo analogue of cladribine was more active against TCL and CLL as compared to both the ribose equivalent and the bromo ribose analogue of cladribine. The compound containing both the bromine atom and a ribose ring was least active among the compounds possessing a primary amino group at the C6 position. Thus, it appears that a free amino group at this location is critical to the activity of cladribine. Interestingly, the C6 piperidinyl analog of cladribine showed low activity. Tests against MDA-MB-231 breast cancer cells showed that only cladribine and its ribose analogue showed some activity. The bromo analogues were about 10 times less active and all others showed no potential. Despite the lack of a major improvement in the activity of cladribine, or the identification of new compounds with activity against breast cancer, this work has provided a route to four doubly functionalizable nucleoside derivatives. The orthogonal reactivities of these compounds, *i.e.*, S_NAr at the C6 position and metal catalysis at the C2 position, can be used for development of novel nucleoside analogues. We anticipate pursuing further work along these lines in the future.

Supplementary Materials: Supplementary can be accessed at: http://www.mdpi.com/1420-3049/20/10/18437/s1.

Acknowledgments: This work was supported by National Science Foundation Grant CHE-1265687 to MKL, in part by National Institutes of Health Grant SC3GM111171 to MMMM from the National Institute on General Medical Sciences, and Title V PPOHA US Department of Education P031M105050 to Universidad Central del Caribe. Infrastructural support at CCNY was provided by National Institute on Minority Health and Health Disparities Grant G12MD007603, while that at UCC was provided by G12MD007583. Siva Subrahmanyam Relangi, Venkateshwarlu Gurram, Muralidharan Kaliyaperumal, Somesh Sharma, and Narender Pottabathini gratefully acknowledge support from GVKBIO and Siva Subrahmanyam Relangi thanks Anna Venkateswara Rao (Koneru Lakshmaiah University, Guntur (Andhra Pradesh), India) for his support.

Author Contributions: Sakilam Satishkumar, Prasanna K. Vuram, and Siva Subrahmanyam Relangi had equal contributions to the synthesis portion of this work. Sakilam Satishkumar and Prasanna K. Vuram (City College of New York) performed optimization of the synthetic procedures, executed syntheses of the deoxyribose series, performed spectroscopic analyses of the compounds, and produced a part of the experimental section. Messrs. Relangi and Venkateshwarlu Gurram (GVKBIO) executed syntheses of the ribose series, performed spectroscopic analyses of the compounds, and produced a part of the experimental section. Hong Zhou and Robert J. Kreitman (National Cancer Institute) tested the compounds against HCL, TCL, and CLL. Michelle M. Martínez Montemayor tested the compounds against breast cancer cell lines. Lijia Yang (City College of New York) performed HRMS analysis of all compounds synthesized in Lakshman's laboratories. Narender Pottabathini (GVKBIO) was responsible for the oversight of the work performed in his laboratories. Mahesh K. Lakshman (City College of New York) conceived and designed the research, assisted with data analysis, and wrote a significant portion of this manuscript.

Conflicts of Interest: The authors declare no conflict of interest.

References and Notes

1. Venner, H. Synthese der den natürlichen entsprechenden 2-desoxy-nucleoside des adenins, guanins, und hypoxanthins. *Chem. Ber.* **1960**, *63*, 100–110. [CrossRef]
2. Carson, D.A.; Kaye, J.; Seegmiller, J.E. Lymphospecific toxicity in adenosine deaminase deficiency and purine nucleoside phosphorylase deficiency: Possible role of nucleosidase kinase(s). *Proc. Natl. Acad. Sci. USA* **1977**, *74*, 5677–5681. [CrossRef] [PubMed]
3. Carson, D.A.; Wasson, D.B.; Kaye, J.; Ullman, B.; Martin, D.W., Jr.; Robins, R.K.; Montgomery, J.A. Deoxycytidine kinase-mediated toxicity of deoxyadenosine analogs toward malignant human lymphoblasts *in vitro* and toward murine L1210 leukemia *in vivo. Proc. Natl. Acad. Sci. USA* **1980**, *77*, 6865–6869. [CrossRef] [PubMed]
4. Wang, L.; Karlsson, A.; Arnér, E.S.J.; Eriksson, S. Substrate specificity of mitochondrial 2'-deoxyguanosine kinase. Efficient phosphorylation of 2-chlorodeoxyadenosine. *J. Biol. Chem.* **1993**, *268*, 22847–22852. [PubMed]
5. Sjöberg, A.H.; Wang, L.; Eriksson, S. Substrate specificity of human recombinant mitochondrial deoxyguanosine kinase with cytostatic and antiviral purine and pyrimidine analogs. *Mol. Pharmacol.* **1998**, *53*, 270–273. [PubMed]
6. Genini, D.; Adachi, S.; Chao, Q.; Rose, D.W.; Carrera, C.J.; Cottam, H.B.; Carson, D.A.; Leoni, L.M. Deoxyadenosine analogs induce programmed cell death in chronic lymphocytic leukemia cells by damaging the DNA and by directly affecting the mitochondria. *Blood* **2000**, *96*, 3537–3543. [PubMed]
7. Marzo, I.; Pérez-Galán, P.; Giraldo, P.; Rubio-Félix, D.; Anel, A.; Naval, J. Cladribine induces apoptosis in human leukaemia cells by caspase-dependent and -independent pathways acting on mitochondria. *Biochem. J.* **2001**, *359*, 537–546. [CrossRef] [PubMed]

8.   Klöpfer, A.; Hasenjäger, A.; Belka, C.; Schulze-Osthoff, K.; Dörken, B.; Daniel, P.T. Adenine deoxynucleotides fludarabine and cladribine induce apoptosis in a CD95/Fas receptor, FADD and caspase-8-independent manner by activation of the mitochondrial cell death pathway. *Oncogene* **2004**, *23*, 9408–9418. [CrossRef] [PubMed]

9.   Goodman, G.R.; Burian, C.; Koziol, J.A.; Saven, A. Extended follow-up of patients with hairy cell leukemia after treatment with cladribine. *J. Clin. Oncol.* **2003**, *21*, 891–896. [CrossRef] [PubMed]

10.   Sigal, D.S.; Miller, H.J.; Schram, E.D.; Saven, A. Beyond hairy cell: The activity of cladribine in other hematologic malignances. *Blood* **2010**, *116*, 2884–2896. [CrossRef] [PubMed]

11.   The US National Institutes of Health Clinical Trials. Available online: https://www.clinicaltrials.gov/ct2/results?term=cladribine&Search=Search (accessed on 18 June 2015).

12.   Ikehara, M.; Tada, H. Studies of nucleosides and nucleotides.   XVIV. Purine cyclonucleosides.   I.   8,2′-Cyclonucleoside derived from 2-chloro-8-mercapto-9-β-D-xylofuranosyladenine. *J. Am. Chem. Soc.* **1965**, *87*, 606–610. [CrossRef] [PubMed]

13.   Christensen, L.F.; Broom, A.D.; Robins, M.J.; Bloch, A. Synthesis and biological activity of selected 2,6-disubstituted-(2-deoxy-α- and -β-D-*erythro*-pentofuranosyl)purines. *J. Med. Chem.* **1972**, *15*, 735–739. [CrossRef]

14.   Kazimierczuk, Z.; Cottam, H.B.; Revankar, G.R.; Robins, R.K. Synthesis of 2′-deoxytubercidin, 2′-deoxyadenosine, and related 2′-deoxynucleosides via a novel direct stereospecific sodium salt glycosylation procedure. *J. Am. Chem. Soc.* **1984**, *106*, 6379–6382. [CrossRef]

15.   Lioux, T.; Gosselin, G.; Mathé, C. Azido/tetrazole tautomerism in 2-azidoadenine β-D-pentofuranonucleoside derivatives. *Eur. J. Org. Chem.* **2003**, 3997–4002. [CrossRef]

16.   Zhong, M.; Nowak, I.; Robins, M.J. Regiospecific and highly stereoselective coupling of 6-(substituted-imidazol-1-yl)purines with 2-deoxy-3,5-di-O-(p-toluoyl)-α-D-*erythro*-pentofuranosyl chloride. Sodium-salt glycosylation in binary solvent mixtures: Improved synthesis of cladribine. *J. Org. Chem.* **2006**, *71*, 7773–7779. [CrossRef] [PubMed]

17.   Yang, F.; Zhu, Y.; Yu, B. A dramatic concentration effect on the stereoselectivity of N-glycosylation for the synthesis of 2′-deoxy-β-ribonucleosides. *Chem. Commun.* **2012**, *48*, 7097–7099. [CrossRef] [PubMed]

18.   Henschke, J.P.; Zhang, X.; Huang, X.; Mei, L.; Chu, G.; Hu, K.; Wang, Q.; Zhu, G.; Wu, M.; Kuo, C.; Chen, Y. A stereoselective process for the manufacture of a 2′-deoxy-β-D-ribonucleoside using the Vorbrüggen glycosylation. *Org. Process Res. Dev.* **2013**, *17*, 1419–1429. [CrossRef]

19.   Xu, S.; Yao, P.; Chen, G.; Wang, H. A new synthesis of 2-chloro-2′-deoxyadenosine (cladribine), CdA). *Nucleosides Nucleotides Nucleic Acids* **2011**, *30*, 353–359. [CrossRef] [PubMed]

20.   Sakakibara, N.; Kakoh, A.; Maruyama, T. First synthesis of [6-^{15}N]-cladribine using ribonucleoside as a starting material. *Heterocycles* **2012**, *85*, 171–182.

21.   Mikhailopulo, I.A.; Zinchenko, A.I.; Kazimierczuk, Z.; Barai, V.N.; Bokut, S.B.; Kalinichenko, E.N. Synthesis of 2-chloro-2′-deoxyadenosine by microbiological transglycosylation. *Nucleosides Nucleotides* **1993**, *12*, 417–422. [CrossRef]

22.   Barai, V.N.; Zinchenko, A.I.; Eroshevskaya, L.A.; Kalinichenko, E.N.; Kulak, T.I.; Mikhailopulo, I.A. A universal biocatalyst for the preparation of base- and sugar-modified nucleosides *via* an enzymatic transglycosylation. *Helv. Chim. Acta* **2002**, *85*, 1901–1908. [CrossRef]

23.   Komatsu, H.; Araki, T. Efficient chemo-enzymatic syntheses of pharmaceutically useful unnatural 2′-deoxynucleosides. *Nucleosides Nucleotides Nucleic Acids* **2005**, *24*, 1127–1130. [CrossRef] [PubMed]

24.   Taran, S.A.; Verevkina, K.N.; Esikova, T.Z.; Feofanov, S.A.; Miroshnikov, A.I. Synthesis of 2-chloro-2′-deoxyadenosine by microbiological transglycosylation using a recombinant *Escherichia coli* strain. *Appl. Biochem. Microbiol.* **2008**, *44*, 162–166. [CrossRef]

25.   Janeba, Z.; Francom, P.; Robins, M.J. Efficient syntheses of 2-chloro-2′-deoxyadenosine (cladribine) from 2′-deoxyguanosine. *J. Org. Chem.* **2003**, *68*, 989–992. [CrossRef] [PubMed]

26.   Peng, Y. A practical synthesis of 2-chloro-2′-deoxyadenosine (cladribine) from 2′-deoxyadenosine. *J. Chem. Res.* **2013**, *37*, 213–215. [CrossRef]

27.   Wan, Z.-K.; Wacharasindhu, S.; Binnun, E.; Mansour, T. An efficient direct amination of cyclic amides and cyclic ureas. *Org. Lett.* **2006**, *8*, 2425–2428. [CrossRef] [PubMed]

28.   Wan, Z.-K.; Binnun, E.; Wilson, D.P.; Lee, J. A highly facile and efficient one-step synthesis of N^6-adenosine and N^6-2′-deoxyadenosine derivatives. *Org. Lett.* **2005**, *7*, 5877–5880. [CrossRef] [PubMed]

29. Bae, S.; Lakshman, M.K. O^6-(Benzotriazol-1-yl)inosine derivatives: Easily synthesized, reactive nucleosides. *J. Am. Chem. Soc.* **2007**, *129*, 782–789. [CrossRef] [PubMed]

30. Wan, Z.-K.; Wacharasindhu, S.; Levins, C.G.; Lin, M.; Tabei, K.; Mansour, T.S. The scope and mechanism of phosphonium-mediated S_NAr reactions in heteocyclic amides and ureas. *J. Org. Chem.* **2007**, *72*, 10194–10210. [CrossRef] [PubMed]

31. Ashton, T.D.; Scammells, P.J. Microwave-assisted direct amination: Rapid access to multi-functionalized N^6-substituted adenosines. *Aust. J. Chem.* **2008**, *61*, 49–58. [CrossRef]

32. Bae, S.; Lakshman, M.K. Unusual deoxygenation and reactivity studies related to O^6-(benzotriazol-1yl) inosine derivatives. *J. Org. Chem.* **2008**, *73*, 1311–1319. [CrossRef] [PubMed]

33. Bae, S.; Lakshman, M.K. A novel polymer supported approach to nucleoside modification. *J. Org. Chem.* **2008**, *73*, 3707–3713. [CrossRef] [PubMed]

34. Bae, S.; Lakshman, M.K. Synthetic utility of an isolable nucleoside phosphonium salt. *Org. Lett.* **2008**, *10*, 2203–2206. [CrossRef] [PubMed]

35. Lakshman, M.K.; Frank, J. A simple method for C-6 modification of guanine nucleosides. *Org. Bimol. Chem.* **2009**, *7*, 2933–2940. [CrossRef] [PubMed]

36. Kokatla, H.P.; Lakshman, M.K. One-pot etherification of purine nucleosides and pyrimidines. *Org. Lett.* **2010**, *12*, 4478–4881. [CrossRef] [PubMed]

37. Lakshman, M.K.; Singh, M.K.; Parrish, D.; Balachandran, R.; Day, B.W. Azide-tetrazole equilibrium of C-6 azidopurine nucleosides and their ligation reactions with alkynes. *J. Org. Chem.* **2010**, *75*, 2461–2473. [CrossRef] [PubMed]

38. Lakshman, M.K.; Kumar, A.; Balachandran, R.; Day, B.W.; Andrei, G.; Snoeck, R.; Balzarini, J. Synthesis and biological properties of C-2 triazolylinosine derivatives. *J. Org. Chem.* **2012**, *77*, 5870–5883. [CrossRef] [PubMed]

39. Ghosh, S.; Greenberg, M.M. Synthesis of cross-linked DNA containing oxidized abasic site analogues. *J. Org. Chem.* **2014**, *79*, 5948–5957. [CrossRef] [PubMed]

40. Devine, S.M.; May, L.T.; Scammells, P.J. Design, synthesis, and evaluation of N^6-substituted 2-aminoadenosine-5′-N-methylcarboxamides as A_3 adenosine receptor agonists. *Med. Chem Commun.* **2014**, *5*, 192–196. [CrossRef]

41. Krüger, S.; Meier, C. Synthesis of site-specific damaged DNA strands by 8-(acetylarylamino)-2′-deoxyguanosine adducts and effects on various DNA polymerases. *Eur. J. Org. Chem.* **2013**, *2013*, 1158–1169. [CrossRef]

42. Van der Henden van Noort, G.J.; Overkleeft, H.S.; van der Marel, G.A.; Filippov, D.V. Synthesis of nucleotidylated poliovirus VPg proteins. *J. Org. Chem.* **2010**, *75*, 5733–5736. [CrossRef] [PubMed]

43. Printz, M.; Richert, C. Pyrenylmethyldeoxyadenosine: A 3′-cap for universal DNA hybridization probes. *Chem. Eur. J.* **2009**, *15*, 3390–3402. [CrossRef] [PubMed]

44. Devine, S.M.; Scammells, P.J. Synthesis and utility of 2-halo-O^6-(benzotriazol-1-yl)-functionalized purine nucleosides. *Eur. J. Org. Chem.* **2011**, 1092–1098. [CrossRef]

45. Sampath, U.; Bartlett, L.; (Reliable Biopharmaceutical, Inc., St. Louis, MO, USA). Process for the Production of 2-halo-6-aminopurine Derivatives. US Patent 6,252,061 B1, 26 June 2001.

46. Francom, P.; Robins, M.J. Nucleic acid related compounds. 118. Nonaqueous diazotization of aminopurine derivatives. Convenient access to 6-halo and 2,6-dihalopurine nucleosides and 2′-deoxynucleosides with acyl or silyl halides. *J. Org. Chem.* **2003**, *68*, 666–669. [CrossRef] [PubMed]

47. Pottabathini, N.; Bae, S.; Pradhan, P.; Hahn, H.-G.; Mah, H.; Lakshman, M.K. Synthesis and reactions of 2-chloro and 2-tosyloxy 2′-deoxyinosine derivatives. *J. Org. Chem.* **2005**, *70*, 7188–7195. [CrossRef] [PubMed]

48. Robins, M.J.; Uznański, B. Nucleic acids related compounds. 34. Non-aqueous diazotization with *tert*-butyl nitrite. Introduction of fluorine, chlorine, and bromine at C-2 of purine nucleosides. *Can. J. Chem.* **1981**, *59*, 2608–2611. [CrossRef]

49. Francom, P.; Janeba, Z.; Shibuya, S.; Robins, M.J. Nucleic acid related compounds. 116. Nonaqueous diazotization of aminopurine nucleosides. Mechanistic considerations and efficient procedures with *tert*-butyl nitrite or sodium nitrite. *J. Org. Chem.* **2002**, *67*, 6788–6796. [CrossRef] [PubMed]

50. Kazimierczuk, Z.; Vilpo, J.A.; Seela, F. Base-modified nucleosides related to 2-chloro-2′-deoxyadenosine. *Helv. Chim. Acta* **1992**, *75*, 2289–2297. [CrossRef]

51. Martínez-Montemayor, M.M.; Rosario-Acevedo, R.; Otero-Franqui, E.; Cubano, L.A.; Dharmawardhane, S.F. *Ganoderma lucidum* (Reishi) inhibits cancer cell growth and expression of key molecules in inflammatory breast cancer. *Nutr. Cancer* **2011**, *63*, 1085–1094.

52. Suarez-Arroyo, I.J.; Rosario-Acevedo, R.; Aguilar-Perez, A.; Clemente, P.L.; Cubano, L.A.; Serrano, J.; Schneider, R.J.; Martinez-Montemayor, M.M. Anti-tumor effects of *Ganoderma lucidum* (Reishi) in inflammatory breast cancer in *in vivo* and *in vitro* models. *PLoS ONE* **2013**, *8*, e57431. [CrossRef] [PubMed]

53. GraphPad Software. Available online: http://www.graphpad.com.

Sample Availability: *Sample Availability*: Limited amounts of compounds **12a–18a** and **12b–18b** may be available from the authors.

![molecules logo]

molecules

Article

Site-Selective Ribosylation of Fluorescent Nucleobase Analogs Using Purine-Nucleoside Phosphorylase as a Catalyst: Effects of Point Mutations

Alicja Stachelska-Wierzchowska [1,*], Jacek Wierzchowski [1], Agnieszka Bzowska [2] and Beata Wielgus-Kutrowska [2]

[1]   Department of Physics and Biophysics, University of Varmia & Masuria in Olsztyn, 4 Oczapowskiego St., 10-719 Olsztyn, Poland; jacek.wie@uwm.edu.pl

[2]   Division of Biophysics, Institute of Experimental Physics, University of Warsaw, Zwirki i Wigury 93, 02-089 Warsaw, Poland; abzowska@biogeo.uw.edu.pl (A.B.); beata@biogeo.uw.edu.pl (B.W.-K.)

*   Correspondence: alicja.stachelska@uwm.edu.pl; Tel.: +48-89-523-3406; Fax: +48-89-523-3861

Academic Editor: Mahesh K. Lakshman
Received: 13 November 2015 ; Accepted: 9 December 2015 ; Published: 28 December 2015

Abstract: Enzymatic ribosylation of fluorescent 8-azapurine derivatives, like 8-azaguanine and 2,6-diamino-8-azapurine, with purine-nucleoside phosphorylase (PNP) as a catalyst, leads to N9, N8, and N7-ribosides. The final proportion of the products may be modulated by point mutations in the enzyme active site. As an example, ribosylation of the latter substrate by wild-type calf PNP gives N7- and N8-ribosides, while the N243D mutant directs the ribosyl substitution at N9- and N7-positions. The same mutant allows synthesis of the fluorescent N7-β-D-ribosyl-8-azaguanine. The mutated form of the *E. coli* PNP, D204N, can be utilized to obtain non-typical ribosides of 8-azaadenine and 2,6-diamino-8-azapurine as well. The N7- and N8-ribosides of the 8-azapurines can be analytically useful, as illustrated by N7-β-D-ribosyl-2,6-diamino-8-azapurine, which is a good fluorogenic substrate for mammalian forms of PNP, including human blood PNP, while the N8-riboside is selective to the *E. coli* enzyme.

Keywords: fluorescent nucleosides; enzymatic ribosylation; 8-azapurines; purine nucleoside phosphorylase

1. Introduction

Purine-nucleoside phosphorylase (PNP, E.C.2.4.2.1) is a key enzyme of purine metabolism, important, *inter alia*, for the proper activity of the immune system in mammals [1–3]. Potent inhibitors of PNP, like immucillin H (forodesine) and some analogs [4], are clinically approved to treat lymphomas [4–6], and others are considered as potential anti-parasitic and anti-malarial drugs [4,7]. Bacterial forms of PNP are of interest because they can be used as a suicidal-gene in cancer chemotherapy [8]. Besides this, PNP is used as a catalyst in the gram-scale preparative ribosylation of purines and purine analogs [9–11], thanks to the reverse (synthetic) pathway of the phosphorolytic process. Nucleoside analogues are widely applied as pharmaceuticals and as biochemical probes for enzymological studies [5–7,9–13].

In the preceding papers [14,15], we have demonstrated that enzymatic ribosylation of some 8-azapurines leads not only to the canonical nucleoside analogs, but also to non-typical, and highly fluorescent ribosides, ribosylated at the N7- and N8-positions (see Figure 1 below). Now we present a kinetic analysis of these processes, with application of several wild and mutated forms of PNP as catalysts. We will demonstrate a considerable selectivity of the 8-azapurine ribosylation sites with various PNP forms, and a remarkable sensitivity of the ribosylation process to point mutations at the critical active site residue. Finally, we present an example of potential analytical application of the obtained compounds to blood PNP determination.

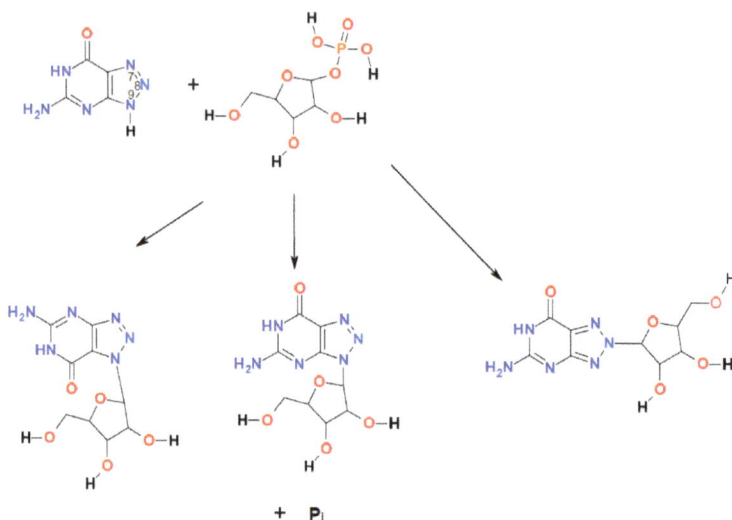

Figure 1. Ribosylation of 8-azaguanine with α-D-ribose-1-phosphate: possible reaction products. The purine numbering is maintained for simplicity. Only the major tautomeric form of 8-azaguanine (N(9)H) is shown.

2. Results and Discussion

In comparison with natural purines, their 8-aza analogues are not as good PNP substrates, but their ribosides are highly fluorescent and therefore can be utilized as probes in enzymology or clinical investigations [11]. Our aim was to identify those forms of PNP which can be used as catalyst in the effective and selective enzymatic syntheses of these ribosides.

We have investigated bacterial (*E. coli*) and mammalian (calf) forms of PNP, as the most widely accessible, and representing two main classes of the enzyme, as well as their mutated forms, which has been previously shown to express altered activity of the phosphorolytic process [1]. In particular, the N243D mutant of the calf and human enzymes is known to catalyze the phosphorolysis not only of Guo and Ino (as do the wild forms), but also of Ado [1,16]. We expected that the analogous mutant of the *E. coli* PNP, D204N, obtained recently [17], will be also interesting as a potential catalyst.

In the ribosylation reactions, we used α-D-ribose-1-phosphate, prepared enzymatically (see Section 3 for details), as a ribosyl donor [14]. Additionally, we have also measured kinetics of phosphorolysis of 8-azapurine ribosides in the phosphate buffer.

2.1. Ribosylation of 8-Azaguanine and 2,6-Diamino-8-azapurine

Ribosylation of 8-azaguanine (8-azaGua) and 2,6-diamino-8-azapurine (8-azaDaPur) was followed spectrophotometrically, and the reaction products analyzed using HPLC separation coupled with spectrophotometric and fluorimetric detectors. The products were identified by comparing their spectral properties with those published earlier [18–20]. The non-typical ribosides of both substrates are spectrally distinct from the N9-nucleosides in that the UV spectra of the formers are quite red-shifted, allowing relatively easy detection at 315 nm. Typical elution profile from the C8-column is shown on Figure 2.

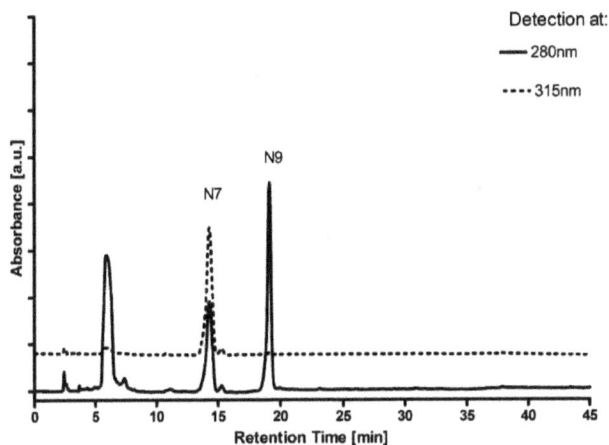

Figure 2. The HPLC elution profile for the mixture obtained from the ribosylation of 8-azaguanine with α-D-ribose-1-phosphate as a ribose donor, catalyzed by the N243D mutant of the calf PNP. The Kromasil C8 column (5 µm × 250 × 10 mm) was used, and the reaction mixture eluted with water (10 min) followed by 0%–30% methanol gradient. The first peak (~6 min retention time) was identified as unreacted 8-azaguanine.

As pointed earlier, enzymatic ribosylation of 8-azaguanine goes fairly rapidly, when catalyzed by calf PNP [11], and the only product of this process is N9-riboside. By contrast, the mutated (N243D) form of the enzyme gives as a major product N7-riboside, although the overall rate of this process is quite slow. Similar qualitative differences can be observed for 8-azaDaPur ribosylation: while the wild-type enzyme gives a mixture of N8 and N7 forms, the application of the N243D mutant gives mainly N7, some N9- and virtually no N8-riboside (see Table 1).

Table 1. Kinetic parameters for enzymatic ribosylation of selected 8-azapurines in 25 mM HEPES buffer, pH 6.6, by α-D-ribose-1-phosphate, using various forms of PNP (nd = not determined). Standard errors are estimated to be ~15%.

Substrate/Enzyme	K_m (µM)	V_{max} (Relative) *	Approximate Product Ratio: N9-riboside: N8-riboside:N7-riboside
8-azaGua/calf PNP-wt	~90	21	1:0:0
8-azaGua/calf PNP-N243D	>100	>0.2 ***	1:0:2
8-azaGua/E. coli PNP-wt	>200	~1 ***	20:1:0
8-azaGua/E. coli PNP-D204N	nd	traces	predominantly N9
8-azaDaPur/calf PNP-wt	60	~1	0:1:1 **
8-azaDaPur/calf PNP-N243D	35	0.6	1:0:3
8-azaDaPur/E. coli PNP-wt	>200	~3 ***	1:2:0
8-azaDaPur/E. coli PNP-D204N	>200	~0.4 ***	10:1:0

* relative to guanine ribosylation under the same conditions (=100); ** this ratio is dependent on the reaction progress; *** reaction rates measured at 200 µM substrate.

The *E. coli* PNP (wild type) gives predominantly N9-riboside of 8-azaguanine, with some participation of a highly fluorescent N8-form. Similar specificity was obtained using the D204N mutant, although the reaction was much slower. With 8-azaDaPur as a substrate, the N8-riboside becomes a dominant ribosylation product for the wild-type enzyme, while the mutant gave mainly N9-nucleoside. Ribosylation rates are typically lower for the mutated enzymes, comparing with the wild types. The K_m values, determined by standard methods, are lower for the calf enzymes (Table 1), while the *E. coli* PNP was not saturated under the applied conditions.

The *E. coli* PNP is known to ribosylate adenine and 8-azaadenine [1,11]. In both cases, the N9-ribosides were the only detectable ribosylation products. We have found that the D204N mutant reacting with the latter substrate does produce also a non-typical product with red-shifted UV spectra, tentatively identified as N8-riboside (data not shown).

The altered specificity of the PNP mutants towards some of the purine analogs has been observed in other laboratories [21]. It can be related to differences in binding modes of 8-azapurine bases in the PNP active site, including binding in the "upside-down" position (rotation by 180 deg when compared with standard binding mode). In fact, such an "upside-down" binding mode was already observed in the crystal structure of calf PNP complexed with N7-acycloguanosine inhibitor [22]. The altered binding geometry can possibly be associated also with prototropic tautomerism or proton affinity alterations [23–26]. This illustrates remarkable plasticity of the mammalian and bacterial PNP active site, which may be profited from in designing new more potent and better membrane-permeable PNP inhibitors—potential pharmaceutical agents [1–7].

There is a considerable progress in enzymatic synthesis of nucleoside analogues on a gram scale using PNP as a catalyst [9,10,27–30]. This work demonstrates that in some substrates there is a possibility to alter ribosylation specificity of PNP by point mutations, an effect noted also by other authors [21]. Although the non-typical ribosides are less biologically active than their N9-β isomers, they sometimes exhibit interesting spectral properties and could be applied as fluorescent probes [14,15,21].

2.2. Phosphorolysis of 8-Azaguanine and 2,6-Diamino-8-azapurine Ribosides

On the basis of the micro-reversibility principle, it can be expected that mutations in the active site of PNP can alter the phosphorolytic processes as well. We have therefore purified the 8-azapurine ribosides and examined them as potential PNP substrates in the phosphate buffer, with results summarized in Table 2. The phosphorolytic reactions were followed spectrophotometrically and, if possible, also fluorimetrically, and the results are summarized in Table 2.

Table 2. Kinetic parameters for enzymatic phosphorolysis of selected ribosides in 25 mM phosphate buffer, pH 6.5, at 25 °C, catalyzed by various forms of PNP. Errors about 15%.

Substrate	Enzyme (wt = Wild Type)	K_m (µM)	V_{max} (Relative) *
N7-β-D-ribosyl-Gua **	calf PNP-wt	27	0.6
N7-β-D-ribosyl-Gua **	*E. coli*-wt	~450	33
N7-ribosyl-8-azaGua	calf PNP-wt	nd	>0.13
N7-ribosyl-8-azaGua	calf PNP-N243D	nd	>1.5
N7-ribosyl-8-azaDaPur	calf PNP-wt	52	~20
N7-ribosyl-8-azaDaPur	calf PNP-N243D	>50	>60
N7-ribosyl-8-azaDaPur	*E. coli* PNP-wt	~80	~1.7
N7-ribosyl-8-azaDaPur	*E. coli* PNP-D204N	nd	~0.7 ***
N8-ribosyl-8-azaDaPur	*E. coli* PNP-wt	7	1.1
N8-ribosyl-8-azaDaPur	*E. coli* PNP-D204N	nd	~1.6 ***
N9-ribosyl-8-azaDaPur	*E. coli* PNP-wt	~20	~0.02

* relative to guanosine phosphorolysis under the same conditions (=100); ** data from ref. [1]; *** rate with 40 µM substrate.

8-Azaguanosine was reported to be a very weak substrate for mammalian PNP [11]. Similarly, only traces of activity of the calf PNP towards N7-riboside were detected. One possible reason can be unfavorable equilibrium of the phosphorolytic process, which for N9-riboside was estimated to be ~300 in favor of nucleoside synthesis, compared to ~50 for natural purines [11]. The mutated form of the calf enzyme is somewhat more active (Table 2).

As mentioned earlier, there is a considerable specificity in the phosphorolytic pathway in both calf and *E. coli* PNP in relation to 8-azaDaPur ribosides [15]. There was no detectable activity towards

N9-riboside with the calf PNP, and only residual with the *E. coli* enzyme. By contrast, the N8-riboside is quite effectively phosphorolysed by the wild type *E. coli* PNP, but not by the mutant (see Table 2). Worth noting is low K_m value for this process (Table 2), contrasting with very high K_m observed in the synthetic reaction (see Table 1).

The N7-riboside of 8-azaDaPur seems to be quite rapidly and specifically phosphorolysed by calf PNP [15]. The mutation at Asn243 markedly accelerates this process (see Table 2), while virtually no activity towards N8- or N9-ribosides was detected using both wild and mutated types of PNP. This finding is somewhat surprising in view of the fact that N8-ribosides are produced in the reverse (synthetic) reaction catalyzed by wild (but not the mutated) form of calf PNP (see previous sections).

Although the spectrum of non-typical substrates of mammalian, and especially bacterial forms of PNP is broad [1,9,11], recent years brought much more examples of applications of these enzymes to synthesis and phosphorolysis of biologically interesting nucleoside analogs [10,27–30]. But the exact prediction of the substrate preferences of various forms is at present difficult [31], we think that large variability of natural PNP and possibility of mutations in the active site can offer new potentially useful catalyst for synthetic procedures in nucleoside chemistry.

2.3. Fluorescence of 8-Azaguanine and 2,6-Diamino-8-azapurine Ribosides and Potential Applications

Fluorescent nucleoside analogues are widely studied because of potential applications in enzymology [11,32,33]. It has been reported that many 8-azapurines and their nucleosides exhibit measurable fluorescence in aqueous medium at room temperature, which was in several instances applied to mechanistic and analytical studies [11,34,35]. The highest yields of fluorescence are observed in 8-azaxanthine and 2,6-diamino-8-azapurine, as well as in some N-alkyl derivatives of these [11,36]. Nucleosides of adenine and hypoxanthine are more fluorescent than the parent bases [13], and N8-ribosides of many 8-azapurines, including N8-ribosyl-8-azaguanine and N8-ribosyl-8-azaDaPur, exhibit yields comparable to the best fluorophores known [11].

2.3.1. Fluorescence of 2,6-Diamino-8-azapurine Ribosides

Ribosylation of 2,6-diamino-8-azapurine leads to at least three fluorescent ribosides (Table 3). The N9-riboside is very highly fluorescent, but its fluorescence is generally similar, in terms of λ_{max} and decay time, to that of the free base [11]. By contrast, fluorescence of N7- and N8-ribosides is red-shifted by ~70 nm and thanks to this shift their cleavage via phosphorolysis (or hydrolysis) leads to a high fluorogenic effect at ~360 nm, which can be analytically useful (see next section).

Table 3. Ionization constants (pK_a values) and spectral parameters for neutral and ionic forms of the selected 8-azapurine ribosides and free bases. The UV spectral data are compiled from the literature [11,18,20] and fluorescence parameters determined in this and previous [11,15] works.

Compound	pK_a	Form * (pH)	UV Absorption		Fluorescence		
			λ_{max} (nm)	ε_{max} ($M^{-1} \cdot cm^{-1}$)	λ_{max} (nm)	ϕ	τ (ns)
9-β-D-ribofuranosyl-8-azaDaPur	2.9	n (7)	285	10,800	368	0.9	6
		c (2)	283	8100	360	nd **	nd **
8-β-D-ribofuranosyl-8-azaDaPur	4.9	n (7)	313	8200	430	0.41	10.6
		c (2.7)	264	13,200	430	nd **	nd **
7-β-D-ribofuranosyl-8-azaDaPur	3.95	n (7)	314	~5500	420	0.063	1.5; 0.45
		c (2)	258	~12,000	420	nd **	nd **
8-azaDaPur (free base)	3.7; 7.7	n (6)	280	8500	365	0.40	7.5; 0.2
9-β-D-ribofuranosyl-8-azaGua	8.05	n (5)	256	12,900	347	~0.01	~0.1
		ma (10)	278	11,700	362	0.55	5.6
7-β-D-ribofuranosyl-8-azaGua	7.4	n (5)	302	4900	410	~0.04	nd **
		ma (10)	304	5100	420	~0.03	nd **
8-azaGua (free base)	6.5	n (4.5)	249	11,200	395	0.05–0.33	6.2

* n—neutral form; c—cation; ma—monoanion; ** nd—not determined.

It is of interest that protonation of N7- and N8-ribosides, which leads to a significant blue shift in the UV spectra (Table 3), apparently does not alter the observed fluorescence band. This is undoubtedly due to a large shift in the acid-base equilibrium in the excided state (pK*), as compared to ground-state (pK$_a$), so upon excitation the protonated 8-azapurine moiety undergoes rapid deprotonation. This process is observed also in acidified alcohols, where dual fluorescence can be observed, especially for the N8-riboside (data not shown), an analogous effect reported earlier for the N8-methyl derivative [36].

It must be stressed that the strong fluorescence of 8-azaDaPur and its ribosides is sensitive to buffer concentration, isotope exchange and other environmental factors [36], partially related to excited-state proton transfer reactions and relatively long fluorescence decay times (Table 3). This creates some difficulty in analytical applications, which can be overcome by using internal concentration standards (e.g., purified products of enzymatic reactions at standardized concentrations).

2.3.2. Blood PNP Determination Using Fluorogenic Ribosides

It has been reported earlier that 7-β-D-ribofuranosyl-8-azaDaPur is a highly fluorogenic substrate for calf PNP in phosphate buffer [15]. Human PNP is similar to the calf enzyme [1], and we have therefore verified a possibility to detect PNP activity in human blood using the same substrate. The experiment was run with 1000-fold diluted, lysed human blood in 25 mM phosphate, and in a phosphate-free 20 mM HEPS buffer, pH 6.6, as a control. *Ca.* 16 µM N7-riboside was used as a substrate. Virtually no fluorescence change was observed in the phosphate-free buffer (data not shown), but in the presence of phosphate the appearance of an emission band at 365 nm indicated free base production (Figure 3, Table 3).

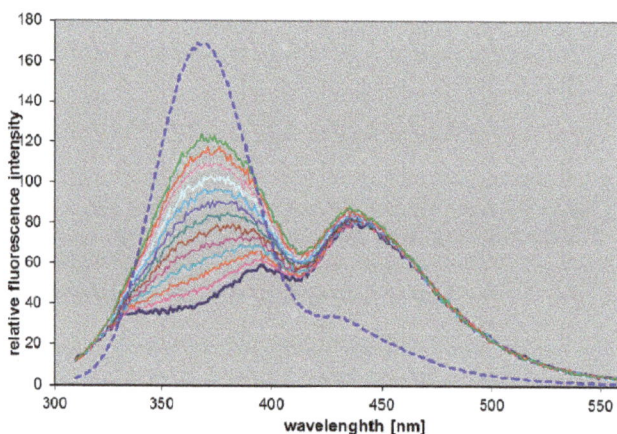

Figure 3. Fluorescence changes during the incubation of 16 µM N7-ribosyl-8-azaDaPur in 1000-diluted blood in the presence of 25 mM phosphate. Spectra, excited at 300 nm, were recorded every 5 min, and after 60 min an aliquot of the purified calf PNP was added, and fluorescence recorded after next 3 min (dotted line, 5-fold diminished relative to remaining curves). Note that the first (lowest) curve reflects blood fluorescence background, and minimum at 415 nm is due to the re-absorption by hemoglobin.

The apparent K$_m$ for this process was rather high, ~90 µM (data not shown), and the presented fluorogenic effect (Figure 3) can undoubtedly be enhanced by optimizing assay conditions. Although phosphorolysis of 7-β-D-ribofuranosyl-8-azaDaPur is slow in comparison to the best PNP substrates, like guanosine or 7-methylguanosine [11], the fluorogenic effect is large thanks to high fluorescence yield of the product (~0.4), and the present assay, in its optimized version, may be more sensitive and simpler than other assays proposed for this enzyme [1,37].

Of special interest is high selectivity of the fluorogenic ribosides of 8-azaDaPur to various forms of PNP, namely, mammalian (N7-ribosides) and bacterial (N8-ribosides), particularly in view of recent applications of bacterial (*E. coli*) PNP as a suicidal-gene in cancer chemotherapy [8]. These two forms of PNP can likely be assayed selectively in the same blood sample, with no pre-purification necessary.

3. Experimental Section

Recombinant *E. coli* and calf spleen PNP, as well as mutated form of these, were obtained and purified as described elsewhere [17,38]. Stock solutions (0.1–1.7 mM per subunit) were stored frozen at −20 °C and diluted prior to experiments. 2,6-Diamino-8-azapurine sulfate, N7-methylguanosine (m⁷Guo) and 8-azaguanine were from Sigma-Aldrich (St. Louis, MO, USA), the latter was re-crystallized as a monosodium salt. N7-methylguanosine was used without further purification.

Fluorescence was measured on a Varian Eclipse instrument (Varian Corp., Palo Alto, CA, USA), and UV absorption kinetic experiments on a Cary 300 (Varian). All buffers were of analytical grade and displayed no fluorescence background.

α-D-Ribose-1-phosphate (100 mM solution in 100 mM HEPES buffer, pH ~7.2) was prepared enzymatically from the N7-methylguanosine and inorganic phosphate, using the modified procedure of Krenitsky *et al.* [39]. The recombinant calf PNP (~2 μg per 1 mL reaction volume) was used as catalyst, and the reaction progress was monitored fluorimetrically [40]. Alternatively, the N243D mutant can be used as a more selective towards m⁷Guo [41]. The second reaction product, N7-methylguanine, was removed in nearly 97% by spontaneous crystallization and filtration. The phosphorylated ribose solution was stored at −20 °C and assayed using previously described fluorimetric method [37]. It was found that 1-year storage caused hydrolysis of not more than 20% of the compound.

Synthetic reactions were carried out on a milligram scale as previously described [14], typically in 1 mL volume, in HEPES buffer. Enzyme concentrations were 1–10 μg/mL, and ribose-1-phosphate ~5 mM. After 24 h reaction mixtures were frozen.

Reaction products were analyzed by the analytical reverse-phase HPLC on a UFLC system from Shimadzu (Kyoto, Japan) equipped with UV (diode-array) detection at 280 nm and 315 nm. The column used was a Kromasil reversed-phase analytical C8 column (250 × 4.6 mm, 5-μm particle size). For product separation, an analogous semi-preparative column was used. The eluent was deionized water (10 min), followed by a linear gradient from 0% to 30% methanol (60 min). The reactions were carried out at pH ~6.5, and samples containing 8-azaguanine ribosides were acidified to ~5 prior to the HPLC analysis.

Kinetic parameters of the enzymatic reactions were calculated using linear regression analysis of the double-reciprocal plots. Substrate concentration in the kinetic experiments ranged typically from 1 to ~200 μM, and enzyme concentrations were adjusted so that the reaction rates were in the range 0.1 to ~5 uM/min. Fluorescence measurements were conducted in semi-micro cuvettes (pathlength 0.4 cm, volumen 1 mL) to diminish the inner filter effect.

Acknowledgments: This work has been supported by the Ministry of Higher Education of Poland, grant #N N301 044939 and by the University of Varmia & Masuria in Olsztyn, internal grant #17.610.006-110. The authors are indebted to Goran Mikleusevic for a sample of the D204N mutated form of the *E. coli* PNP and to Mariusz Szabelski for determination of fluorescence lifetimes.

Author Contributions: A.S.-W. and J.W. are responsible for kinetic and spectroscopic measurements, and A.S.-W. additionally for HPLC analysis. Enzyme cloning and purifications has been done by A.M.B. and B.W.K. This paper was written jointly by all the authors.

Conflicts of Interest: The authors declare no conflict of interest.

Abbreviations

PNP, purine-nucleoside phoshorylase; 8-azaGua, 8-azaguanine; 8-azaDaPur, 2,6-diamino-8-azapurine.

References

1. Bzowska, A.; Kulikowska, E.; Shugar, D. Purine nucleoside phosphorylases: Properties, functions, and clinical aspects. *Pharmacol. Ther.* **2000**, *88*, 349–425. [CrossRef]
2. Grunebaum, E.; Cohen, A.; Roifman, C.M. Recent advances in understanding and managing adenosine deaminase and purine nucleoside phosphorylase deficiencies. *Curr. Opin. Allergy Clin. Immunol.* **2013**, *13*, 630–638. [CrossRef] [PubMed]
3. Edwards, P.N. A kinetic, modeling and mechanistic re-analysis of thymidine phosphorylase and some related enzymes. *J. Enzym. Inhib. Med. Chem.* **2006**, *21*, 483–499. [CrossRef] [PubMed]
4. Taylor Ringia, E.A.; Schramm, V.L. Transition states and inhibitors of the purine nucleoside phosphorylase family. *Curr. Top. Med. Chem.* **2005**, *5*, 1237–1258. [CrossRef] [PubMed]
5. Robak, P.; Robak, T. Older and new purine nucleoside analogs for patients with acute leukemias. *Cancer Treat. Rev.* **2013**, *39*, 851–861. [CrossRef] [PubMed]
6. Al-Kali, A.; Gandhi, V.; Ayoubi, M.; Keating, M.; Ravandi, F. Forodesine: Review of preclinical and clinical data. *Future Oncol.* **2010**, *6*, 1211–1217. [CrossRef] [PubMed]
7. De Jersey, J.; Holý, A.; Hocková, D.; Naesens, L.; Keough, D.T.; Guddat, L.W. 6-oxopurine phosphoribosyltransferase: A target for the development of antimalarial drugs. *Curr. Top. Med. Chem.* **2011**, *11*, 2085–2102. [CrossRef] [PubMed]
8. Portsmouth, D.; Hlavaty, J.; Renner, M. Suicide genes for cancer therapy. *Mol. Asp. Med.* **2007**, *28*, 4–41. [CrossRef] [PubMed]
9. Mikhailopulo, I.A.; Miroshnikov, A.I. Biologically important nucleosides: Modern trends in biotechnology and application. *Mendeleev Commun.* **2011**, *21*, 57–68. [CrossRef]
10. Zhou, X.; Mikhailopulo, I.A.; Cruz Bournazou, M.N.; Neubauer, P. Immobilization of thermostable nucleoside phosphorylases on MagReSyn® epoxide microspheres and their application for the synthesis of 2,6-dihalogenated purine nucleosides. *J. Mol. Catal. B Enzym.* **2015**, *115*, 119–127. [CrossRef]
11. Wierzchowski, J.; Antosiewicz, J.M.; Shugar, D. 8-Azapurines as isosteric purine fluorescent probes for nucleic acid and enzymatic research. *Mol. BioSyst.* **2014**, *10*, 2756–2774. [CrossRef] [PubMed]
12. Giorgi, I.; Scartoni, V. 8-Azapurine nucleus: A versatile scaffold for different targets. *Mini Rev. Med. Chem.* **2009**, *9*, 1367–1378. [CrossRef] [PubMed]
13. Wierzchowski, J.; Wielgus-Kutrowska, B.; Shugar, D. Fluorescence emission properties of 8-azapurines and their nucleosides, and application to the kinetics of the reverse synthetic reaction of PNP. *Biochim. Biophys. Acta* **1996**, *1290*, 9–17. [CrossRef]
14. Stachelska-Wierzchowska, A.; Wierzchowski, J.; Wielgus-Kutrowska, B.; Mikleušević, G. Enzymatic synthesis of highly fluorescent 8-azapurine ribosides using purine-nucleoside phosphorylase reverse reaction: Variable ribosylation sites. *Molecules* **2013**, *18*, 12587–12598. [CrossRef] [PubMed]
15. Wierzchowski, J.; Stachelska-Wierzchowska, A.; Wielgus-Kutrowska, B.; Mikleušević, G. Two fluorogenic substrates for purine-nucleoside phosphorylase, selective for mammalian and bacterial forms of the enzyme. *Anal. Biochem.* **2014**, *446*, 25–27. [CrossRef] [PubMed]
16. Stoeckler, J.D.; Poirot, A.F.; Smith, R.M.; Parks, R.E., Jr.; Ealick, S.E.; Takabayashi, K.; Erion, M.D. Purine nucleoside phosphorylase. 3. Reversal of purine base specificity by site-directed mutagenesis. *Biochemistry* **1997**, *36*, 11749–11756. [CrossRef] [PubMed]
17. Mikleušević, G.; Štefanić, Z.; Narczyk, M.; Wielgus-Kutrowska, B.; Bzowska, A.; Luić, M. Validation of the catalytic mechanism of *Escherichia coli* purine nucleoside phosphorylase by structural and kinetic studies. *Biochimie* **2011**, *93*, 1610–1622. [CrossRef] [PubMed]
18. Montgomery, J.A.; Shortnacy, A.T.; Secrist, J.A., III. Synthesis and biological evaluation of 2-fluoro-8-azaadenosine and related compounds. *J. Med. Chem.* **1983**, *26*, 1483–1489. [CrossRef] [PubMed]
19. Seela, F.; Lampe, S. 8-Aza-2'-deoxyguanosine and related 1,2,3-triazolo[4,5-*d*]pyrimidine 2'-deoxyribofuranosides. *Helv. Chim. Acta* **1993**, *72*, 2388–2397. [CrossRef]
20. Elliott, R.D.; Montgomery, J.A. Analogues of 8-azaguanosine. *J. Med. Chem.* **1976**, *19*, 1186–1191. [CrossRef]
21. Ye, W.; Paul, D.; Gao, L.; Seckute, J.; Sangaiah, R.; Jayaraj, K.; Zhang, Z.; Kaminski, P.A.; Ealick, S.E.; Gold, A.; *et al.* Ethenoguanines undergo glycosylation by nucleoside 2'-deoxyribosyltransferases at non-natural sites. *PLoS ONE* **2014**, *9*, e115082. [CrossRef] [PubMed]

22. Luić, M.; Koellner, G.; Shugar, D.; Saenger, W.; Bzowska, A. Calf spleen purine nucleoside phosphorylase: Structure of its ternary complex with an N(7)-acycloguanosine inhibitor and a phosphate anion. *Acta Crystallogr. D57* **2001**, *57*, 30–36. [CrossRef]

23. Kierdaszuk, B.; Modrak-Wójcik, A.; Wierzchowski, J.; Shugar, D. Formycin A and its N-methyl analogues, specific inhibitors of *E. coli* purine nucleoside phosphorylase: Induced tautomeric shift on binding to enzyme, and enzyme → ligand fluorescence resonance energy transfer. *Biochim. Biophys. Acta* **2000**, *1476*, 109–128. [CrossRef]

24. Pyrka, M.; Maciejczyk, M. Theoretical study of tautomeric equillibria of 2,6–diamino-8-azapurine and 8-aza-isoguanine. *Chem. Phys. Lett.* **2015**, *627*, 30–35. [CrossRef]

25. Wierzchowski, J.; Bzowska, A.; Stępniak, K.; Shugar, D. Interactions of calf spleen purine nucleoside phosphorylase with 8-azaguanine, and a bisubstrate analogue inhibitor: Implications for the reaction mechanism. *Z. Naturforsch.* **2004**, *59*, 713–725. [CrossRef]

26. Gasik, Z.; Shugar, D.; Antosiewicz, J.M. Resolving differences in substrate specificities between human and parasite phosphoribosyltransferases via analysis of functional groups of substrates and receptors. *Curr. Pharm. Des.* **2013**, *19*, 4226–4240. [CrossRef] [PubMed]

27. Zhou, X.; Szeker, K.; Jiao, L.Y.; Oestreich, M.; Mikhailopulo, I.A.; Neubauer, P. Synthesis of 2,6-dihalogenated purine nucleosides by thermostable nucleoside phosphorylases. *Adv. Synth. Catal.* **2015**, *357*, 1237–1244. [CrossRef]

28. Fateev, I.V.; Kharitonova, M.I.; Antonov, K.V.; Konstantinova, I.D.; Stepanenko, V.N.; Esipov, R.S.; Seela, F.; Temburnikar, K.W.; Seley-Radtke, K.L.; Stepchenko, V.A.; *et al.* Recognition of Artificial Nucleobases by *E. coli* Purine Nucleoside Phosphorylase *versus* its Ser90Ala Mutant in the Synthesis of Base-Modified Nucleosides. *Chem. Eur. J.* **2015**, *21*, 13401–13419. [CrossRef] [PubMed]

29. Calleri, E.; Cattaneo, G.; Rabuffetti, M.; Serra, I.; Bavaro, T.; Massolini, G.; Speranza, G.; Ubiali, D. Flow-Synthesis of Nucleosides Catalyzed by an Immobilized Purine Nucleoside Phosphorylase from Aeromonas hydrophila: Integrated Systems of Reaction Control and Product Purification. *Adv. Synth. Catal.* **2015**, *357*, 2520–2528. [CrossRef]

30. Calleri, E. Immobilized purine nucleoside phosphorylase from Aeromonas hydrophila as an on-line enzyme reactor for biocatalytic applications. *J. Chromatogr. B* **2014**, *968*, 79–86. [CrossRef] [PubMed]

31. Štefanić, Z.; Mikleušević, G.; Narczyk, M.; Wielgus-Kutrowska, B.; Bzowska, A.; Luić, M. Still a long way to fully understanding the molecular mechanism of *Escherichia coli* Purine Nucleoside Phosphorylase. *Croat. Chem. Acta* **2013**, *86*, 117–127. [CrossRef]

32. Sinkeldam, R.W.; Greco, N.J.; Tor, Y. Fluorescent analogs of biomolecular building blocks: Design, properties, and applications. *Chem. Rev.* **2010**, *110*, 2579–2619. [CrossRef] [PubMed]

33. Pollum, M.; Martínez-Fernández, L.; Crespo-Hernández, C.E. Photochemistry of nucleic acid bases and their thio- and aza-analogues in solution. *Top. Curr. Chem.* **2015**, *355*, 245–355. [PubMed]

34. Liu, L.; Cottrell, J.W.; Scott, L.G.; Fedor, M.J. Direct measurement of the ionization state of an essential guanine in the hairpin ribozyme. *Nat. Chem. Biol.* **2009**, *5*, 351–357. [CrossRef] [PubMed]

35. Cottrell, J.W.; Scott, L.G.; Fedor, M.J. The pH dependence of hairpin ribozyme catalysis reflects ionization of an active site adenine. *J. Biol. Chem.* **2011**, *286*, 17658–17664. [CrossRef] [PubMed]

36. Wierzchowski, J.; Mędza, G.; Szabelski, M.; Stachelska-Wierzchowska, A. Properties of 2,6-diamino-8-azapurine, a highly fluorescent purine analog and its N-alkyl derivatives: Tautomerism and excited-state proton transfer reactions. *J. Photochem. Photobiol. A* **2013**, *265*, 49–57. [CrossRef]

37. Wierzchowski, J.; Ogiela, M.; Iwańska, B.; Shugar, D. Selective fluorescent and fluorogenic substrates for purine-nucleoside phosphorylases from various sources, and direct fluorimetric determination of enzyme levels in human and animal blood. *Anal. Chim. Acta* **2002**, *472*, 63–74. [CrossRef]

38. Breer, K.; Girstun, A.; Wielgus-Kutrowska, B.; Staron, K.; Bzowska, A. Overexpression, purification and characterization of functional calf purine nucleoside phosphorylase (PNP). *Protein Expr. Purif.* **2008**, *61*, 122–130. [CrossRef] [PubMed]

39. Krenitsky, T.A.; Koszalka, G.W.; Tuttle, J.V. Purine nucleoside synthesis, an efficient method employing nucleoside phosphorylases. *Biochemistry* **1981**, *20*, 3615–3621. [CrossRef] [PubMed]

40. Kulikowska, E.; Bzowska, A.; Wierzchowski, J.; Shugar, D. Properties of two unusual, and fluorescent substrates of purine-nucleoside phosphorylase: 7-methylguanosine and 7-methylinosine. *Biochim. Biophys. Acta* **1986**, *874*, 355–366. [CrossRef]

41. Wielgus-Kutrowska, B. *Division of Biophysics, Institute of Experimental Physics*; University of Warsaw: Warsaw, Poland, Unpublished Work.

Sample Availability: Samples of the compounds are available from the authors.

MDPI AG

St. Alban-Anlage 66

4052 Basel, Switzerland

Tel. +41 61 683 77 34

Fax +41 61 302 89 18

http://www.mdpi.com

Molecules Editorial Office

E-mail: molecules@mdpi.com

http://www.mdpi.com/journal/molecules

www.ingramcontent.com/pod-product-compliance
Lightning Source LLC
Chambersburg PA
CBHW051842210326
41597CB00033B/5740